# Backscattering and RF Sensing for Future Wireless Communication

# Backscattering and RF Sensing for Future Wireless Communication

*Edited by*

*Qammer H. Abbasi*
James Watt School of Engineering,
University of Glasgow,
Glasgow, United Kingdom

*Hasan T. Abbas*
James Watt School of Engineering,
University of Glasgow,
Glasgow, United Kingdom

*Akram Alomainy*
School of Electronic Engineering and
Computer Science, Queen Mary University of London,
London, United Kingdom

*Muhammad Ali Imran*
James Watt School of Engineering,
University of Glasgow,
Glasgow, United Kingdom

This edition first published 2021


*Registered Office(s)*
John Wiley & Sons, Inc., 111 River Street, Hoboken, NJ 07030, USA
John Wiley & Sons Ltd, The Atrium, Southern Gate, Chichester, West Sussex, PO19 8SQ, UK

*Editorial Office*
The Atrium, Southern Gate, Chichester, West Sussex, PO19 8SQ, UK

For details of our global editorial offices, customer services, and more information about Wiley products visit us at www.wiley.com.

Wiley also publishes its books in a variety of electronic formats and by print-on-demand. Some content that appears in standard print versions of this book may not be available in other formats.



*Library of Congress Cataloging-in-Publication Data*

Name: Abbasi, Qammer H., editor.
Title: Backscattering and RF sensing for future wireless communication / edited by Qammer H. Abbasi, James Watt School of Engineering, University of Glasgow, UK, Hasan Tahir Abbas, James Watt School of Engineering, University of Glasgow, UK, Akram Alomainy, School of Electronic Engineering and Computer Science, Queen Mary University of London, UK, Muhammad Ali Imran, James Watt School of Engineering, University of Glasgow, UK.
Description: Hoboken, NJ, USA : Wiley, 2021. | Includes bibliographical references and index.
Identifiers: LCCN 2020055879 (print) | LCCN 2020055880 (ebook) | ISBN 9781119695653 (hardback) | ISBN 9781119695660 (adobe pdf) | ISBN 9781119695684 (epub)
Subjects: LCSH: Backscatter communication. | Wireless communication systems–Technological innovations. | Radio resource management (Wireless communications)
Classification: LCC TK5103.256 .B334 2021 (print) | LCC TK5103.256 (ebook) | DDC 621.382/4–dc23
LC record available at https://lccn.loc.gov/2020055879
LC ebook record available at https://lccn.loc.gov/2020055880

Cover Design: Wiley
Cover Images: © metamorworks/iStock/Getty Images Plus/Getty Images

Set in 9.5/12.5pt STIXTwoText by SPi Global, Chennai, India

Printed and bound in Great Britain by CPI Group (UK) Ltd, Croydon CR0 4YY

C9781119695653_160321

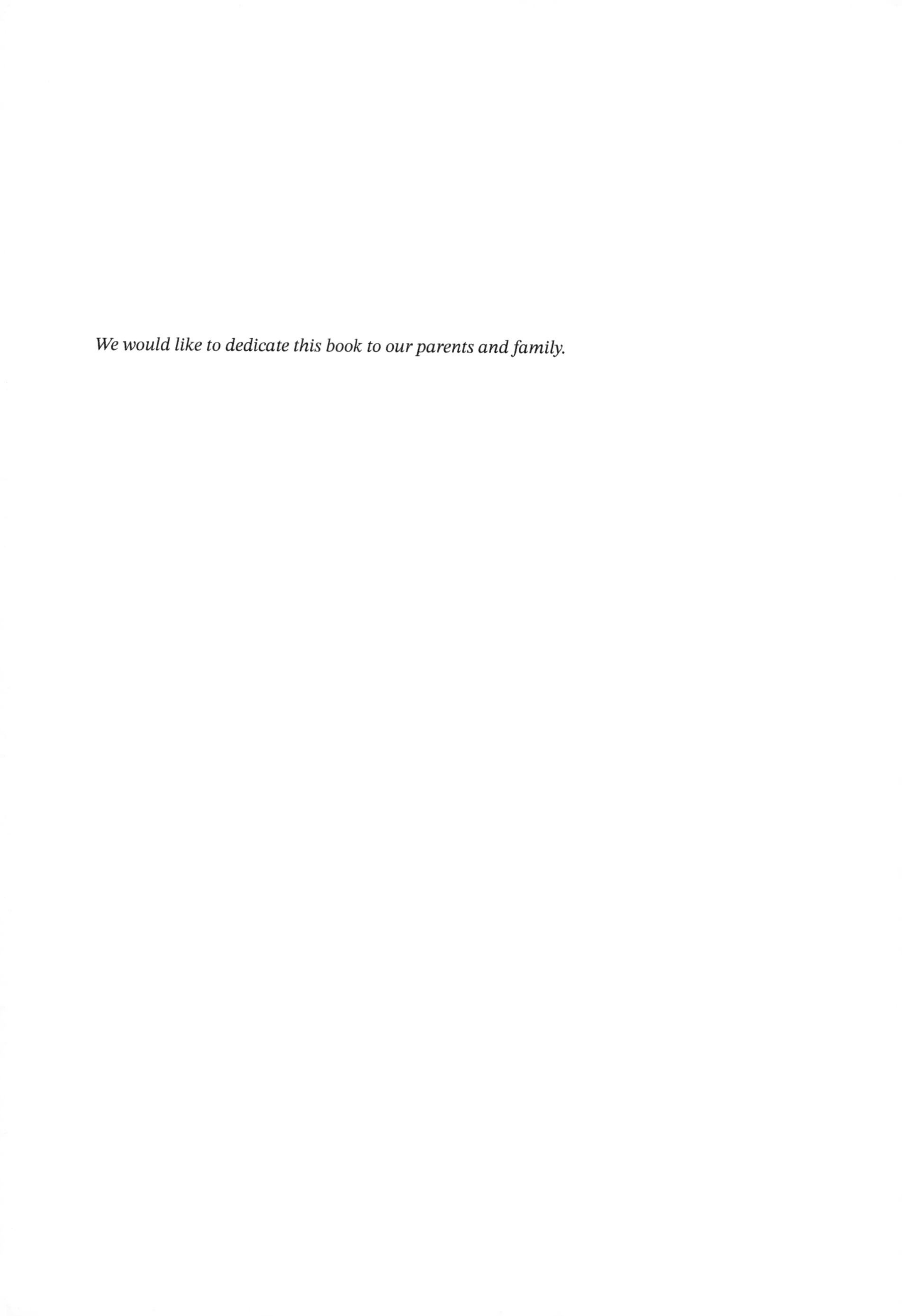

*We would like to dedicate this book to our parents and family.*

# Contents

**Preface** *xii*
**List of Contributors** *xv*

1 **Intelligent Reflective Surfaces – State of the Art** *1*
*Jalil ur Rehman Kazim, Hasan T. Abbas, Muhammad Ali Imran, and Qammer H. Abbasi*
1.1 Background *1*
1.2 Introduction to Reflective Surfaces *1*
1.3 Intelligent Reflective Surface *2*
1.3.1 Theory of Operation *3*
1.4 Potential Use Cases *5*
1.4.1 Application of RIS in Communication *5*
1.4.2 Application of RIS in Health Care *6*
1.5 RIS as an Alternative to Relay *6*
1.6 Coded and Programmable RIS *7*
1.7 Functionalities of RIS *8*
1.7.1 Beam Focusing *8*
1.7.2 Beam Scattering/Diffusion *8*
1.7.3 Multi-Beam Generation *9*
1.7.4 Anomalous Reflection *9*
1.7.5 RIS as a Single RF Chain Transmitter *10*
1.8 Architecture *10*
1.8.1 Electromagnetic Layer *11*
1.8.2 DC Control Layer *11*
1.8.3 Controller Board Interfacing *12*
1.9 Summary *14*
References *14*

2 **Signal Modulation Schemes in Backscatter Communications** *19*
*Yuan Ding, George Goussetis, Ricardo Correia, Nuno Borges Carvalho, Romwald Lihakanga, and Chaoyun Song*
2.1 Introduction *19*
2.2 Principles of Backscatter Modulations and Some Common Schemes *20*

2.2.1 Amplitude Shift Keying (ASK) *20*
2.2.2 Frequency Shift Keying (FSK) *22*
2.2.3 Phase Shift Keying (PSK) *24*
2.2.4 Quadrature Amplitude Modulation (QAM) *25*
2.3 Chirp Spread Spectrum (CSS) Modulation *26*
2.4 Multicarrier Backscatter Transmission *34*
2.5 Summary *37*
References *38*

**3 Electromagnetic Waves Scattering Characteristics of Metasurfaces** *41*
*Muhammad Ali Babar Abbasi, Dmitry E. Zelenchuk, and Abdul Quddious*
3.1 Introduction *41*
3.1.1 General Classifications of Metasurfaces *42*
3.1.2 Characterization Approaches of Metasurfaces *44*
3.2 Metasurface Applications and Practical Examples *46*
3.2.1 Absorptive, Reflective, and Diffusion–Type Metasurfaces *47*
3.2.2 Refractive and Transmission–Type Metasurfaces *48*
3.2.3 Digitally Encoded Metasurfaces *49*
3.2.4 Polarization-Sensitive Metasurface Spectral Filtering *50*
3.2.5 Beamforming with Polarization-Controlling Metasurfaces *53*
3.2.6 Beamforming with Reflective-Type Metasurfaces *55*
3.2.7 Circular Polarization-Selective Metasurfaces *56*
3.2.8 Passive Lossless Huygens' Metasurfaces *59*
3.2.9 Cavity–Excited Huygens' Metasurface Antenna *61*
3.2.10 Passive Lossless Omega–Type Bianisotropic Metasurface *63*
3.3 Conclusion *65*
References *65*

**4 Metasurfaces Based on Huygen's Wavefront Manipulation** *71*
*Abubakar Sharif, Jun Ouyang, Kamran Arshad, Muhammad Ali Imran, and Qammer H. Abbasi*
4.1 Introduction *71*
4.2 Huygens' Metasurfaces (HMSs) *72*
4.3 HMS Applications *76*
4.3.1 HMS Refraction *76*
4.3.2 Antennas Beamforming *77*
4.3.3 Perfect Reflections and Focusing *77*
4.3.4 Beam Scanning *77*
4.4 Conclusion and Key Scientific Issues to Be Addressed *78*
4.5 Future Trends *79*
References *81*

**5 Metasurface: An Insight into Its Applications** *85*
*Fahad Ahmed and Nosherwan Shoaib*
5.1 Polarization *86*

5.2 Polarizers *86*
5.2.1 Basic Principle *86*
5.2.2 Geometrical Configuration *95*
5.2.2.1 Cross-Polarizer – HWP *97*
5.2.2.2 Circular Polarizer – QWP *101*
5.2.2.3 HWP and QWP in Transmission Mode *104*
5.3 Beam Splitter *106*
5.3.1 Design Example *107*
5.3.2 Mathematical Background *109*
5.4 Absorbers *111*
5.4.1 Mathematical Background *111*
5.4.2 Design Examples *112*
5.5 Summary *116*
References *117*

**6 The Role of Smart Metasurfaces in Smart Grid Energy Management** *119*
*Islam Safak Bayram, Muhammad Ismail, and Raka Jovanovic*
6.1 Introduction *119*
6.2 Smart Metasurfaces *120*
6.3 Communication Support in Smart Power Grids *122*
6.3.1 Demand Response and Energy Management *122*
6.3.2 Plug-In Electric Vehicle Load Management *124*
6.3.3 Grid Monitoring and State Estimation *124*
6.3.4 Peer-to-Peer Energy Trading *126*
6.3.5 Potential Applications of Intelligent Surfaces in Smart Energy Grids *127*
6.4 Case Study: Communication System Performance Improvement in Vehicle-to-Grid Networks *127*
6.5 Conclusions *130*
References *130*

**7 Passive UHF RFID Tag Antennas-Based Sensing for Internet of Things Paradigm** *133*
*Abubakar Sharif, Jun Ouyang, Kamran Arshad, Muhammad Ali Imran, and Qammer H. Abbasi*
7.1 Introduction: Healthcare Provision and Radar Technology *133*
7.2 UHF RFID Fundamentals and Performance Metrics *136*
7.2.1 UHF RFID Microchips Insight *137*
7.2.2 Performance Parameters for Passive UHF RFID Tag Antennas *137*
7.2.2.1 Input Impedance and Bandwidth *137*
7.2.2.2 Radar Cross Section (RCS) *140*
7.2.2.3 Read Range *140*
7.3 Sensing Methodology and Techniques *143*
7.3.1 Single Tag Antenna-Based Sensing *143*
7.3.1.1 UHF RFID Backscattering-Based Sensing Methodology *144*
7.3.2 Single Tag Antenna with On-chip Circuitry-Based Sensing *146*

7.3.3 Multiport Tag Antenna-Based Sensing *148*
7.3.4 Reference Multiple Tag-Based Sensing *149*
7.4 Conclusion and Outstanding Challenges *150*
7.5 Future Trends *151*
References *152*

**8 RF Sensing for Healthcare Applications** *157*
*Syed Aziz Shah, Hasan Abbas, Muhammad Ali Imran, and Qammer H. Abbasi*
8.1 Introduction *157*
8.2 Basics of RF Sensing in Remote Healthcare *158*
8.3 Challenges in RF Sensing Technologies *159*
8.3.1 Robust RF System for Healthcare *159*
8.3.2 Reliability of RF Sensing for Remote Patient Monitoring *160*
8.3.3 Affordability in RF Systems for Healthcare *160*
8.3.4 Ethical Approval and Consent of Patients for Data Acquisition *160*
8.4 Wireless Sensing Technologies for Healthcare Applications *160*
8.4.1 Radar Sensing Technologies for Healthcare Applications *161*
8.4.2 Active Radar Sensing in Remote Healthcare *161*
8.4.3 Passive Radar in Remote Healthcare *161*
8.5 RF Sensing Signal Processing for Patient Monitoring *162*
8.5.1 Feature Extraction from Single RF Sensor *162*
8.5.2 Radar Features for Machine Learning Algorithms *163*
8.5.3 Automatic Feature Selection in Machine Learning *163*
8.5.4 Multiple Sensor Combination *164*
8.5.5 Working Function of Multiple Sensors *164*
8.5.6 Multiple Sensor Architectures *164*
8.5.7 Classification of Active Radar Sensor Node *164*
8.6 Active Radar Sensing in Digital Healthcare *166*
8.7 Posture Recognition on Bed *166*
8.7.1 RF Sensing for Patients with Sleep Disorders *166*
8.7.2 Radio Frequency Identification for Patients *167*
8.7.3 Radar Sensing for Occupancy Monitoring in Healthcare Sector *167*
8.7.4 Activities of Daily Livings and Critical Events *168*
8.7.5 Noncontact Wi-Fi Sensing for Patient Monitoring *169*
8.7.6 Wi-Fi-Based Activities of Daily Livings *170*
8.7.7 Vital Signs Monitoring using Wi-Fi Signals *170*
8.7.8 Sleep Attack Detection using CSI Wi-Fi Technologies *172*
8.8 Radio Frequency Identification Sensing for Patient Monitoring *172*
8.8.1 Radio Frequency Identification Sensing for Patient for Patient Tracking *172*
8.8.2 Radio Frequency Identification Sensing for Patient for Disease Detection *172*
8.8.3 Radio Frequency Identification Sensing for Patient to Identify Falls *173*
8.9.4 Radio Frequency Identification Sensing for Patient for Intricate Body Movement Observations *174*
References *174*

**9 Electromagnetic Wave Manipulation with Metamaterials and Metasurfaces for Future Communication Technologies** *179*
*Muhammad Qasim Mehmood, Junsuk Rho, and Muhammad Zubair*
9.1 Introduction *179*
9.2 Meta-Atoms for Optical Frequencies *185*
9.2.1 Nano Half-Wave Plates *186*
9.2.2 Step-Indexed Nano-Waveguides *189*
9.2.3 Supercell *190*
9.3 Applications *192*
9.4 Summary *200*
References *200*

**10 Conclusion** *205*
*Qammer H. Abbasi, Hasan T. Abbas, Akram Alomainy, and Muhammad Ali Imran*
10.1 Future Hot Topics *205*
10.1.1 Signal Modulation *205*
10.1.2 Channel Estimation *205*
10.1.3 Low-Throughput *206*
10.1.4 Network Security *206*
10.1.5 Energy Management *206*
10.2 Concluding Remarks *206*

**Index** *207*

# Preface

Since the early 1990s, we have witnessed a wireless communication revolution that has had an enormous impact on our lives. Breakthroughs in the radiofrequency (RF) and microwave (MW) technologies have propelled the mass proliferation of cell phone, Internet, and digital data streaming technologies that have drastically changed our social interactions.

Despite the massive technological advances, the wireless communication infrastructure is on the verge of saturation due to increased demand. Moreover, with the imminent arrival of a new frontier in the form of device-to-device (D2D) communications in the form of Internet of Things (IoT) technologies, it is expected that soon billions of small computing devices will go online. To meet the data and power requirements, the engineers need to come up with innovative network technologies and deployment schemes. Although RF backscatter communication technology introduced was long time ago in 1948, interest in it has been recently revitalized due to potentially many low-powered IoT applications particularly that are related to healthcare. Backscatter communication allows a device such as an IoT node to not only transmit data through reflection and modulation of an incoming electromagnetic wave, but wirelessly receive power as well. In the last few years, a new frontier of intelligent reflective surfaces (IRS) has opened with the help of which the performance of a backscatter communication system by modifying the wireless propagation medium itself.

This book is an effort to provide an overarching perspective of the latest technological developments in the backscatter communication with a focus on its applications in RF sensing enabled by intelligent reflective metasurfaces. The book chapters contributed to this book are written by experts in the field with many years devoted in the development and analysis of solving engineering problems related to wireless communications and sensing. A particular emphasis is given to the notion of reconfiguration of a wireless channel with the help of the IRS. We aim the book to be an indispensable resource for wireless communication industry engineers, academics, students, and technologists. We have designed the book in a way that only an introductory knowledge of wireless communications and electromagnetic fields is required to comprehend the contents. The book has been planned for a board audience, and we hope that this book will provide a concise picture of the future trends of increasing the efficiency and performance of a wireless communication network. The book contents will assist the research and the associated stakeholder communities to narrow down potential proposal ideas for next-generation wireless communication technologies.

## Organization

A remarkable feature of this book is the detailed discussion of the fundamental physics of backscatter communication enabled by the IRS. Chapter 2 gives an encyclopedic insight to the theory of operation of IRS using an excellent analogy with the aperture antennas. A unique perspective of the state-of-the-art of techniques used for reconfiguring the electromagnetic beams is also given.

Chapter 3 introduces the reader to an in-depth treatment of the signal modulation schemes that are essential in establishing a reliable, low-cost backscatter communication framework. The discussion is devoted to the IoT-centric applications for which a backscatter communication can be exploited to harvest the RF energy as a power source.

Chapters 4 and 5 present some basic concepts of the electromagnetic theory of artificially engineered, sub-wavelength structures known as metasurfaces with the help of which the behavior of electromagnetic waves for wireless communication can be controlled in a way that is not possible through naturally occurring materials. Specifically, Chapter 4 breaks down the metasurface into unit cells formed by a combination of electric and magnetic sources that can be modeled by an analogous electric circuit. Various metasurface designs are analyzed in the microwave frequency range using full-wave electromagnetic simulations are provided.

Chapter 5 covers a special class of metasurfaces that take a leaf from the field of optics and use Huygens's principle of wave propagation to manipulate the electromagnetic waves. An extensive presentation of the Huygens metasurface is provided with the help of the surface equivalence theorem. Toward the end, some critical questions that could pave the path for future research are raised.

Chapter 6 continues the discussion on metasurfaces by delving into the working principles and mathematical foundations of devices such as polarizers, beam splitters, and absorbers operating not only at the microwave frequency, but from a future perspective, millimeter-wave frequencies as well. The design of the devices in this chapter is backed by an extensive full-wave electromagnetic analysis and various unit-cell configurations, implemented with the help of a commercial electromagnetic solver.

Chapters 7–10 follow the common theme of the applications of metasurfaces in various aspects of our lives. Chapter 7 discusses a very interesting application of metasurfaces in improving the energy efficiency and management smart grids for power distribution. Some of the early research endeavors to minimize the reliance of IoT devices on the power grid are also touched upon in which the ambient RF energy is harvested to charge the device batteries.

Chapter 8 surveys the use of RF identification that can arguably be regarded as the most significant application of backscatter communication. The chapter discusses how RFID tags have evolved from simple product labeling to wireless, pervasive sensors that can power up the IoT devices.

Chapter 9 describes various use cases of RF sensing for healthcare, and more importantly, how techniques discussed in the chapter can revolutionize the assisted living by reducing the human involvement. The chapter discusses noncontact and highly accurate human activity and detection systems that only require a single RF source. Some of the challenges that may arise due to privacy concerns are also addressed that have prevented the discussed technologies from going mainstream. Finally, the use of machine learning for intelligent and predictive analytics is also covered in detail.

Chapter 10 provides an extensive review of the applications of metasurfaces for wireless communication and sensing not only at the optical frequency but microwave and terahertz frequencies as well. Some of the applications discussed deal with meta-lens for the creation of computer-generated holograms and optical vortices for ultra-high bandwidth communications.

Finally, the book concludes with Chapter 11 in which we present some of the future directions in the field of RF sensing and backscatter communications and discuss some of the open issues related to the real-world applications of the technologies.

*Qammer H. Abbasi*
*School of Engineering*
*University of Glasgow*
*Glasgow*
*United Kingdom*

*Hasan T. Abbas*
*School of Engineering*
*University of Glasgow*
*Glasgow*
*United Kingdom*

*Akram Alomainy*
*School of Electronic Engineering*
*and Computer Science*
*Queen Mary University of London*
*London*
*United Kingdom*

*Muhammad Ali Imran*
*School of Engineering*
*University of Glasgow*
*Glasgow*
*United Kingdom*

# List of Contributors

***Hasan T. Abbas***
James Watt School of Engineering
University of Glasgow
Glasgow
United Kingdom

***Muhammad Ali Babar Abbasi***
The Centre for Wireless Innovation (CWI)
Institute of Electronics
Communications and Information
Technology (ECIT)
Queen's University of Belfast
Belfast
United Kingdom

***Qammer H. Abbasi***
James Watt School of Engineering
University of Glasgow
Glasgow
United Kingdom

***Fahad Ahmed***
National University of Sciences and
Technology (NUST)
Islamabad
Pakistan

***Akram Alomainy***
School of Electronic Engineering and
Computer Science
Queen Mary University of London
London
United Kingdom

***Ayman Althuwayb***
Jouf University
Sakaka
Aljouf
Kingdom of Saudi Arabia

***Kamran Arshad***
College of Engineering and IT
Ajman University
Ajman
United Arab Emirates

***Islam Safak Bayram***
University of Strathclyde
Glasgow
United Kingdom

***Nuno Carvalho***
Universidade de Aveiro
Aveiro
Portugal

***Ricardo Correia***
Universidade de Aveiro
Aveiro
Portugal

***Yuan Ding***
Herriot-Watt University
Edinburgh
United Kingdom

***George Goussetis***
Herriot-Watt University
Edinburgh
United Kingdom

***Muhammad Ali Imran***
James Watt School of Engineering
University of Glasgow
Glasgow
United Kingdom

***Muhammad Ismail***
Tennessee Tech University
Cookeville
TN
USA

***Raka Jovanovic***
Hamad bin Khalifa University
Doha
Qatar

***Jalil Kazim***
University of Glasgow
Glasgow
United Kingdom

***Romwald Lihakanga***
Herriot-Watt University
Edinburgh
United Kingdom

***Muhammad Qasim Mehmood***
Information Technology University
Lahore
Pakistan

***Jun Ouyang***
School of Electronic Science and Engineering
University of Electronic Science and Technology of China
Chengdu
China

***Abdul Quddious***
KIOS Research Center
University of Cyprus
Nicosia
Cyprus

***Junsuk Rho***
Pohang University of Science and Technology
Pohang
South Korea

***Syed Aziz Shah***
Faculty Research Centre for Intelligent Healthcare
Coventry University
Coventry
United Kingdom

***Abubakar Sharif***
School of Electronic Science and Engineering
University of Electronic Science and Technology of China
Chengdu
China

and

Department of Electrical Engineering
Government College University Faisalabad
Faisalabad
Pakistan

***Nosherwan Shoaib***
National University of Sciences and Technology (NUST)
Islamabad
Pakistan

***Chaoyan Song***
Herriot-Watt University
Edinburgh
United Kingdom

***Dmitry Zelenchuk***
The Centre for Wireless Innovation (CWI)
Institute of Electronics
Communications and Information Technology (ECIT)
Queen's University of Belfast
Belfast
United Kingdom

***Muhammad Zubair***
Information Technology University
Lahore
Pakistan

# 1

# Intelligent Reflective Surfaces – State of the Art

*Jalil ur Rehman Kazim, Hasan T. Abbas, Muhammad Ali Imran, and Qammer H. Abbasi*

*James Watt School of Engineering, University of Glasgow, Glasgow, United Kingdom*

## 1.1 Background

The era of wireless connectivity of individuals from connected laptops and smartphones has moved further beyond to connected "things." We are going toward connected cars, smart homes, smart cities, and connected wearables. This ubiquitous connectivity aims at improving the quality of life and enhancing productivity. Any entity connected to the Internet is referred to be a part of "Internet of Things" (IoT). This includes a simple temperature sensor that uploads information on a weather server to a more sophisticated smartwatch, monitoring the vital signs of patients. A subclass of the IoT family is labeled as IoT sensors. The IoT sensor network plays a key role in gathering important data that are used for analysis and predictive purposes. These are designed to be low cost and low power, hence sometimes referred to as battery-free devices. The lower power is achieved via ambient backscatter communication that does not include active radio frequency (RF), components such as mixers and/or power amplifiers. Information is transmitted via reflected waves by modulating the incident RF signal [1]. On the contrary, backscatter communication suffers from direct-link interference and might not be suitable in a noisy environment. The information could be lost if the signal-to-noise ratio (SNR) is below the threshold due to round-trip attenuation. As a result, the ambient backscatter communication may be restricted to operate in a limited range. To extend the communication range of the IoT sensors, a potential solution is to place active relays in their vicinity. Relays have active components that receive weak signals, amplify them, and retransmit them to the destination. But this would increase the overall network cost and utilize more energy. A novel solution will be to use a system with low cost, high gain but with no active components.

## 1.2 Introduction to Reflective Surfaces

The field of electromagnetics describes the spatial and temporal behavior of the electric and magnetic field. In free space, the temporal aspect of the electromagnetics gives rise to the classification of the electromagnetic spectrum such as RF, microwave, millimeter

*Backscattering and RF Sensing for Future Wireless Communication,* First Edition.
Edited by Qammer H. Abbasi, Hasan T. Abbas, Akram Alomainy, and Muhammad Ali Imran.

waves, infrared, visible light, ultraviolet, X-rays, and nuclear waves, i.e., alpha, beta, and gamma rays. Similarly, the spatial behavior is studied when the electromagnetic (EM) waves interact with conductors and dielectrics, giving rise to phenomena such as reflection, refraction, diffraction, dispersion, and absorption. These properties have always fascinated human beings, e.g., the dispersion of light from a dielectric surface such as a water droplet in the air making a rainbow, similarly, the bending of a solid object in water due to refraction of light and so on. The phenomena of total internal reflection of light in a dielectric medium resulted in the invention of optical fiber, which is used as a backbone in the optical communication system. At the lower end of the spectrum involving radio waves and microwaves, the study of reflection of these waves from conductors led to the invention of parabolic dish antennas. These curved surfaces are now being used for the transmission and reception of waves over long distances. Hence, reflective surfaces whether conductive or dielectric play vital role in wireless communication system till date.

## 1.3 Intelligent Reflective Surface

A novel idea of a smart reflector that is gaining a lot of interest is the reconfigurable intelligent surface (RIS), which operates by reflecting the waves in the same way as a backscattering device. It consists of planar, low cost, nearly passive elements that would smartly manipulate EM waves in specific directions. As a result, this would boost the signal at the receiver with no added hardware cost. As compared to existing complex transceiver architectures with multiple RF chains and complex signal processing hardware, RIS practically has no RF chain, i.e., no power amplifier, phase shifters, attenuators, mixers, and analog-to-digital converter (ADC) components. This, in turn, eliminates the hardware complexity and the exorbitant cost of the RF equipment. Hence, the overall potential impact of RIS is reduced deployment and hardware costs, network complexity, and power consumption. Researchers in [2–6] have envisioned a smart radio environment, which will be able to use the RIS as a software-controlled entity to recycle the existing signals instead of generating a new one. An RIS-assisted backscattering communication network is depicted in Figure 1.1, which reflects the incoming waves from the source toward the IoT device without an additional RF source.

In various published studies, these smart reflectors have been addressed by various names, i.e., smart reflect-arrays [7], large intelligent surface (LIS) [8], large intelligent metasurface (LIM) [9], reconfigurable metasurface [10], reconfigurable intelligent surface [11], software-defined surface (SDS) [12], software-defined metasurfaces (SDMs) [13], passive intelligent surface (PIS) [14], and passive intelligent mirrors [15]. The initial concept of an active reflective surface was proposed in [16] using a computer simulation. The reflective surface that was modeled as a frequency-selective surface was controlled via positive-intrinsic-negative (PIN) diodes. It was noted that the active surface was efficient in enhancing coverage and quality of service. In [7], a reflective surface was proposed using varactor diodes to provide a continuous phase shift. A disadvantage of varactor diodes is that the response time is very large and achieving phase accuracy is extremely challenging.

In the aforementioned research works, the main focus was to electronically steer the beam of the incoming signal. With the desire to achieve multifunctionality from the

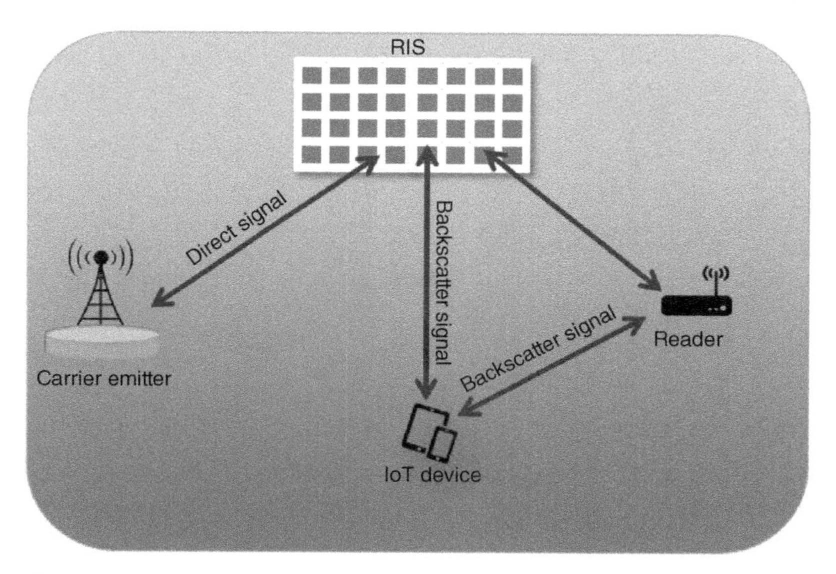


**Figure 1.1** RIS-assisted backscatter communication. RIS, reconfigurable intelligent surface.

reflective surface, the concept of digitally controlled metasurfaces [17, 18] has been proposed and gained significant attention from the research community. This emerging concept has eventually converged the information science domain with EM domain giving rise to programmable metasurfaces [17].

### 1.3.1 Theory of Operation

The incoming wave upon incidence on the medium is reflected from a perfect conductor and at the same time transmitted into the medium creating a refracted wave if it is a dielectric medium. A uniform metallic plate has a constant surface impedance. Hence, the reflection from a metallic surface is explained by Snell's law [19]. In case of a uniform, infinite conductive planar surface, the reflected wave has a 180° phase difference with the incident wave [20]. Hence, any uniformly large conducting surface will invert the phase of an incoming wave by 180°. An alteration to the surface impedance of conductor results in wave reflected in a different direction. For instance, the RIS consists of "$N$" unit cells arranged periodically in a two-dimensional (2D) manner. The metasurface properties of RIS are exploited by keeping the interelement spacing among the unit cells less than $\lambda/2$ and greater than $\lambda/8$, where "$\lambda$" is the wavelength of the signal. Each unit cell independently acts as a scatterer and/or a point source by itself, thus capturing the incoming wave and retransmitting it with some controllable phase. The reflector is "intelligent" in the sense that the phase can be controlled using an external circuit. The adjustment of the phase profile on the surface of the RIS depends upon the EM source being near or far from the RIS. An EM source generates spherical waves when excited. The spherical wave upon propagation converts into a plane wave at a distance of $2D^2/\lambda$ from the source [21], where "$D$" is the dimension of the RIS. The boundary beyond the $2D^2/\lambda$ is called the far-field distance, while the area inside the far-field is regarded as the near-field.

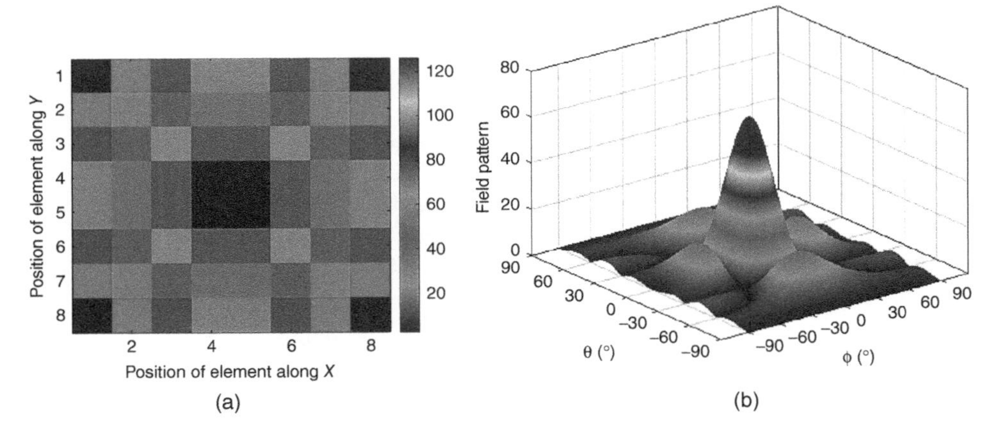


**Figure 1.2** (a) Phase profile for an 8 × 8 metasurface and (b) 3D view of the magnitude of the field pattern. 3D, three-dimensional.

In case of a reflector illuminated by a far-field source the pattern function of the reflector obtained is given in equation 1 from [22] The phase profile and magnitude of the field pattern is plotted in Figure 1.2

$$F(\theta, \varphi) = \sum_{m=1}^{M}\sum_{n=1}^{N} A_{mn} \exp\{j[k_o D_x(m - 1/2)\sin\theta\cos\varphi + k_o D_y(n - 1/2)\sin\theta\sin\varphi + \phi_{m,n}]\} \tag{1.1}$$

where "$D_x$" and "$D_y$" are the interelement distances along the $x$- and $y$-axes. The position of the unit cell is represented by "$m$" and "$n$" along the $x$- and $y$-axes, where "$m$" and "$n$" are greater or equal to 1. Moreover, "$k_o$" is the free space wavenumber; "$\varphi$" and "$\theta$" the azimuth and the elevation angles, respectively, which are specified for steering the beam. "$A_{mn}$" is the reflection amplitude. Similarly, in the scenario where the RIS is placed in the near-field of the transmitter, the pattern function for even number of unit cells becomes

$$F(\theta, \varphi) = \sum_{m=1}^{M}\sum_{n=1}^{N} A_{mn} \exp\{j[\varphi'(m, n) + kD_x(m - 1/2)\sin\theta\cos\varphi + kD_y(n - 1/2)\sin\theta\sin\varphi]\} \tag{1.2}$$

where $\varphi'(m, n)$ is the spatial phase delay required for the phase compensation from the near-field source placed at a distance "$Z$" and expressed as

$$\varphi'(m, n) = k\left[\sqrt{(Z^2 + (D_x(m - 1/2))^2 + (D_y(n - 1/2))^2)}\right] - Z^2 \tag{1.3}$$

The aperture reflector due to its parabolic shape acts as a lens and focuses the incoming beams onto a single point. As the beams converge to that point, the energy of the signal is amplified. The RIS enhances the signal in the same way as the conventional aperture reflector. The gain provided by the RIS is proportional to the effective aperture size. For simplicity, let us assume an RIS operating in millimeter-wave band, e.g., 30 GHz with an area of 1 m$^2$; hence, the theoretical gain achieved for maximum aperture efficiency will be

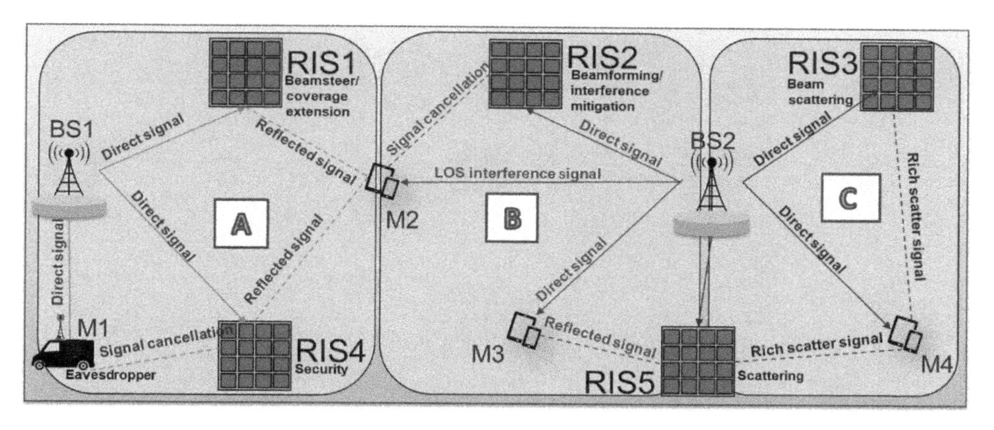


**Figure 1.3** RIS deployment in B5G scenarios. RIS, reconfigurable intelligent surface.

around 50 dB. Similarly, it was revealed in [4] that the received power is proportional to the square of the number of elements on the metasurface, i.e., a 1 $m^2$. RIS with 100 elements will enhance the received power by approximately 40 dB.

## 1.4 Potential Use Cases

The ability of the RIS to improve the received signal strength at the receivers was shown in an indoor environment using simulation [3]. It was shown that the average power received over 12 receivers without the RIS was −75 dBm with the minimum received power at −250 dBm, while in the presence of RIS, the average power received was around 20.6 dBm with a minimum received power of 12.4 dBm. Hence, the RIS can find many potential scenarios where it can be deployed in an indoor environment. As an assisted technology in 5G and/or beyond 5G (B5G), it can be deployed indoor and outdoor with emphasis to communication scenario and/or future health care.

### 1.4.1 Application of RIS in Communication

The RIS, due to its ability to manipulate the EM signals, can be used in different application scenarios in wireless communication. In Figure 1.3, four different application scenarios are shown. In scenario A, the RIS1 is responsible to enhance the coverage of the EM signal from the base station BS1 to the mobile node M2. The RIS1 can collect random signals from the BS1 and focus it toward the M2. In scenario B, the RIS2 acts as an interference cancellation surface. It can send a phase-inverted signal toward M2 to cancel the unwanted BS2 signal reaching M2. Similarly, in scenario A, the RIS4 can use the signal inversion technique to avoid the eavesdropper M1 from hacking the signal of BS1. In scenario C, the RIS3 and RIS5 can act as scattering surfaces to generate a rich scattering environment for M3 and M4 to use multiple-input-multiple-output (MIMO) communication.

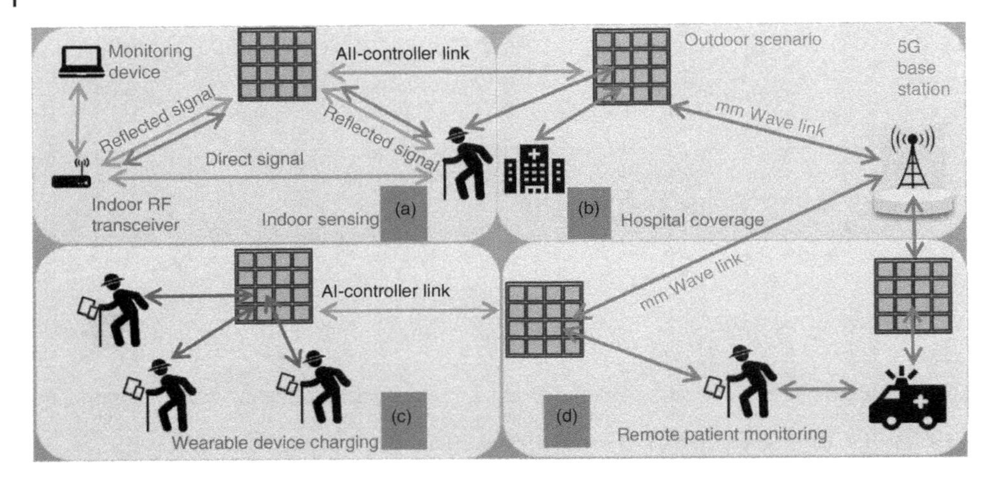


**Figure 1.4** RIS deployment in assistive health care scenarios. RIS, reconfigurable intelligent surface.

### 1.4.2 Application of RIS in Health Care

Apart from a communication perspective, the RIS can also contribute to future health care scenarios. This is illustrated in Figure 1.4. With low-latency and high-speed 5G links, the RIS can be effectively used for indoor sensing applications to monitor various activities of patients that require constant assistive care service. As the RIS can reflect the signals from the RF transceivers, vital signs monitoring of these patients could be easily performed in a noninvasive way [23]. Similarly, in scenario D, outdoor monitoring of elderly patients can be carried out using direct 5G signals. In case of an emergency, the RIS could be able to send beacon signals to the main control room and an ambulance could be called for without any human-assisted call. An important case using the RIS can be the wireless charging of devices [24] as shown in scenario C. The focused signals from the RIS could be directed toward the wearable device worn on the patient wrists or the on-body sensor which are usually low-powered. This would be useful for dementia patients, which involve constant human presence for their care.

## 1.5 RIS as an Alternative to Relay

Similar technologies to the RIS include an active relay station. The job of the relay station is to receive, process, amplify, and retransmit the information. This comes with two major problems. First, a typical active relay usually processes the received signal before transmitting and hence requires complex front-end signal-processing circuits. Furthermore, the relay contains power amplifiers that amplify the captured signal before retransmission, thereby increasing the overall network cost. Thus, the relay station must have a dedicated power source to drive all the active components. Second, the active relay can be operated in two duplexing modes. A typical relay is operated in a half-duplex mode to avoid the extra complex processing. A drawback in the half-duplex mode is that it becomes less spectral efficient. On the contrary, the RIS does not employ any transmitter components and uses the

**Table 1.1** Performance comparison of various technologies compared to RIS.

| Technology | Reflection mechanism | Hardware cost | Energy consumption | Duplex | RF chain | Role |
|---|---|---|---|---|---|---|
| Backscatter | Passive | Low | Low | Full | 0 | Source |
| MIMO relay | Active | Very high | High | Half/full | $M$ | Assist |
| Massive MIMO | Active | Very high | Very high | Half/full | $M$ | Source |
| RIS | Passive/active | Low | Low | Full | 0 | Source/assist |

MIMO, multiple-input-multiple-output; RIS, reconfigurable intelligent surface.

aperture size to achieve the high gain and operates in full-duplex mode. On the other hand, the full-duplex mode of operation in a relay is accompanied with the loop-back interference at the relay node and co-channel interference problems at the receivers. A simulation model comparing the RIS and the relay is discussed in [25], where it is shown that for a higher spectrum efficiency rate "$R$," the transmitted power required is less in the RIS under the decode and forward (DF) relaying protocol over a distance of 80 m. In terms of energy efficiency, the DF relay protocol outperforms RIS. But the author in [11] has shown that RIS outperforms the amplify–forward (AF) relay protocol in terms of energy efficiency. An experimental result of the comparison between RIS and relay, however, is yet to take place. A general comparison of RIS with other technologies is given in Table 1.1.

## 1.6 Coded and Programmable RIS

The concept of digital and programmable metamaterials was presented in [17, 18]. Existing metamaterial-based structures are characterized based on the macroscopic medium parameters such as the effective permeability and permittivity, which can be homogeneous or inhomogeneous. Hence, they can be regarded as analogue metamaterials that target the bulk property of the respective surface. In contrast, coded metasurfaces assign each meta-unit cell a binary digit. The phase of a metasurface is then quantized and replaced with a coding sequence, e.g., a simplest coding sequence is 1-bit. The phase distribution on the surface is rounded to either 0° or 180° and is digitally represented with two states, i.e., 0 and 1. An example is shown in Table 1.2 for 2-bit and 3-bit sequences. Various 1-bit elements have been investigated in [26–31]. Using 1-bit elements, only two-phase states can be achieved, i.e., 0° and 180°. However, such a low-phase quantization value results in an almost 3 dB loss in the antenna gain [32]. Increasing the phase quantization increases the overall hardware complexity but enhances the gain and vice versa. For instance, to implement an $n$-bit coding pattern on the metasurface, "$n$" number of routing lines will be required to the unit cell, e.g., 3-bit coding pattern would require three input bias line for each cell. If the metasurface consists of 100 unit cells, the total routing lines to the control circuit would be 300 making the PCB implementation highly complex and multilayer, especially at higher frequencies. Thus, the author in [33] has shown that a 2-bit phase quantization

**Table 1.2** Digital phase mapping for a coded metasurface.

| Bits/phase | 0° | 45° | 90° | 135° | 180° | 225° | 270° | 315° |
|---|---|---|---|---|---|---|---|---|
| 1-bit | 0 | | | | 1 | | | |
| 2-bit | 00 | | 01 | | 10 | | 11 | |
| 3-bit | 00 | 001 | 010 | 011 | 100 | 101 | 110 | 111 |

is a balance between the power loss and the hardware complexity of the RIS. Further contributions regarding 2-bit elements can be found in [34–36]. A vector synthesis approach to achieve higher bit states using only 2-bit phase quantization is to implement time-varying sequences along the metasurface [37], where it is demonstrated that by switching the phase states in the time domain, 16 phase states can be achieved, thereby eliminating the need for multi-bit implementation on the physical level.

## 1.7 Functionalities of RIS

With regard to a simple reconfigurable surface which only performs a single operation such as beam focusing and/or beam steering, a key feature of the RIS technology is a multi-functional operation which can be realized using different coding sequences to efficiently manipulate the EM waves [38]. Some key functionalities are discussed in the following subsections.

### 1.7.1 Beam Focusing

One of the primary features of the RIS is the ability to focus EM waves at a specified point in space by specifying the azimuth and the elevation angle. If the RIS is placed in the near-field, a phase compensation will be required using equations in Section 3.1 to obtain a far-field pattern. It is important to note that the RIS acts as a lens to focus the beam at any given point and not as a mirror. A plane wave reflected from a mirror also results in a plane wave. In contrast, each element of the RIS reflects a signal with a controlled phase shift to focus the wave at a specific point behaving as a lens.

### 1.7.2 Beam Scattering/Diffusion

Since the World War II, immense research effort has been put into the idea of stealth and radar cross-section (RCS) reduction. The idea is to achieve a surface which is transparent to EM waves and not detected by the enemy radar. Conventionally, this is achieved by either coating the object with a material that can absorb the EM waves or optimizing the object dimension to scatter the energy from its surface. Different coded metasurfaces for RCS reduction have been reported such as the chessboard configuration [39, 40], coding sequences obtained from ergodic algorithm [41], and a 1-bit coding sequence generated using a random optimization algorithm [42]. However, the aforementioned works do not provide real-time control over the coding sequences, and coding patterns cannot be altered once the surface is fabricated. The application of RIS for RCS reduction has been demonstrated in [17]. Instead of fixing the coding sequence, the coded sequences can be altered in real time by programming the metasurface. The optimized coding patterns provided an RCS reduction of −10 dB in the frequency band of 7.8–12 GHz.

### 1.7.3 Multi-Beam Generation

Existing antenna technology deploy beam steering circuits in the transmitter to steer the beams toward users. With more users in a given area, it is desirable to generate multiple beams to serve different users simultaneously in an efficient way. This is accomplished using transmitters equipped with phased array antenna technology [43]. Phased array antennas can generate single as well as multiple beams with very high gains. This requires multiple RF chains with attenuators, phase shifters, and costly power amplifiers. To reduce the cost, a passive multi-beam antenna technology is also used [44]. The high gain is provided by the parabolic reflector, which is fed by a phased antenna array transceiver with low gain. Multiple beams generated by the phased array antenna are reflected off the reflector toward a specific location. The technology is mostly adopted in satellite communication, but it is not applicable in mobile communication as the multiple beams reflected from the aperture cannot be steered toward moving users. Low-cost and high-gain multiple beams can be achieved using the RIS technology with no power-hungry active components. Lately, it has been demonstrated that reflection phases of the individual field patterns, each with its reflected beam direction when added together, results in multi-beam generated from the RIS [45]. This can be explained by the equation,

$$|A_{mn}^1|e^{j\phi_{mn}^1} + |A_{mn}^2|e^{j\phi_{mn}^2} + \cdots + |A_{mn}^M|e^{j\phi_{mn}^M} = Ae^{j\phi_o} \tag{1.4}$$

where "$A_{mn}$" is the reflection amplitude and "$\phi_{mn}$" is the reflection phase of the $m$-th and $n$-th element corresponding to an individual "$M$-th" field pattern. If only beam focusing and/or beam-steering is desired, the value of "$A_{mn}$" is kept one. Thus, only the control over the reflection phases is enough to achieve the required functionalities. The author in [46] has shown that controlling the reflection amplitude results in different energies being transmitted into the multiple beams. Hence, the RIS may function as a spatial beam splitter with power levels being controlled by changing the reflection amplitude of the individual beams.

### 1.7.4 Anomalous Reflection

Another potential ability of RIS is that it can act as an anomalous reflector. When a plane wave bounces off a uniformly flat surface whose surface impedance is constant, it is reflected at an angle equal to the angle of incidence which is regarded as specular reflection [19]. On the contrary, the RIS can reflect a plane wave in a different direction which is described as anomalous reflection. In this case, the impedance is designed to vary along the surface. A plane wave incident on a periodic metallic surface may generate a (i) specular reflection, i.e., reflected waves bouncing opposite to the direction of source (angle of incidence is equal to the angle of reflection), (ii) retroreflection, i.e., the reflected wave bounces back toward the source, and (iii) parasitic direction (undesired reflection). The anomalous reflection that is not predicted by the Snell's law can be generated in a controlled way. The anomalous reflection from an RIS can be generated along the horizontal plane by applying a linear phase gradient along the surface by using the equation [47],

$$n_i k_o[\sin(\theta_r) - \sin(\theta_i)] = \frac{\Delta\varphi}{\Delta x} = \frac{2\pi}{ND} \tag{1.5}$$

$$\Rightarrow \theta_r = \arcsin\left(\frac{\lambda_o}{ND}\right) \quad \text{where } N > 3 \tag{1.6}$$

where "$n_i$" is the refractive index of the medium and $k_o$ is the free space wavenumber given by "$2\pi/\lambda_o$". Hence, the reflection angle is determined by the dimension of the unit cell "$D$", which is normally taken as interelement spacing among the elements along the $x$-direction and "$N$" is the number of elements in a super unit cell. A typical design scenario would require calculating the number "$N$" to generate anomalous reflection. Also, the azimuthal angle "$\phi_r$" is computed as

$$\varphi_r = \arctan\left[\frac{\Delta\varphi_y/D_y}{\Delta\varphi_x/D_x}\right] \tag{1.7}$$

For simplicity, if $D_x = D_y$ and $\Delta\phi_x = \Delta\phi_y$, then $\phi_r$ is computed to be 45°.

### 1.7.5 RIS as a Single RF Chain Transmitter

One of the remarkable properties of the RIS is that it can be used as a transmitter if operated in the near-field region. A horn antenna connected to a power amplifier band is used to generate a single tone carrier signal to illuminate the RIS [48]. With the capability to control the phase of each unit cell, a phase-modulated reflected wave can be generated by the RIS. A binary frequency shift keying (BFSK) system was presented in [49]. A first successful prototype that implements an 8PSK modulation scheme is presented in [50]. A prototype of RIS for QPSK is given in [51]. The conventional architecture of the RIS can inherently vary the phase of the unit cell through control voltages. This, in turn, affects the amplitude of the unit cells also. Consequently, the implementation of high-order modulations such as QAM and OFDM requires complex unit cells with independent control of amplitudes as well. Normally, varying the phase also changes the amplitude of the unit cell as they are decoupled to one another. A remarkable method presented in [52] uses a constant envelope, nonlinear modulation technique to solve this problem. An experimental prototype developed by implementing 16 QAM modulation using the RIS was demonstrated in [53]. Furthermore, another experimental prototype of RIS transmitter operating as a 2*2 MIMO is given in [48].

## 1.8 Architecture

The proposed architecture of RIS consists of three main important parts as given in Figure 1.5. This includes the EM layer, direct current (DC) control layer, and a controller unit. All the three components of the RIS must be modeled, integrated, and optimized separately to obtain full the functionality of the hardware. The subparts of the different layers of RIS are briefly explained in the sections below.

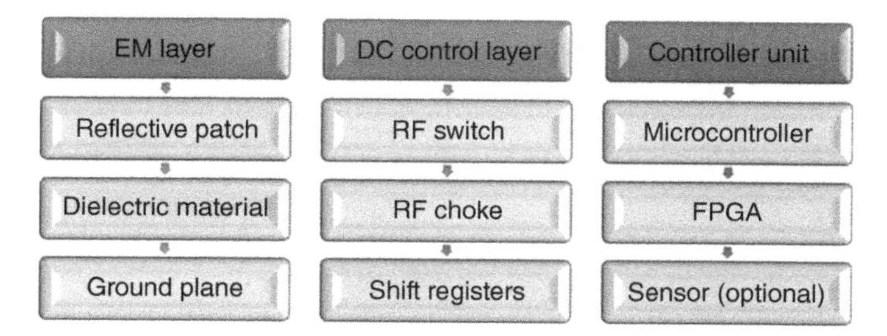


**Figure 1.5** Typical hardware layout of the RIS. RIS, reconfigurable intelligent surface.

**Figure 1.6** EM layer consisting of reflective patches on the top layer backed by DC control layer at the back.

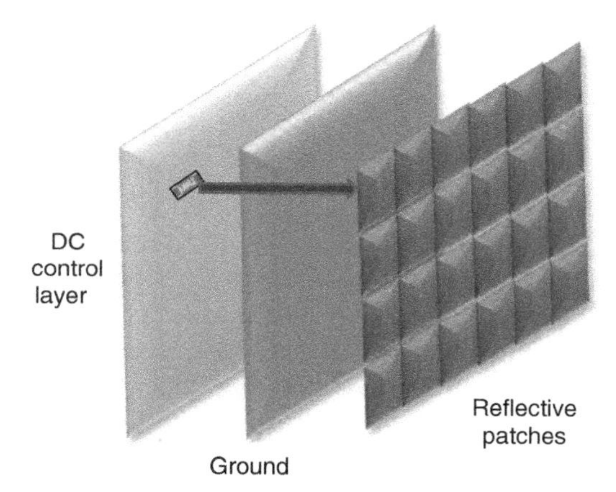


### 1.8.1 Electromagnetic Layer

This is the topmost layer in the RIS. The EM layer is illustrated in Figure 1.6. It consists of metallic patches made of copper. As compared to conventional reflectors elements that have half-wavelength spacing, the RIS elements will have sub-wavelength dimensions and interelement spacing. Hence, they are nonresonant elements and are, therefore, called meta-elements. The meta-elements will be placed in a periodic 2D manner on the surface. These reflective elements are backed by a ground plane that is responsible to reflect the incoming EM wave.

The important aspect of the EM layer design is the selection of the dielectric material. A thicker dielectric substrate with the lower dielectric constant is considered suitable as it provides enhanced impedance bandwidth [54]. On the contrary, the reflection efficiency tends to decrease by using a thicker dielectric. This can be increased by using interelement spacing below the half-wavelength [55].

### 1.8.2 DC Control Layer

An overview of DC control layer is given in Figure 1.7. The reconfigurability of the reflective surface is achieved by changing the amplitude and phase properties of the meta-unit cells. In existing literature, two types of control methods have been implemented, i.e., analogue and digital. An analogue control, which normally consists of varactor diodes [56],

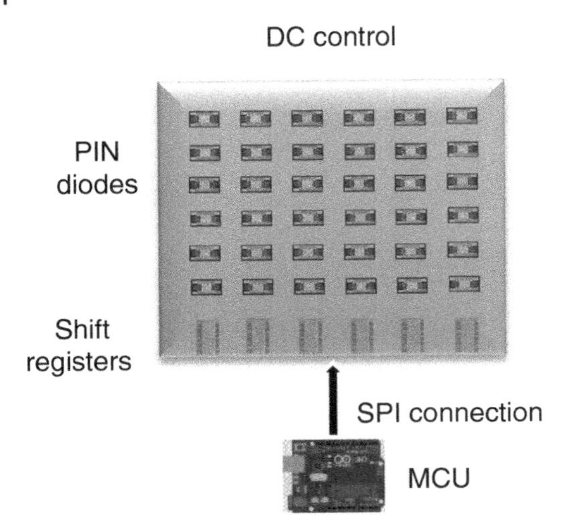


**Figure 1.7** The layout of the DC control layer.

barium–strontium–titanate (BST) capacitors [57], and/or liquid crystal technology [58], has high-resolution tuning ability to control the phase of the unit cell in the EM layer. However, BST capacitor and liquid crystal are not commonly available technologies and require complex fabrication technique to be embedded in the reflective surface. The varactor diode, on the other hand, is easily available but requires higher biasing voltages. Furthermore, separate digital-to-analogue chips are also required to control it via the controller unit. On the other hand, a digital control method using PIN diodes [38], micro-electro-mechanical systems switches [59], and/or relays [60] could be adapted for discrete control of the EM layer and are commercially available. The field-effect transistors are seldom used in for RF switching especially in the reflective surface as they have a very high insertion loss at higher frequencies as compared to the PIN diode [61]. Similarly, relay switches having higher footprint cannot be integrated on RIS printed circuit board (PCB) if multi-bit control is required. PIN diode has the advantage of a lower RF insertion loss, and as a two-terminal device, it is widely used in existing switching technologies. Performance comparison of various switches is given in Table 1.3.

### 1.8.3 Controller Board Interfacing

The intelligence to the reconfigurable surface will be provided using the controller board that can be a simple microcontroller unit or field programmable gate arrays. As proposed in the previous section, a programmable RIS can be achieved with the help of PIN diodes controlled by the controller unit. The controller unit will send digital signals to switch the PIN diodes changing the phase properties of the surface. An efficient algorithm that might incorporate machine learning can be embedded in the RIS to reduce the real-time computational complexity of the hardware. The controller board could be integrated with off-the-shelf sensor platforms, such as temperature sensors, acoustic sensors, optical sensor and so on, to make the RIS self-adaptable and more aware of the environment in which the RIS is operated.

**Table 1.3** Performance comparison of digital switching devices [62].

| Parameters | PIN-diode-based RF switch | FET-based RF switch | Ferrite-based RF switch | MMIC-based RF switch | RF MEMS switch |
|---|---|---|---|---|---|
| Weight (oz) | Light (0.5–1) | Light (<1) | Heavy (1–9) | Light (~0.01) | Light |
| Size ($mm^2$) | Small (1–5) | Small (0.1) | Large (~few cm) | Small (2–3) | Small (~nm/μm) |
| Cost | Low | Low | Very high | Low | High cost |
| Availability | Commercially available | Commercially available | Complex in realization | Commercially not available | Commercially not available |
| DC power consumption (mW) | 5–100 | 0.05–0.1 | 10–50 | 0.05–0.1 | 0.05–0.1 |
| Insertion loss (dB) dB (1 GHz) | 0.3–1 | 0.4–2.5 | 0.5–1.5 | 0.8–2 | 0.05–0.2 |
| Power-handling capability (W) | ~kW in pulse mode; ~200 W in CW mode | <10 | >100 | <1 | <1 |
| Switching time (s) | 200–800 ns | 1–100 ns | 1–20 μs | 25–100 ns | 1–300 μs |

FET, field-effect transistor; MEMS, micro-electro-mechanical systems; MMIC, monolithic microwave integrated circuit; RF, radio frequency.
Source: Chakraborty and Gupta [62]. © 2016, IEEE.

## 1.9 Summary

In this chapter, a comprehensive overview of the RIS was presented. The theory of operation of RIS was given, which is similar to the aperture antennas. Various potential application scenarios of the RIS in future wireless communication and health care technologies were shown. The differences between RIS with relay technology were also discussed, and a comparison of RIS with existing technologies was shown. A new class of emerging metasurface, i.e., coded and programmable metasurface, will consequently bridge the gap between the digital domain and the electromagnetic domain. Some of the published works in this area have been briefly discussed. Various important functionalities associated with RIS are also discussed in detail. A significant application of the RIS operated in the near-field as a transmitter is presented. Finally, we have introduced a typical architecture of the RIS with three basic layers, and role and key features of every sublayer have been discussed in detail.

## References


1 Boyer, C. and Roy, S. (2013). Invited paper—backscatter communication and RFID: coding, energy, and MIMO analysis. *IEEE Transactions on Communications* 62 (3): 770–785.

2 Subrt, L. and Pechac, P. (2012). Intelligent walls as autonomous parts of smart indoor environments. *IET Communications* 6 (8): 1004–1010.

3 Liaskos, C., Nie, S., Tsioliaridou, A. et al. (2018). A new wireless communication paradigm through software-controlled metasurfaces. *IEEE Communications Magazine* 56 (9): 162–169.

4 Basar, E., Di Renzo, M., De Rosny, J. et al. (2019). Wireless communications through reconfigurable intelligent surfaces. *IEEE Access* 7: 116753–116773.

5 Wu, Q. and Zhang, R. (2020). Towards smart and reconfigurable environment: intelligent reflecting surface aided wireless network. *IEEE Communications Magazine* 58 (1): 106–112.

6 Gacanin, H. and Di Renzo, M. (2020). *Wireless 2.0: towards an intelligent radio environment empowered by reconfigurable meta-surfaces and artificial intelligence.* arXiv preprint. arXiv:2002.11040.

7 Tan, X., Sun, Z., Jornet, J.M., and Pados, D. (2016). Increasing indoor spectrum sharing capacity using smart reflect-array. *IEEE International Conference on Communications (ICC)*, Kuala Lumpur, Malaysia (22–27 May 2016), 1–6.

8 Huang, C., Zappone, A., Alexandropoulos, G.C., Debbah, M., and Yuen, C. (2018). *Large intelligent surfaces for energy efficiency in wireless communication.* arXiv preprint. arXiv:1810.06934 v1.

9 He, Z. and Yuan, X. (2020). Cascaded channel estimation for large intelligent metasurface assisted massive MIMO. *IEEE Wireless Communications Letters* 9 (2): 210–214.

10 Di Renzo, M. and Song, J. (2019). Reflection probability in wireless networks with metasurface-coated environmental objects: an approach based on random spatial processes. *EURASIP Journal on Wireless Communications and Networking* 2019 (1): 99.

**11** Huang, C., Zappone, A., Alexandropoulos, G.C. et al. (2019). Reconfigurable intelligent surfaces for energy efficiency in wireless communication. *IEEE Transactions on Wireless Communications* 18 (8): 4157–4170.

**12** Basar, E. (2019). *Large intelligent surface-based index modulation: a new beyond MIMO paradigm for 6G*. arXiv preprint. arXiv:1904.06704.

**13** Liaskos, C., Tsioliaridou, A., Nie, S. et al. (2019). An interpretable neural network for configuring programmable wireless environments. *IEEE 20th International Workshop on Signal Processing Advances in Wireless Communications (SPAWC)*, Cannes, France (2–5 July 2019), 1–5.

**14** Mishra, D. and Johansson, H. (2019). Channel estimation and low-complexity beamforming design for passive intelligent surface assisted MISO wireless energy transfer. *ICASSP 2019 IEEE International Conference on Acoustics, Speech and Signal Processing (ICASSP)*, Brighton, United Kingdom (12–17 May 2019), 4659–4663.

**15** Huang, C., Zappone, A., Debbah, M., and Yuen, C. (2018). Achievable rate maximization by passive intelligent mirrors. *2018 IEEE International Conference on Acoustics, Speech and Signal Processing (ICASSP)*, Calgary, AB, Canada (15–20 April 2018), 3714–3718.

**16** Subrt, L. and Pechac, P. (2012). Controlling propagation environments using intelligent walls. *2012 6th European Conference on Antennas and Propagation (EUCAP)*, Prague, Czech Republic (26–30 March 2012), 1–5.

**17** Cui, T.J., Qi, M.Q., Wan, X. et al. (2014). Coding metamaterials, digital metamaterials and programmable metamaterials. *Light: Science & Applications* 3 (10): e218–e218.

**18** Della Giovampaola, C. and Engheta, N. (2014). Digital metamaterials. *Nature Materials* 13 (12): 1115–1121.

**19** Young, H.D., Freedman, R.A., and Ford, A.L. (2013). *University Physics with Modern Physics Technology Update*. Pearson Education.

**20** Sadiku, M.N. (2014). *Elements of Electromagnetics*. Oxford university Press.

**21** Balanis, C.A. (2016). *Antenna Theory: Analysis and Design*. John Wiley & Sons.

**22** Wan, X., Qi, M.Q., Chen, T.Y., and Cui, T.J. (2016). Field-programmable beam reconfiguring based on digitally-controlled coding metasurface. *Scientific Reports* 6: 20663.

**23** Shah, S.A. and Fioranelli, F. (2019). RF sensing technologies for assisted daily living in healthcare: a comprehensive review. *IEEE Aerospace and Electronic Systems Magazine* 34 (11): 26–44.

**24** Tran, N.M., Amri, M.M., Park, J.H. et al. (2019). A novel coding metasurface for wireless power transfer applications. *Energies* 12 (23): 4488.

**25** Björnson, E., Özdogan, Ö., and Larsson, E.G. (2019). Intelligent reflecting surface vs. decode-and-forward: how large surfaces are needed to beat relaying? *IEEE Wireless Communications Letters* 9 (2): 244–248.

**26** Kamoda, H., Iwasaki, T., Tsumochi, J. et al. (2011). 60-GHz electronically reconfigurable large reflectarray using single-bit phase shifters. *IEEE Transactions on Antennas and Propagation* 59 (7): 2524–2531.

**27** Bayraktar, O., Civi, O.A., and Akin, T. (2012). Beam switching reflectarray monolithically integrated with RF MEMS switches. *IEEE Transactions on Antennas and Propagation* 60 (2) PART 2: 854–862.

28 Carrasco, E., Barba, M., and Encinar, J.A. (2012). X-band reflectarray antenna with switching-beam using PIN diodes and gathered elements. *IEEE Transactions on Antennas and Propagation* 60 (12): 5700–5708.

29 Montori, S., Cacciamani, F., Gatti, R.V. et al. (2015). A transportable reflectarray antenna for satellite Ku-band emergency communications. *IEEE Transactions on Antennas and Propagation* 63 (4): 1393–1407.

30 Zhang, M.T., Gao, S., Jiao, Y.C. et al. (2016). Design of novel reconfigurable reflectarrays with single-bit phase resolution for Ku-band satellite antenna applications. *IEEE Transactions on Antennas and Propagation* 64 (5): 1634–1641.

31 Yang, H., Yang, F., Cao, X. et al. (2017). A 1600-element dual-frequency electronically reconfigurable reflectarray at X/Ku-band. *IEEE Transactions on Antennas and Propagation* 65 (6): 3024–3032.

32 Wu, Q. and Zhang, R. (2020). Beamforming optimization for wireless network aided by intelligent reflecting surface with discrete phase shifts. *IEEE Transactions on Communications* 68 (3): 1838–1851.

33 Yang, H., Yang, F., Xu, S. et al. (2017). A study of phase quantization effects for reconfigurable reflectarray antennas. *IEEE Antennas and Wireless Propagation Letters* 16: 302–305.

34 Pereira, R., Gillard, R., Sauleau, R. et al. (2010). Four-state dual polarisation unit-cells for reflectarray applications. *Electronics Letters* 46 (11): 742–743.

35 Pereira, R., Gillard, R., Sauleau, R. et al. (2012). Dual linearly-polarized unit-cells with nearly 2-bit resolution for reflectarray applications in X-band. *IEEE Transactions on Antennas and Propagation* 60 (12): 6042–6248.

36 Yang, X., Xu, S., Yang, F., and Li, M. (2017). A novel 2-bit reconfigurable reflectarray element for both linear and circular polarizations. *Proceedings of the 2017 IEEE Antennas and Propagation Society International Symposium*, San Diego, CA, USA (9–14 July 2017), vol. 2017, 2083–2084.

37 Zhang, L., Wang, Z.X., Shao, R.W. et al. (2020). Dynamically realizing arbitrary multi-bit programmable phases using a 2-bit time-domain coding metasurface. *IEEE Transactions on Antennas and Propagation* 68 (4): 2984–2992.

38 Yang, H., Cao, X., Yang, F. et al. (2016). A programmable metasurface with dynamic polarization, scattering and focusing control. *Scientific Reports* 6 (1): 35692.

39 Esmaeli, S. and Sedighy, S. (2015). Wideband radar cross-section reduction by AMC. *Electronics Letters* 52 (1): 70–71.

40 Paquay, M., Iriarte, J., Ederra, I. et al. (2007). Thin AMC structure for radar cross-section reduction. *IEEE Transactions on Antennas and Propagation* 55 (12): 3630–3638.

41 Liu, X., Gao, J., Xu, L. et al. (2016). A coding diffuse metasurface for RCS reduction. *IEEE Antennas and Wireless Propagation Letters* 16: 724–727.

42 Ali, L., Li, Q., Ali Khan, T. et al. (2019). Wideband RCS reduction using coding diffusion metasurface. *Materials* 12 (17): 2708.

43 Mailloux, R.J. (2017). *Phased Array Antenna Handbook*. Artech House.

44 Regier, F.A. (1992). The ACTS multibeam antenna. *IEEE Transactions on Microwave Theory and Techniques* 40 (6): 1159–1164.
45 Wu, R.Y., Shi, C.B., Liu, S. et al. (2018). Addition theorem for digital coding metamaterials. *Advanced Optical Materials* 6 (5): 1701236.
46 Bao, L., Wu, R.Y., Fu, X. et al. (2019). Multi-beam forming and controls by metasurface with phase and amplitude modulations. *IEEE Transactions on Antennas and Propagation* 67 (10): 6680–6685.
47 Yu, N., Genevet, P., Kats, M.A. et al. (2011). Light propagation with phase discontinuities: generalized laws of reflection and refraction. *Science* 334 (6054): 333–337.
48 Tang, W., Dai, J.Y., Chen, M.Z. et al. (2019). *MIMO transmission through reconfigurable intelligent surface: system design, analysis, and implementation*. arXiv preprint. arXiv:1912.09955.
49 Zhao, J., Yang, X., Dai, J.Y. et al. (2019). Programmable time-domain digital-coding metasurface for non-linear harmonic manipulation and new wireless communication systems. *National Science Review* 6 (2): 231–238.
50 Tang, W., Dai, J.Y., Chen, M. et al. (2019). Programmable metasurface-based RF chain-free 8PSK wireless transmitter. *Electronics Letters* 55 (7): 417–420.
51 Tang, W., Li, X., Dai, J.Y. et al. (2019). Wireless communications with programmable metasurface: transceiver design and experimental results. *China Communications* 16 (5): 46–61.
52 Dai, J.Y., Zhao, J., Cheng, Q., and Cui, T.J. (2018). Independent control of harmonic amplitudes and phases via a time-domain digital coding metasurface. *Light: Science and Applications* 1 (7): 90.
53 Dai, J.Y., Tang, W., Yang, L.X. et al. (2020). Realization of multi-modulation schemes for wireless communication by time-domain digital coding metasurface. *IEEE Transactions on Antennas and Propagation* 68 (3): 1618–1627.
54 Nayeri, P., Yang, F., and Elsherbeni, A.Z. (2018). *Reflectarray antennas: theory*. In: *Designs and Applications*. Wiley Online Library.
55 Pozar, D. (2007). Wideband reflectarrays using artificial impedance surfaces. *Electronics Letters* 43 (3): 148–149.
56 Luo, Z., Long, J., Chen, X., and Sievenpiper, D. (2016). Electrically tunable metasurface absorber based on dissipating behavior of embedded varactors. *Applied Physics Letters* 109 (7): 071107.
57 Sazegar, M., Zheng, Y., Maune, H. et al. (2011). Low-cost phased-array antenna using compact tunable phase shifters based on ferroelectric ceramics. *IEEE Transactions on Microwave Theory and Techniques* 59 (5): 1265–1273.
58 Zhang, F., Zhao, Q., Zhang, W. et al. (2010). Voltage tunable short wire-pair type of metamaterial infiltrated by nematic liquid crystal. *Applied Physics Letters* 97 (13): 134103.
59 Arbabi, E., Arbabi, A., Kamali, S.M. et al. (2018). MEMS-tunable dielectric metasurface lens. *Nature Communications* 9 (1): 812.

**60** Tan, X., Sun, Z., Koutsonikolas, D., and Jornet, J.M. (2018). Enabling indoor mobile millimeter-wave networks based on smart reflect-arrays. *IEEE INFOCOM 2018-IEEE Conference on Computer Communications*, Honolulu, HI, USA (16–19 April 2018), 270–278.

**61** McSpadden, J.O., Fan, L., Chang, K., and Huang, J. (1999). Ka-band beam steering reflectarray study. *IEEE Antennas and Propagation Society International Symposium. 1999 Digest. Held in Conjunction With: USNC/URSI National Radio Science Meeting (Cat. No. 99CH37010)*, Orlando, FL, USA (11–16 July 1999), vol. 3, 1662–1665.

**62** Chakraborty, A. and Gupta, B. (2016). Paradigm phase shift: RF MEMS phase shifters: an overview. *IEEE Microwave Magazine* 18 (1): 22–41.

# 2

# Signal Modulation Schemes in Backscatter Communications

*Yuan Ding[1], George Goussetis[1], Ricardo Correia[2], Nuno Borges Carvalho[2], Romwald Lihakanga[1], and Chaoyun Song[1]*

[1] *Heriot-Watt University, Edinburgh, United Kingdom*
[2] *Universidade de Aveiro, Aveiro, Portugal*

## 2.1 Introduction

Electromagnetic waves are inevitably scattered by various objects when they propagate through the environment, creating a complex, and usually unpredictable, multipath communication channel. Many techniques, such as channel equalization and multicarrier signal modulation schemes, have been developed to combat this adverse channel condition. On the other hand, the scattering phenomenon has also been exploited for a number of important applications. For example, Radar, rapidly developed during the WWII, detects the scattered energy (or signal waveforms) by a target object in the backward direction (namely backscatter), so that the information about the target, including direction, distance, speed, etc., can be derived. The backscatter was later employed by the Radio Frequency Identification (RFID) technology. For the logistic applications, RFID uses backscattered signals, which contain the identification of the tags, to automatically identify and track tags that are attached to various objects. Unlike the barcode, the RFID tags do not need to be within the line of sight of the reader, so they can be embedded within the tracked object. RFID has become a mature technology [1] and has been standardized for different frequency operations and applications, such as widely adopted ISO/IEC 18000 and EPC Gen2.

Apart from being utilized for Radar and RFID, the backscatter has now been employed for communication purpose, especially under the trend of booming Internet of Things (IoT). The world around us is becoming increasingly "intelligent," featuring smart homes, smart buildings, smart factories, smart cities, etc. A key underpinning technology is the IoT networks that are able to sense the environment, convey the information (usually in a wireless fashion), and, in some applications, take actions [2]. Some IoT standards have emerged to accommodate the demands, such as Narrowband IoT (NB-IoT) built upon cellular infrastructure, and standalone LoRa and Sigfox. All these technologies are capable of greatly reducing the transmission power and enhancing the receiving sensitivity (resulting in long communication ranges up to kilometers), but at the cost of low data rates due to ultra-narrow frequency bandwidth occupied. From the power consumption's

*Backscattering and RF Sensing for Future Wireless Communication,* First Edition.
Edited by Qammer H. Abbasi, Hasan T. Abbas, Akram Alomainy, and Muhammad Ali Imran.

perspective, even though the nodes in low-power wide-area network, including NB-IoT, LoRa and Sigfox, typically consume tens of mW (e.g., SODAQ Mbili ATmega 1284P), at least an order of magnitude lower than the power consumption in the current WLAN and cellular networks, this remains the biggest challenge of massive deployment of IoT nodes as the cost associated with the maintenance, e.g., replacing batteries, rockets up. In addition, the manufacturing of the huge amount of batteries contributes a lot to the global carbon footprint and the disposal of wasted batteries is a great concern to the environment since only a fraction of them are able to be recycled. Under this background, backscatter communication technology has become a promising solution to enable wireless data exchange between IoT nodes/tags and centralized readers [3], as it removes the costly and power-hungry RF frontends, e.g., mixers and power amplifiers, which exist in every conventional wireless transmitter. Instead, a backscatter tag relies on modulating the scattered incoming electromagnetic waves. Since backscatter tags do not actively radiate, they generally consume three orders of magnitude lower than the common conventional IoT nodes, more precise tens of μW [4]. This ultra-low-power consumption allows the backscatter tags to operate in passive or semi-passive fashion. Passive tags normally harvest RF energies from a dedicated power source, e.g., from readers in commercial RFID systems, or from ambient RF signals [5, 6], and in some outdoor applications, the solar energy harvester could be included [7]. The semi-passive tags are usually powered by batteries, which are only consumed for control/modulation purpose, instead of power hungry radiation, the battery life can normally last for years.

Although the backscatter communication drastically reduces the power consumption, it suffers from the low data rate and short communication distance. These shortcomings are the results of extremely low power scattered by the tags, in most cases even below the noise floor. Many recent academic researches are focusing on addressing these two issues. One obvious solution is to scatter more energy by increasing the number of tag antennas, and each individual scatter is coordinated so that the signals are combined in-phase at the reader. A passive Van Atta retrodirective array was developed in [8] to achieve the beam-forming gain in the backscatter link, demonstrating an extended communication range to hundreds of meters. The other solution relies on the signal modulations, as it can provide a flexible trade-off among power, data rates, and communication range.

This book chapter is dedicated to the signal modulations in backscatter communications and their associated hardware realization. In Section 2.2, the basic principle of backscatter modulation is introduced and some commonly adopted schemes are discussed, leaving two recently developed more advanced modulation schemes, i.e., Chirp Spread Spectrum (CSS) and multicarrier modulation, in Sections 2.3 and 2.4, respectively. Finally, conclusions are drawn in Section 2.5.

## 2.2 Principles of Backscatter Modulations and Some Common Schemes

### 2.2.1 Amplitude Shift Keying (ASK)

The commercial RFID systems employ the simplest form of the Amplitude Shift Keying (ASK) scheme, which is On-Off-Keying (OOK). For illustration in Figure 2.1, the OOK is

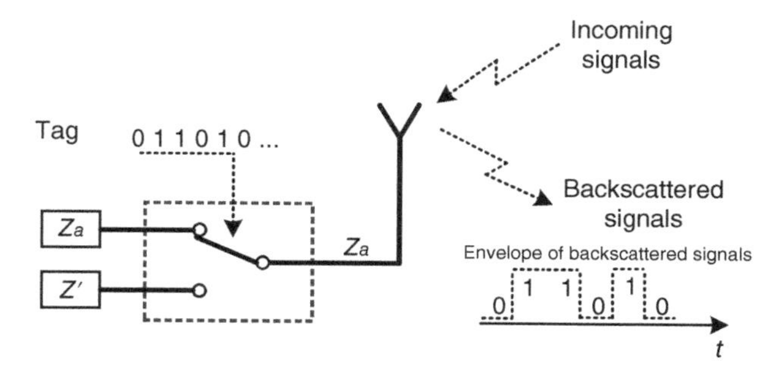


**Figure 2.1** Illustration of a tag backscattering OOK-modulated signals.

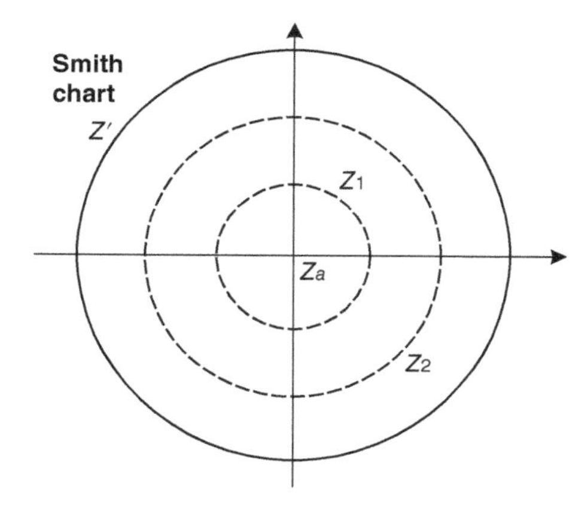


**Figure 2.2** Design guidance on antenna termination impedance for ASK modulation. $Z_a$ and $Z'$ for OOK, and $Z_a$, $Z_1$, $Z_2$, and $Z'$ for 4PAM.

performed by connecting a matching load $Z_a$ or a full-reflection circuit ($Z'$), i.e., open or short, to the tag antenna, so that incoming electromagnetic waves received by the antenna are absorbed or scattered in a sequence that is determined by the data or ID of the tags [9]. It is noted that in this chapter, the structural mode of the backscatter, i.e. the backscatter signals that are independent to the tag antenna loads, is ignored. At the receiving end, the detected waveform is normally processed through an envelope detection circuit, before being digitally sampled and decoded. This 1-bit OOK scheme has been further extended to higher-order ASK modulations, such as the 4PAM in [10]. This higher-order ASK can be implemented using single-pole multi-throw switches or less power consuming devices with controllable impedance, such as transistors. For OOK, the antenna termination impedance, i.e., $Z_a$ and $Z'$, should be designed as shown in Figure 2.2 with maximum separation, while for higher-order ASK, the discrete impedance should be designed to fall onto the reflection coefficient circles in the Smith Chart with equal radius intervals. This design guideline implies that higher order of ASK reduces the Euclidean distance in the magnitude domain, suffering from greater susceptibility to the system noise. Thus, there is a trade-off between data rate and communication distance. One major reason of the popularity of ASK backscatter communication is the simplicity of signal reception, as the envelope detection does not require a frequency and phase reference, i.e., non-coherent detection [11] is sufficient.

For ambient backscatter, namely the incoming signals are not unmodulated continuous waves (CW) but modulated random data [12], ASK modulation can still be applied as long as the ambient signals can be subtracted or be orthogonalized. For example, in [10], the ambient FM signals, upon which 4PAM was applied, were removed by repeatedly averaging in an appropriately selected symbol slot, so that the appended 4PAM signals can be recovered. In [13], the magnitude of wireless channel in commodity OFDM transmissions was backscatter modulated at the packet level, so that the OFDM data and backscatter data are orthogonalized by separating into different domains.

### 2.2.2 Frequency Shift Keying (FSK)

Since modulation in the frequency domain is more resilient to noise compared with modulation in the amplitude domain [14], the Frequency Shift Keying (FSK)-based backscattering modulation was developed for different orders, for example, 2FSK in [15, 16] and 4FSK in [17]. This is commonly achieved by toggling among different loads connected to the tag antenna with a frequency that matches the modulation frequency shifts, see a 2FSK example in Figure 2.3. The selection of loads (say $Z_1$ and $Z_2$ of 2FSK in Figure 2.3), however, is different to the loads for the ASK scheme (say $Z_a$ and $Z'$ of OOK in Figure 2.1). In order to maximize the power of backscattered signals, ideally $Z_1$ and $Z_2$ should be oppositely located on the outmost reflection coefficient circle in the Smith Chart, seen in Figure 2.4, making open and short termination an obvious choice.

Assuming that the incoming signal is a single-frequency ($f_0$) CW, noted as

$$S_{in} = Re(e^{j2\pi f_0 t}), \tag{2.1}$$

the backscattered signal $S_{bs}$, after being modulated by an alternating reflection coefficient $\Gamma$ shown in (2.2), can be expressed as in (2.3).

$$\Gamma(t) = \begin{cases} 1 & \left(\text{when } Z_1 = \infty \text{ is connected, } t \in \left[\frac{n}{\Delta f}, \frac{2n+1}{2\Delta f}\right), \quad n \in \text{integer}\right) \\ -1 & \left(\text{when } Z_2 = 0 \text{ is connected, } t \in \left[\frac{2n+1}{2\Delta f}, \frac{n+1}{\Delta f}\right), \quad n \in \text{integer}\right) \end{cases} \tag{2.2}$$

$$S_{bs} = \Gamma S_{in} \tag{2.3}$$

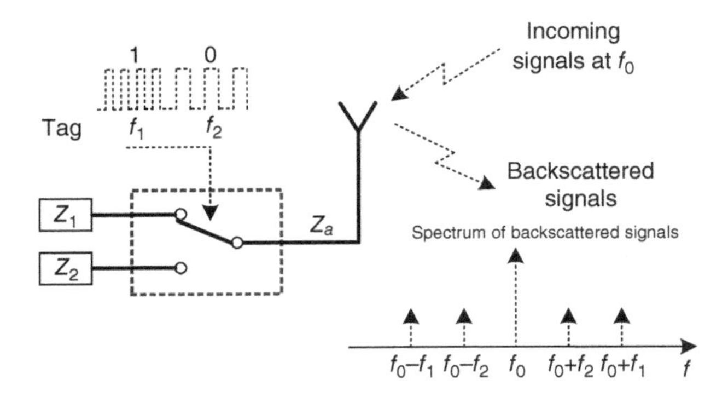


**Figure 2.3** Illustration of a tag backscattering 2FSK-modulated signals.

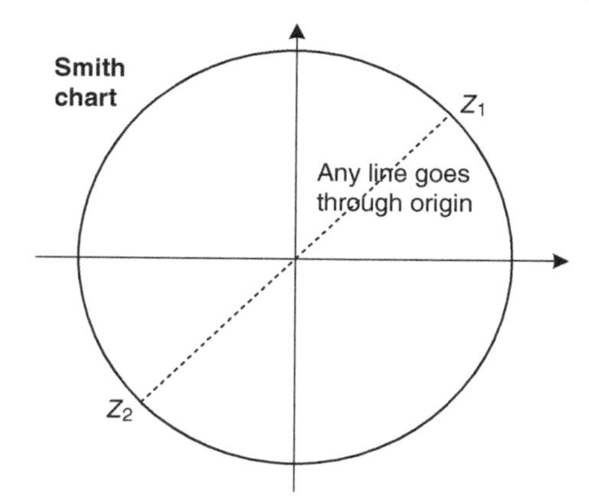


**Figure 2.4** Design guidance on antenna termination impedance for 2FSK modulation.

Since $\Gamma$ in (2.2) is periodic function, it can be expanded as a Fourier series,

$$\Gamma(t) = \frac{4}{\pi} \sum_{m=1,3,5,\ldots}^{\infty} \left[\frac{1}{m} \sin(2m\pi\Delta ft)\right]. \tag{2.4}$$

Thus, substituting (2.1) and (2.4) into (2.3) we get

$$S_{bs} = \frac{4}{\pi} \cos(2\pi f_0 t) \cdot \sum_{m=1,3,5,\ldots}^{\infty} \left[\frac{1}{m} \sin(2m\pi\Delta ft)\right], \tag{2.5}$$

wherein its first harmonic is

$$\begin{aligned} S_{bs}^{(1)} &= \frac{4}{\pi} \cos(2\pi f_0 t) \cdot \sin(2\pi\Delta ft). \\ &= \frac{2}{\pi} \{\sin[2\pi(f_0 + \Delta f)t] - \sin[2\pi(f_0 + \Delta f)t]\} \end{aligned} \tag{2.6}$$

From the derivation in (2.1–2.6), we have the following observations:

- Any periodic switching, at the frequency of $\Delta f$, of a backscatter antenna loading results in the frequency shift of $\Delta f$ of the backscatter signals. For example, this was exploited to shift an out-of-band incoming CW signal to the WiFi band for further modulation [18];
- By mapping the data bits onto the discretely selected $\Delta f$, FSK backscatter modulation is synthesized. For example, by carefully choosing $f_0$ and two $\Delta f$, a Bluetooth Low Energy (BLE) 4.0 compatible (in advertising channels) Gaussian-shaped 2FSK backscatter modulation was generated in [19];
- The baseband component of $S_{bs}$ can be suppressed, e.g. in (2.5), when the load impedance is symmetrically selected in Smith Chart. In this case, the $f_0$ component in backscatter signals is purely contributed by the leakage of the incoming signals through other propagation paths, including the structural scatter mode of the backscatter antenna;
- Backscatter signals generally contain higher harmonics, i.e. $m > 1$, and both sidebands, i.e. harmonics higher than $f_0$ and lower than $f_0$;
- By carefully designing the antenna load impedance, hence the periodic waveforms of the reflection coefficient in time domain, the mirror sideband, and higher-order harmonics of backscatter signals can be purposely manipulated [20].

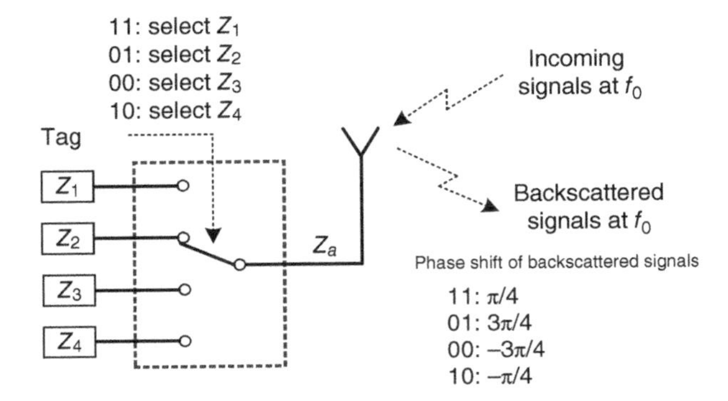


**Figure 2.5** Illustration of a tag backscattering QPSK-modulated signals.

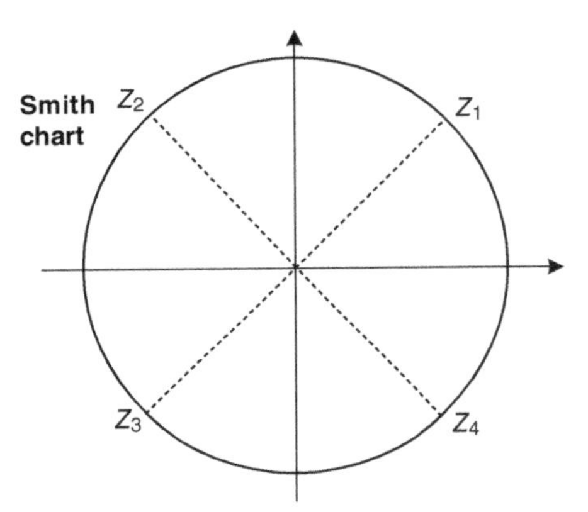


**Figure 2.6** Ideal choice of antenna termination impedance for the example QPSK backscatter modulation in Figure 2.5.

### 2.2.3 Phase Shift Keying (PSK)

The phase of backscatter signals, when the incoming signal is a one-tone CW, is purely determined by the loads terminating the tag antenna. Thus, intuitively, in order to synthesize an $N$-order phase shift keying (PSK) backscatter modulation, $N$ loads uniformly separated on a reflection circle (ideally the outmost circle for maximum power) in the Smith chart are required. A 1-to-$N$ switching circuit, controlled by the data for transmission, is also needed. An example of a QPSK backscatter tag with this arrangement is illustrated in Figure 2.5 with loads selection in Figure 2.6.

However, there are challenges when considering the system and hardware implementation. First, the strong CW signals from the source have to be suppressed at the receiver end before the weak PSK-modulated backscatter signals can be detected and demodulated, since they are all at the same frequency $f_0$, seen in Figure 2.5. Second, low-cost phase coherent detection at receiver end is difficult. Thus, the differential form PSK is commonly used, e.g. DBPSK in [21]; last but not the least, larger $N$ results in higher insertion loss of the 1-to-$N$ switch and the requirement of more loads. Hence, most works have adopted the simplest BPSK (or DBPSK). Though, instead of using switches, a transistor (or multiple transistors) can be used to obtain multiple impedance by varying its bias, it is not a trivial task to maintain a constant envelope of multiple synthesized reflection coefficients [22].

An alternative approach to achieve any $N$-order PSK using only two loads (commonly open and short) was proposed in [22]. With two loads, a square reflection coefficient function $\Gamma(t)$ in (2.2) with frequency of $\Delta f$ is first synthesized. Then, it can be observed that by shifting the time $t$ in (2.2) by $n/(N\Delta f)$, the fundamental component of $\Gamma(t)$ becomes

$$\Gamma(t)\,|_{m=1} = \frac{4}{\pi}\sin\left[2\pi\Delta f\left(t + \frac{n}{N\Delta f}\right)\right] = \frac{4}{\pi}\sin\left(2\pi\Delta ft + \frac{2n\pi}{N}\right). \tag{2.7}$$

As a consequence, the phase of the first upper (or lower) harmonic of the backscatter signal is rotated by $2n\pi/N$ anti-clockwise (or clockwise), achieving the $n$-th constellation symbol in the $N$-order PSK modulation scheme. This method significantly reduces the hardware complexity and associated power consumption, but at the cost of

- Interference to other channels due to the other generated harmonics as the results of the approximation of a square wave to a sinusoidal wave, see (2.4). The interference at the first mirror harmonic frequency should be particularly shifted out of other active channels;
- Wider occupied frequency bandwidth because the backscatter process shifts the frequency to $f_0 \pm \Delta f$ and also generates higher-order harmonics;
- Low data rate since at least $N$ samples are required in a single period of $\Gamma(t)$ function, reducing the data rate by a factor of $N$ when comparing the architecture in Figure 2.5.

### 2.2.4 Quadrature Amplitude Modulation (QAM)

Similar to the PSK modulation, an $M$-ary quadrature amplitude modulation (QAM) scheme can be implemented with $M$ distinct load impedances that are switched across the terminals of the tag's antenna. This can be hardware-consuming, e.g. a 16QAM modulation was implemented with five switches with lumped termination as a 16-to-1 Mux in [23], increasing the associated loss and power consumption (up to mW). In [24], a design compatible with CMOS integration was proposed, wherein the required impedances were synthesized using only resistors and capacitors, reducing the power consumption.

In [25], a backscatter IQ modulator was proposed, consisting of a Wilkinson power splitter and a bias-current-controlled p-i-n diode terminating each branch. Two branches were designed to have a $\pi/4$ phase difference, orthogonalizing the impedance tuning to form $I$ and $Q$ components. The power consumption of this architecture, however, is prohibitively high in the order to tens of mW (e.g. 80 mW in [25]). This is owed to the high current of p-i-n diodes. This issue was addressed by replacing the current-controlled p-i-n diodes with the voltage-controlled E-PHEMT transistors [26], e.g. the ATF-54143 [27] from the Broadcom has a very low static leak current of <0.2 μA when the gate voltage is below the threshold voltage of 0.6 V. This architecture, illustrated in Figure 2.7, is quite appealing for its low power consumption and the capability of achieving any order QAM with only two transistors. This tag architecture is also employed for other backscatter modulation schemes that will be introduced in Sections 2.3 and 2.4.

For the dual-transistor-based IQ backscatter modulator tag architecture shown in Figure 2.7, the incoming signals captured by the tag antenna are firstly 3-dB divided into two paths which are terminated by impedance loads each controlled by a transistor. An identical matching circuit is used to map the transistor impedance onto the real axis, say $R_I$ and $R_Q$. The $I$ and $Q$ paths are characterized by 45° phase difference, achieved by means of

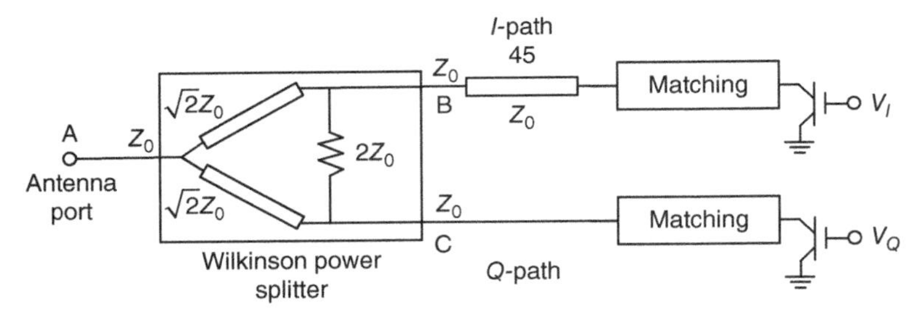


**Figure 2.7** Dual-transistor-based IQ backscatter modulator.

a transmission delay line. With reference to Figure 2.7, the reflected signals at port B and C are, therefore, characterized by phase difference of 90°, since the signal in the $I$-path passes through the delay line twice. This 90° phase difference in two paths makes the impedance manipulation by the two transistors orthogonal in $IQ$ plane. Mathematically, when an ideal Wilkinson power splitter is considered, the reflection coefficient at the antenna port A can be derived in (2.8),

$$\Gamma_A = \frac{1}{2}j\Gamma_I + \frac{1}{2}\Gamma_Q, \tag{2.8}$$

where

$$\Gamma_I = \frac{R_I - Z_0}{R_I + Z_0}; \quad \Gamma_Q = \frac{R_Q - Z_0}{R_Q + Z_0}. \tag{2.9}$$

In practice, the matching networks in Figure 2.7 can be removed without compromising the $IQ$ orthogonality, see (2.8). In this case, the non-real values of the transistor impedance lead to a rotation and a DC offset of the synthesized QAM constellation. These issues, however, can be readily equalized in a modern QAM receiver.

With the measured impedance of the transistor ATF-54143 when its gate voltage is varied within the range between 0 and 0.6 V of a resolution of 1 mV, seen in Figure 2.8, the simulated reflection coefficient at the tag antenna port is plotted as seen in Figure 2.9 (no matching is included). It can be observed that a square-shaped reflection coefficient at the tag antenna port is spanned with only two transistors in an orthogonal fashion. Since no matching circuit is present, the square is rotated and is slightly shifted with respect to the origin. In order to synthesize a QAM backscatter signal, a maximum square is first fitted within the achievable impedance area and then the QAM constellation symbols can be obtained with the maximum magnitude, resulting in maximum backscattered QAM energy. An example of the synthesized 16QAM constellation diagram is illustrated in Figure 2.9. Each synthesized constellation symbol corresponds to a transistor voltage pair ($V_I$, $V_Q$). Thus, a QAM modulation ultimately can be performed by mapping the data onto the corresponding voltage pair.

## 2.3 Chirp Spread Spectrum (CSS) Modulation

CSS is a spread spectrum technique that employs linear frequency-modulated chirp signals to encode information. Similar to the other spread spectrum techniques, it is resilient to the in-band and out-of-band interference. In addition, it does not require fine-grind frequency

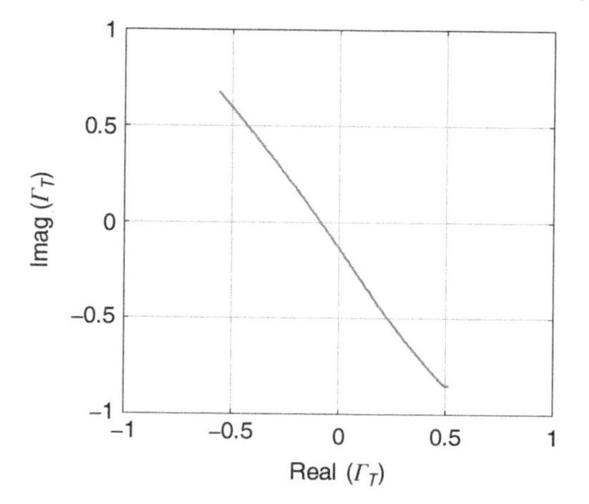


**Figure 2.8** Measured reflection coefficient $\Gamma_T$ of the transistor ATF-54143 when its gate voltage was varied within the range between 0 and 0.6 V of a resolution of 1 mV. It is normalized to 50 Ω.

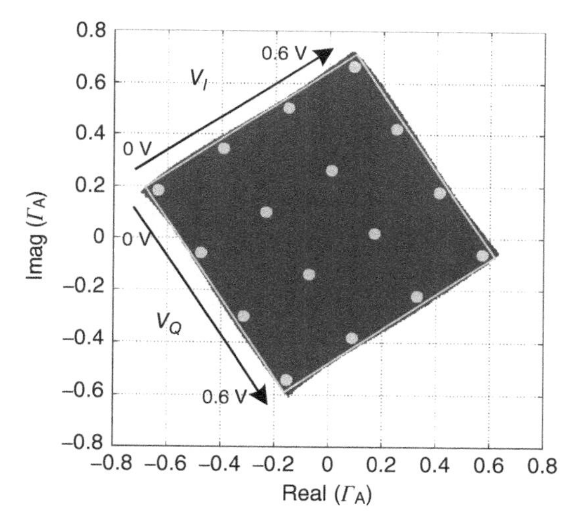


**Figure 2.9** Simulated reflection coefficient $\Gamma_A$ at the tag antenna port with the measured transistor (ATF-54143) impedance.

synchronization at the receiver end. A key characteristic of the CSS is that a time delay in the chirp signal translates to a frequency shift at the output of the fast Fourier transform (FFT) after the correlation with an opposite chirp. The CSS modulation uses this to encode data as cyclic time shifts in the baseline chirp. This CSS modulation is able to achieve very high reception sensitive at the cost of low data rate (or spectrum efficiency). This CSS technique is employed in the LoRa-PHY, which is able to achieve a sensitivity down to −149 dBm, well below noise floor in many receivers.

Aiming to possess this super sensitivity, which means high reliability and long communication range, some researchers have managed to achieve LoRa (or CSS) backscatter signal waveforms. The Passive LoRa, named as PLoRa, was proposed in [28]. Its frequency shifts an ambient LoRa packet to another LoRa channel. The two well-designed frequency shifts result in a different LoRa packet in the new channel, thus, transferring only one bit per LoRa symbol. As expected, an extended communication range up to 1.1 km was achieved. However, from the modulation's perspective, the PLoRa tag actually performs 2FSK modulation on top of an ambient CSS-modulated signals. Hence, it is not categorized as CSS backscatter

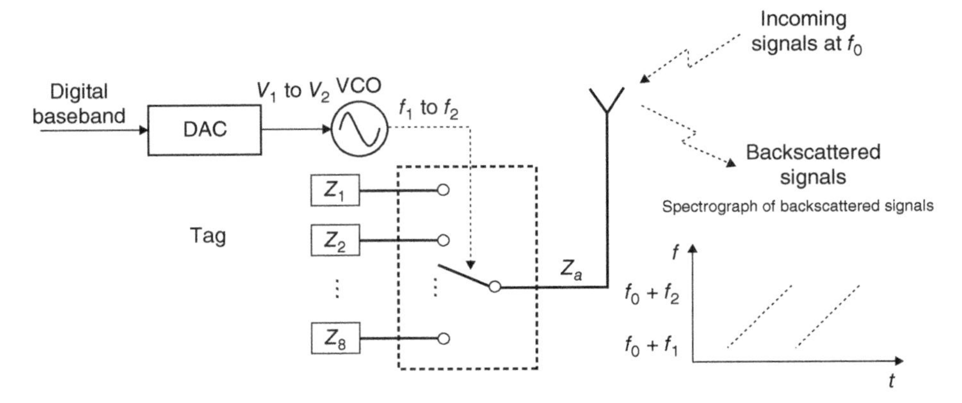


**Figure 2.10** CSS (LoRa-PHY compatible) backscatter tag proposed in [29]. Source: Modified from Talla et al. [29].

modulation in this chapter. This is similar to the passive WiFi tag designed in [13], which is 2PAM modulated, though the backscattered signal is DBPSK modulated (because ambient 802.11b incoming signals are exploited).

This first CSS (LoRa-PHY compatible) backscatter was described in [29], with the proposed tag architecture shown in Figure 2.10. Here, the frequency, with which the tag antenna impedance termination is toggled, is linearly increasing (or decreasing), creating up-chirp (or down-chirp). This can be derived similar to the FSK backscatter modulation discussed in Section 2.2. This linear frequency variation was achieved through a Digital-to-Analog Convertor (DAC)-controlled voltage-controlled oscillator. The required bit for the DAC equals to the spreading factor of the LoRa symbol. Thus, a 12-bit DAC is needed. The number of loads was increased from 2 to 8, in order to suppress the mirror frequency component and the third and fifth harmonics. The experiment conducted using modular commercial off the shelf (COTS) components demonstrated a superior communication range up to 2.8 km when a commercial LoRa node was used as a receiver. It is noted that the switching network between the tag antenna and the eight loads contributes 4 dB loss, and the resistor and capacitor-only impedance for IC compatibility results in 3 dB extra loss [29]. All these indicate further room for improvement.

Based on the QAM tag architecture in [26], a new type of CSS backscatter was proposed in [30, 31]. Different from the approach in [29] wherein the frequency of toggling among loads with an equal phase interval is varied, the method introduced in [30] is to fix the toggling frequency, but, instead, vary the phase intervals among consecutive loads. This exploits the fact that the derivative of the phase change equals the frequency. The implementation of this concept, however, requires much greater number of loads. Thanks to the IQ backscatter tag illustrated in Figure 2.7 which is able to synthesize a nearly continuous impedance coverage shown in Figure 2.9, the required loads can be readily generated with the carefully controlled transistor gate voltage pair ($V_I$, $V_Q$).

The CSS backscatter waveform synthesis procedures and the underpin principle are now elaborated as below [30, 31].

- The CSS modulation has a constant envelope, indicating that the tag antenna termination impedance should be located on a reflection coefficient circle. In order to maximize the

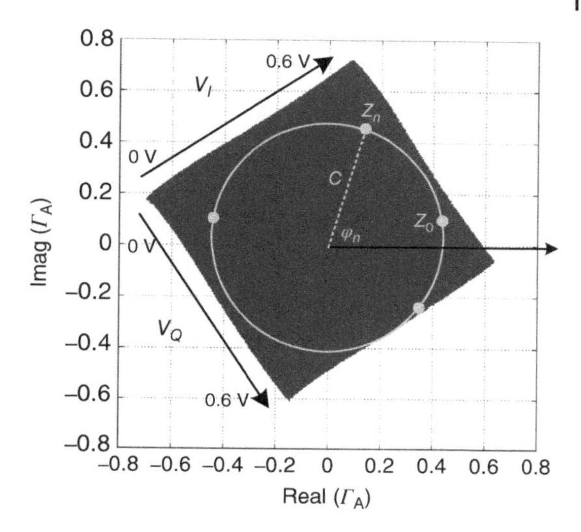


**Figure 2.11** CSS backscatter waveform synthesis based on IQ backscatter modulator in Figure 2.7.

power of backscatter signals, a reflection coefficient circle with the maximum radius is fitted inside the shaded impedance area spanned by the IQ backscatter modulator, see illustration in Figure 2.11.

- Select $N+1$ load impedance on the fitted reflection coefficient circle, denoted as $Z_n$ ($n = 0, 1, 2, \ldots, N$). Thus, the reflection coefficient $\Gamma_n$, when the load $Z_n$ is connected to the tag antenna, can be expressed as

$$\Gamma_n = \frac{Z_n - Z_a}{Z_n + Z_a} = C\cdot \exp(j\varphi_n), \tag{2.10}$$

where the magnitude $C$ and phase $\varphi_n$ of $\Gamma_n$ are labelled in Figure 2.11. It is noted that the shaded impedance area could be shifted toward the center by carefully designing a transistor matching network, in order to further enlarge the reflected signal magnitude, i.e. larger $C$. The choice of $Z_n$, hence $\Gamma_n$, will be derived in the next step.

- When the loads are sequentially selected with the time interval of $\Delta t$, the samples, with the sampling frequency of $\Delta f = 1/\Delta t$, of the backscattered signal can be written as

$$\begin{aligned} S_{bs}(n)|_{t=n/\Delta f} &= \exp(j2\pi f_0 t)\cdot C\exp(j\varphi_n). \\ &= C\cdot \exp\left[j2\pi\left(f_0 + \frac{\varphi_n \Delta f}{2n\pi}\right)t\right] \end{aligned} \tag{2.11}$$

Here, the incoming signal is assumed to be a single-tone CW with frequency $f_0$, i.e. $\exp(j2\pi f_0 t)$.

From (2.11), it can be observed that in order to achieve a chirp signal, the frequency term $(\varphi_n \Delta f)/(2n\pi)$ has to be a linear function with respect to time, which is sampled with equal interval. In other words, $(\varphi_n \Delta f)/(2n\pi)$ needs to be a linear function of $n$, namely

$$\begin{aligned} &\frac{\varphi_n \Delta f}{2n\pi} = A\cdot n + B \\ &\Rightarrow \varphi_n = \frac{2\pi A}{\Delta f}\cdot n^2 + \frac{2\pi B}{\Delta f}\cdot n, \end{aligned} \tag{2.12}$$

**Table 2.1** Design example of an up-chirp.

| $n$ | $\varphi_n$ (°) | $\Gamma_n$ | $Z_n$ (Ω) |
|---|---|---|---|
| 0 | 0 | 0.4 | 116.7 |
| 1 | 43.2 | 0.2916 + $j$0.2738 | 72.8 + $j$47.5 |
| 2 | 172.8 | −0.3968 + $j$0.0510 | 21.5 + $j$2.6 |
| 3 | 28.8 | 0.3505 + $j$0.1927 | 91.5 + $j$42.0 |
| 4 | 331.2 | 0.3505 − $j$0.1927 | 91.5 − $j$42.0 |
| 5 | 0 | 0.4 | 116.7 |
| 6 | 115.2 | −0.1703 + $j$0.3619 | 28.0 + $j$24.1 |
| 7 | 316.8 | 0.2916 − $j$0.2738 | 72.8 − $j$47.5 |
| 8 | 24.8 | −0.1703 − $j$0.3619 | 28.0 − $j$24.1 |
| 9 | 259.2 | −0.0750 − $j$0.3929 | 32.1 − $j$30.0 |
| 10 | 0 | 0.4 | 116.7 |

$C$ is assumed to be 0.4, and $Z_a$ is assumed to be 50 Ω.

where $A$ and $B$ are scaler constants, defining the starting frequency, i.e. $f_0 + B$, and the bandwidth, i.e. $N|A|$, of the chirp signal. When $A > 0$, it is an up-chirp, while when $A < 0$, a down-chirp is synthesized. Thus, from (2.12), the impedance of $N$ loads can be calculated and the corresponding $N$ pairs of $(V_I, V_Q)$ can be obtained to generate required chirp backscatter waveforms.

In order to further elaborate the design procedures summarized above, an example is now presented. Assuming the frequency of the incoming waveform is $f_0$, we want to synthesize an up-chirp that occupies a bandwidth of 512 kHz from $f_0$ + 50 kHz to $f_0$ + 562 kHz, and that lasts for 1 ms. When the updating frequency of the loads is $\Delta f$ = 10 kHz, it can be calculated that $N$ = 1 ms × 10 kHz = 10. Thus, from (2.11) and (2.12), it can be derived that $A = (512 \times 10^3)/10 = 51\,200$, and $B = 50\,000$. With known $A$ and $B$, $\varphi_n$ ($n$ = 0, 1, 2, …, $N$), and hence $\Gamma_n$ and $Z_n$, can be obtained, which are listed in Table 2.1 ($C$ is assumed to be 0.4, and $Z_a$ is assumed to be 50 Ω). When the corresponding 11 pairs of $(V_I, V_Q)$, associated with each $Z_n$, are alternatingly updated at a frequency of 10 kHz, the required up-chirp can be synthesized.

In order to validate the CSS design approach, an IQ backscatter modulator, operating at 2.45 GHz, was fabricated, see photograph in Figure 2.12a. All possible synthesized reflection coefficients measured with a resolution of 1 mV (both $V_I$ and $V_Q$), from 0 to 0.6 V, are depicted in Figure 2.12b (for better visualization, a step of 10 mV was used for plotting). Parallel open stubs were added at the transistors' drains to shift the response toward the center of the Smith Chart. By doing so, a set of impedances can be achieved whose phases vary $2\pi$, on a maximum constant reflection coefficient circle. After the tag circuit characterization, the reflection coefficients circle of VSWR of 1.9 was identified, as well as the corresponding $V_I$ and $V_Q$. Thus, it is possible to create CSS-modulated LoRa symbols by following the design procedures elaborated earlier.

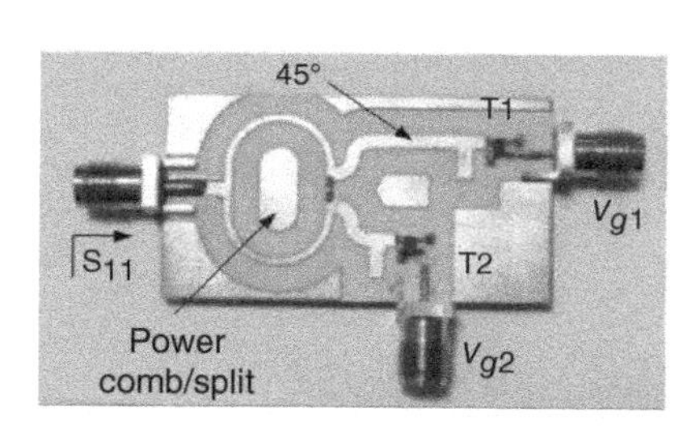


(a)

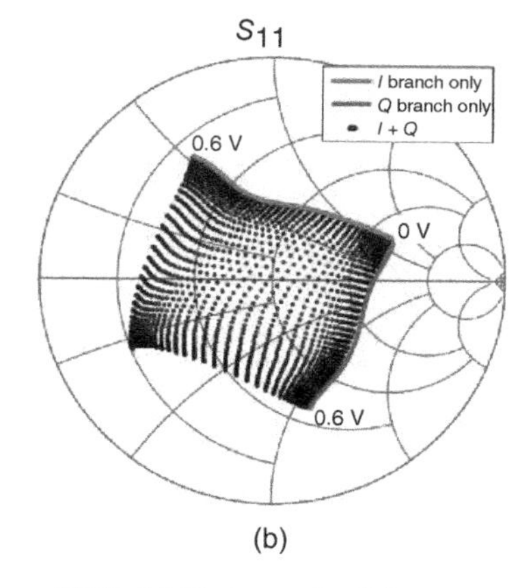


(b)

**Figure 2.12** IQ impedance modulator front-end [31]. (a) Photograph of the tag prototype. Transistors T1 and T2 are controlled by an external baseband source. (b) All possible synthesized impedances measured with a grid of 10 mV step, from 0 to 0.6 V. Source: Belo et al. [31].

From (2.11) and (2.12), a single LoRa symbol was synthesized in order to validate the symbol generation process. For this, a two channel Arbitrary Waveform Generator (AWG) was selected to produce the control voltages required for $V_I$ and $V_Q$. Additionally, a Vector Signal Analyzer (VSA) acquires the backscattered signal and down-converts it to the complex baseband. The captured complex signal is then loaded into the MATLAB and processed.

Considering a LoRa signal with a spreading factor of 7 and bandwidth of 125 kHz, the phase progression required to generate a symbol of, for example, index 44 is depicted in Figure 2.13a. Here, a CW signal generator was used to produce the backscatter carrier, and the AWG was loaded with the control bias voltage waveforms, $V_I$ and $V_Q$, synthesized in advance. A directional coupler was employed to take a fraction of the reflected wave to feed the VSA. Then, the down-converted $I/Q$ signal was loaded into MATLAB, down-sampled to exact 125 kHz and multiplied by the reference down-chirp. Finally, the Discrete Fourier Transform (DFT) was applied. It can be observed that a peak at the fast Fourier transform bin 44 was observed, see Figure 2.13b, indicating the successful decoding of the LoRa symbol.

After symbol coding and decoding validation, a bit stream consisting of 84 000 bits (12 000 symbols) was generated and the corresponding control voltage pairs ($V_I$, $V_Q$) were synthesized and loaded into the AWG. The spectrogram of the first 40 symbols received by the VSA is shown in Figure 2.14a. The represented signal consists of a preamble with 8 reference symbols, 2 synchronization symbols, and 30 data symbols. Figure 2.14b illustrates the spectrogram from the decoded symbols. After multiplying each symbol with the conjugate of the reference symbol, the preamble, synchronization, and transmitted data symbols can be clearly seen. The data symbols are represented by the constant frequencies that result from such multiplication. The frequency represents the symbol, and the length of the data symbol represents the time that it took to be fully transmitted. The synchronization was performed by delaying the received packet until a maximum occurs at the first DFT bin, for the first eight preamble symbols.

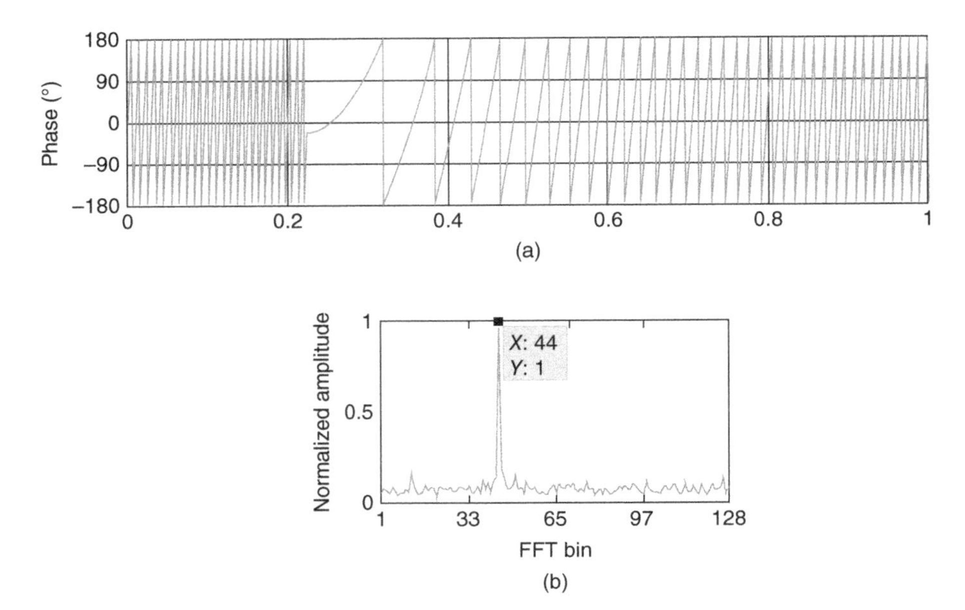


**Figure 2.13** (a) The phase updating profile in time required to synthesize a LoRa symbol of index 44. (b) Successfully decoded LoRa symbol in FFT domain [31] (assuming LoRa signal with a spreading factor of 7 and bandwidth of 125 kHz). Source: Belo et al. [31]. © 2019, IEEE.

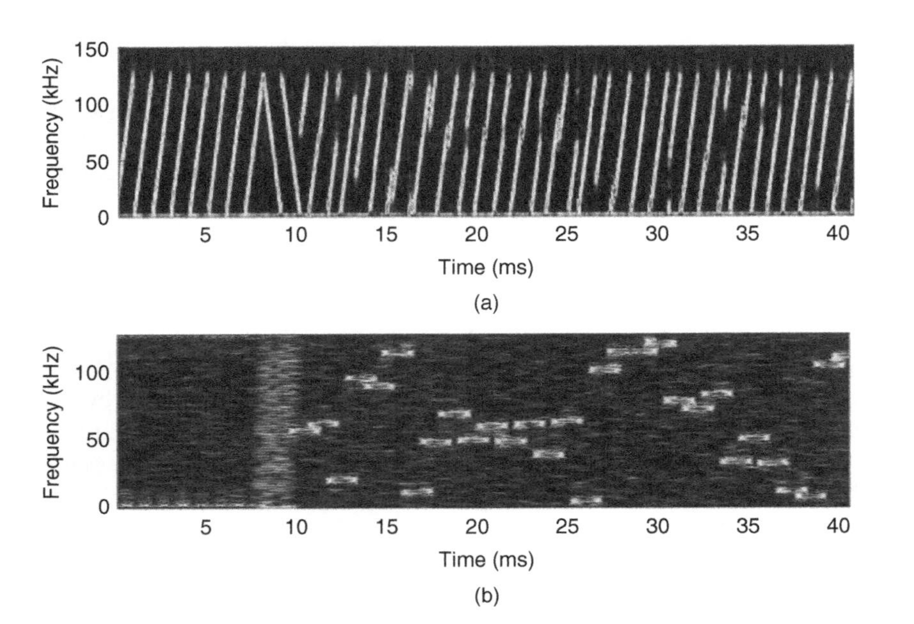


**Figure 2.14** (a) Received LoRa packet spectrogram (measurement). (b) Spectrogram after LoRa decoding. Only the first 40 symbols out of 12 000 are presented. The packet consists of a preamble of 8 reference symbols, 2 synchronization symbols, and 30 data symbols. Source: Belo et al. [31]. © 2019, IEEE.

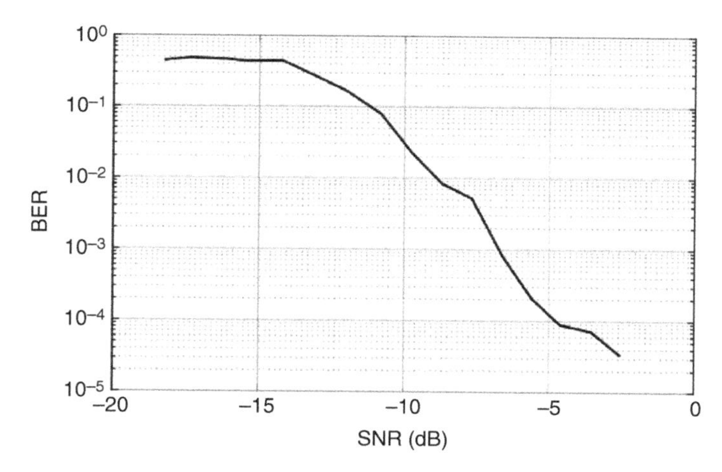


**Figure 2.15** Measured BER versus receive SNR. Source: Belo et al. [31]. © 2019, IEEE.

In order to evaluate the performance of the proposed LoRa backscatter modulator front-end, cabled as well as over-the-air tests were conducted. The cabled laboratory experiment was designed to evaluate the bit error rate (BER) performance versus signal-to-noise ratio (SNR). For this, the 12 000 symbols generated earlier were transmitted repeatedly for each SNR value, sampled by the VSA and processed in MATLAB. The achieved BER was computed by taking an average value of all those measurements, and it is plotted in Figure 2.15. The achieved results are in line with other reported works for the same spreading factor [32, 33].

Additionally, with the same cabled configuration, intentional noise was added to the control voltages to evaluate the required processing unit precision and its robustness against external interferences. Both control voltages were corrupted by random perturbations around the nominal value. The instantaneous phase error is shown to be $<2°$ for perturbations of $\pm 1$ mV, $<10°$ for $\pm 8$ mV, and for $\pm 32$ mV it can get as high as 35°. It was seen that the BER vs SNR curve, under control bias noise, is exactly the same. However, the backscatter input power has to be increased to achieve a same SNR. This aspect will be presented with the over-the-air measurement.

With the laboratory set-up illustrated in Figure 2.16, several over-the-air tests were performed. A 2.45 GHz power source that generates the backscatter carrier is co-located with the proposed device. A circulator is employed to redirect the reflected wave to the antenna

**Figure 2.16** Block diagram of the laboratory set-up used for over-the-air measurements. Source: Belo et al. [31]. © 2019, IEEE.

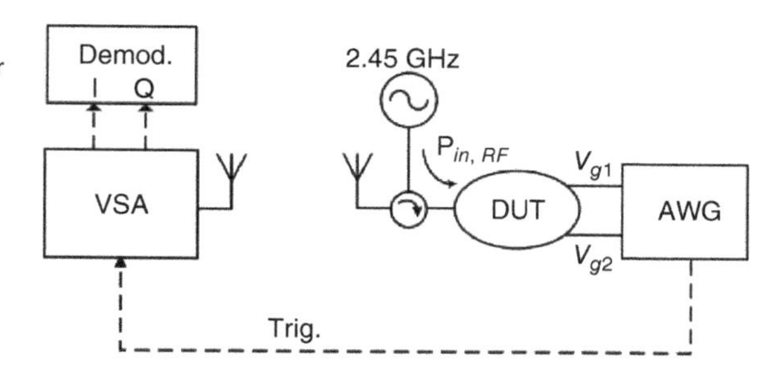

as shown. The receiver consists of a VSA that down-converts the signal to the complex baseband. Then, the acquired signal is loaded into MATLAB and processed to obtain an estimation of the transmitted information. Two similar single element patch antennas were designed to operate at 2.45 GHz, with an estimated gain of 6 dBi.

Three different indoor scenarios within a laboratory environment were targeted for evaluation and are shown in Figure 2.17. The aim is to provide information about how much power the device requires from the backscatter carrier to produce a successful transmission. The first, Figure 2.17a, is a typical indoor scenario with Line-of-Sight (LoS) conditions. An estimation of the distance between the receiver (VSA) and the device is 10 m. In this particular scenario, intentional perturbations were also added to the control voltages, namely, ±8 and ±32 mV. In the second scenario, Figure 2.17b, the device was positioned close to the floor at roughly 7.5 m from the receiver. In this situation, there were desks with laboratorial instruments and other common laboratory hardware in-between the device and the receiver antennas. In the third scenario, Figure 2.17c, the set-up was re-positioned in order to account the effects produced by a wall on the signal path, with the receiver located 10 m away from the device.

In all these indoor experiments, 12 000 symbols were transmitted several times for each backscatter carrier input power level (steps of 1 dB) and the percentage of the overall symbol error was computed and is shown in Figure 2.18. It is shown that in the first scenario, −40 dBm of backscatter carrier input power was required to produce error-free transmissions. Additionally, it is shown that a random perturbation of ±8 mV on the control voltages does not produce noticeable effects, while ±32 mV produces a sensitivity decrease of 2.2 dB. Higher perturbation values require higher backscatter carrier input power to maintain the same receive SNR. In the second scenario, the non-line-of-sight (NLoS) conditions imposed by the desks determine that the required backscatter carrier input power for error-free transmission was −25 dBm. Finally, due to the wall attenuation, the input power required for error-free transmissions was −26 dBm. It should be noted that the measurements were taken during normal operation of the laboratory and under possible heavy 2.45 GHz WiFi network interference.

Backscatter communications are generally associated with wireless power transfer systems. Thus, with the presented approach, it is possible to build a full passive wirelessly powered LoRa backscatter communication device that may operate over larger distances when compared with the conventional SNR needed for ASK, FSK, or PSK backscatter modulation systems. Moreover, it is shown that the proposed device has the prominent versatility to backscatter signals compatible with many other standards, thanks to the IQ impedance modulation.

## 2.4 Multicarrier Backscatter Transmission

A number of recent backscatter developments have focused on extending the communication range, like the CSS modulation discussed in Section 2.3. In this section, another important aspect of the backscatter communication, i.e. increasing data rates, is presented. In addition to the high-order QAM in Section 2.4, the multicarrier backscatter modulation is introduced here for the first time. It is noted that although FSK-based frequency

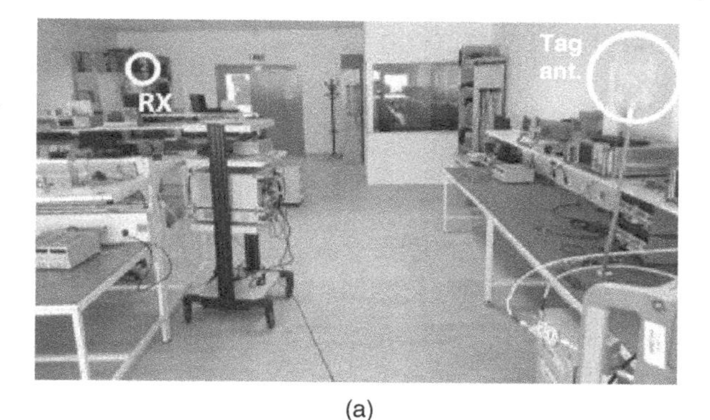


(a)

(b)

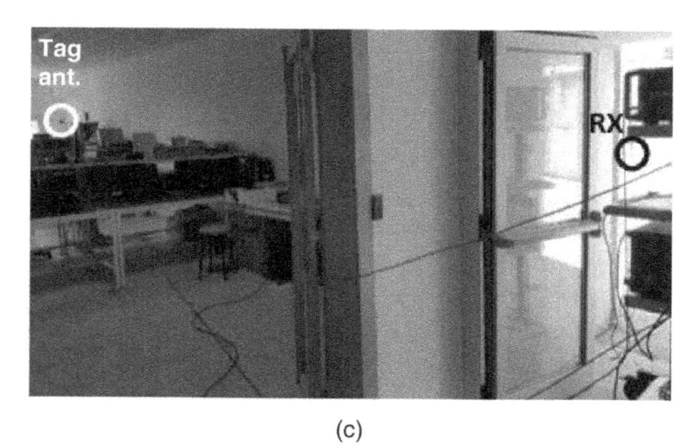


(c)

**Figure 2.17** Three typical over-the-air lab test environments. (a) Typical indoor line-of-sight channel (communication distance of 10 m). (b) Typical indoor non-line-of-sight channel with furniture and instruments in-between (communication distance of 7.5 m). (c) Typical indoor non-line-of-sight channel with a wall in-between (communication distance of 10 m). Source: Belo et al. [31]. © 2019, IEEE.

hopping and the CSS modulation [31] generate backscatter signals that occupy a frequency bandwidth, at any time instant, the signals only have a single frequency component. Instead, a multicarrier signal would involve backscattering multiple frequencies simultaneously. It is worth mentioning that this work intends to generate multicarrier backscatter signals when the incoming electromagnetic wave is a signal carrier. This is distinct to the system presented in [13], wherein the tag is performing ASK upon incoming multicarrier waveforms.

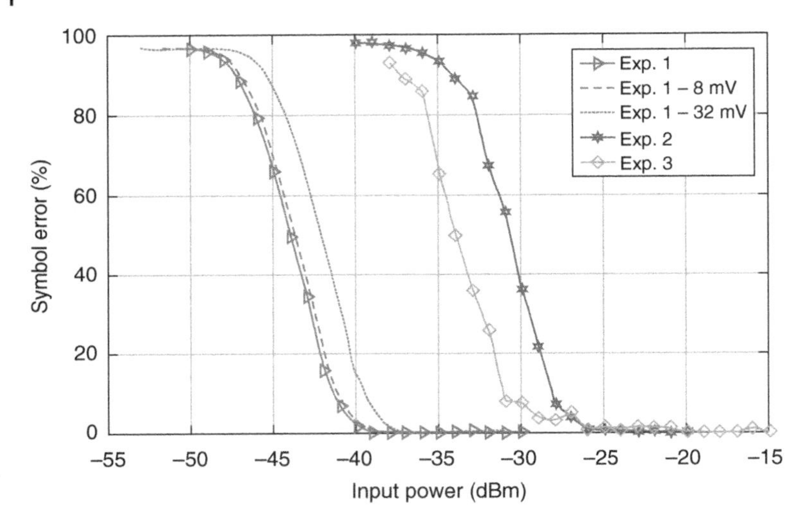


**Figure 2.18** Symbol error percentage versus backscatter carrier input power level measured for all over-the-air test scenarios. Results for perturbations of ±8 and ±32 mV are provided for the first experimental scenario. Source: Belo et al. [31]. © 2019, IEEE.

In this section, we demonstrate the synthesis of multicarrier backscattering waveforms from a CW incidence using the same dual-transistor-based IQ modulator described in Sections 2.2 and 2.3.

As an illustrative example, the authors synthesize the preamble of IEEE 802.11g (WiFi 3), which has 64 OFDM subcarriers, each of whom is BPSK modulated. The preamble lasts for five OFDM symbols, therefore, comprises $64 \times 5 = 320$ samples in the time domain. Considering the baseband signal, each sample is characterized by its magnitude and phase and, therefore, is a complex number. This complex sequence, which consists of 320 samples in the time domain with the sampling period being the inverse of the OFDM subcarrier frequency spacing, is plotted in Figure 2.19a. The magnitude of samples is small due to power normalization. The sample sequence, which contains 64 subcarriers, is then frequency upconverted in order to generate OFDM RF signals, ready to be radiated.

As pointed out, these baseband time-domain samples are used to modulate the OFDM fundamental RF carrier. This, in fact, is the same function that the IQ backscatter modulator applies on the incoming electromagnetic CW. By applying the IQ values associated with this sequence, it is thus possible to backscatter an OFDM signal from an incoming single carrier signal.

In order to illustrate this, we first scale the samples within the shaded area shown in Figure 2.11, see Figure 2.19b, in order to maximize the backscattering signal power. Each sample (circle dot) in Figure 2.19b corresponds to a pair of transistor control voltage ($V_I$, $V_Q$). In other words, when a correct sequence of 320 ($V_I$, $V_Q$) is applied at the IQ backscatter modulator at the OFDM sampling rate, the incoming single carrier electromagnetic wave is converted/backscattered into the OFDM-modulated IEEE 802.11g preamble. It is noted that the pulse-shaping filter should be added following the method presented in [34]. Using the same procedure, the OFDM payloads can be generated, thus, equipping this low-cost low-power tags the capability of communicating with WiFi (or others, e.g. LTE) devices.

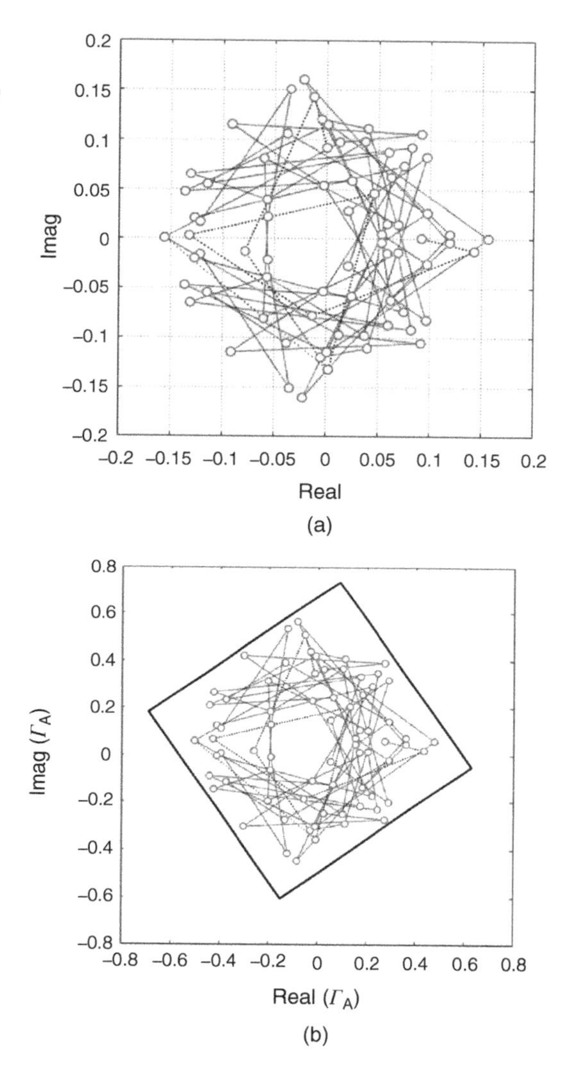


**Figure 2.19** (a) 320 samples in IQ plane in the IEEE 802.11g (some samples are overlapped); (b) the scaled 320 samples to fit into the shaded area in Figure 2.11.

## 2.5 Summary

With the booming of the IoT networks that are the foundation of the future highly intelligent world, the cost- and power-efficient IoT nodes are particularly demanded to stimulate the industrial uptake, and equally important, to ease the pressure of detrimental impacts on the fragile environment. The backscatter communication principle, removing the costly and power hungry RF components in conventional active radios, is a perfect fit to facilitate this trend. While a surge of studies has been conducted to reveal the benefits of backscatter communications on the network topologies and network capacity [35, 36], the fundamental implementation of reliable backscatter tags with preferred signal modulation schemes is still a challenge. In this chapter, the authors have summarized the existing modulation schemes in backscatter communications with their respective hardware realizations. In particular, the two recently introduced CSS and multicarrier modulations were presented in detail, as they made a crucial step to address the current limitations on backscatter links, i.e. short communication range and low transmission data rates. The authors expect further improvement along these two research tracks in order to enable ubiquitous sensing envisaged in our society.

## References


1 Liu, A., Shahzad, M., Liu, X., and Li, K. (2017). *RFID Protocol Design, Optimization, and Security for the Internet of Things*. The Institution of Engineering and Technology. ISBN: 978-1-78561-332-6.

2 Dixit, M., Kumar, J., and Kumar, R. (2015). Internet of Things and its challenges. *Int. Conf. Green Computing Internet of Things (ICGCIoT)*, Noida (8–10 October 2015), 810–814.

3 Philipose, M., Smith, J.R., Jiang, B. et al. (2005). Battery-free wireless identification and sensing. *Pervasive Comput.* 4 (1): 37–45.

4 Van Huynh, N., Hoang, D.T., Lu, X. et al. (2018). Ambient backscatter communications: a contemporary survey. *IEEE Commun. Surv. Tuts.* 20 (4): 2889–2922, Fourthquarter.

5 Assimonis, S., Daskalakis, S., and Bletsas, A. (2016). Sensitive and efficient RF harvesting supply for batteryless backscatter sensor networks. *IEEE Trans. Microwave Theory Tech.* 64 (4): 1327–1338.

6 Song, C., Yela, A., Huang, Y. et al. (2019). A novel Quartz clock with integrated wireless energy harvesting and wireless sensing for smart home applications. *IEEE Trans. Ind. Electron.* 66 (5): 4042–4053.

7 Zhang, Y., Shen, S., Chiu, C., and Murch, R. (2019). Hybrid RF-solar energy harvesting systems utilizing transparent multiport micromeshed antennas. *IEEE Trans. Microwave Theory Tech.* 67 (11): 4534–4546.

8 Hester, J. and Tentzeris, M. (2016). Inkjet-printed flexible mm-Wave Van-Atta reflectarrays: a solution for ultralong-range dense multitag and multisensing chipless RFID implementations for IoT smart skins. *IEEE Trans. Microwave Theory Tech.* 64 (12): 4763–4773.

9 Kimionis, J., Bletsas, A., and Sahalos, J.N. (2014). Increased range bistatic scatter radio. *IEEE Trans. Commun.* 62 (3): 1091–1104.

10 Daskalakis, S.N., Correia, R., Goussetis, G. et al. (2018). 4-PAM modulation of ambient FM backscattering for spectrally efficient low power applications. *IEEE Trans. Microwave Theory Tech.* 66 (12): 5909–5921.

11 Qian, J., Gao, F., Wang, G. et al. (2017). Noncoherent detections for ambient backscatter system. *IEEE Trans. Wirel. Commun.* 16 (3): 1412–1422.

12 Memon, M.L., Saxena, N., Roy, A., and Shin, D.R. (2019). Backscatter communications: inception of the battery-free era—a comprehensive survey. *Electronics* 8 (2): 1–69.

13 Kellogg, B., Parks, A., Gollakota, S. et al. (2014). Wi-Fi backscatter: internet connectivity for RF-powered devices. *Proc. ACM SIGCOMM*, Chicago, IL (17–22 August 2014), 1–12.

14 Rappaport, T.S. (2002). *Wireless Communications: Principles and Practice*, 2e. Prentice Hall.

15 Varshney, A., Pérez-Penichet, C., Rohner, C., and Voigt, T. LoRea: a backscatter architecture that achieves a long communication range. *Proc. ACM Embedded Netw. Sensor Syst. (SenSys)*, Delft (5–8 November 2017), 1–14.

16 Vougioukas, G., Daskalakis, S.N., and Bletsas, A. (2016). Could battery-less scatter radio tags achieve 270-meter range? *Proc. IEEE Wireless Power Transfer Conf. (WPTC)*, Aveiro (5–6 May 2016), 1–3.

17 Liu, V., Parks, A., Talla, V. et al. (2013). Ambient backscatter: wireless communication out of thin air. *ACM SIGCOMM Comput. Commun. Rev.* 43 (4): 39–50.

18 Kellogg, B., Talla, V., Gollakota, S., and Smith, J.R. (2016). Passive Wi-Fi: bringing low power to Wi-Fi transmissions. *Proc. USENIX Symp. Netw. Syst. Design Implement.* (NSDI), Santa Clara, CA (16–18 March 2016), 151–164.

19 Ensworth, J.F. and Reynolds, M.S. (2017). BLE-backscatter: ultralow-power IoT nodes compatible with Bluetooth 4.0 Low Energy (BLE) smartphones and tablets. *IEEE Trans. Microwave Theory Tech.* 65 (9): 3360–3368.

20 Iyer, V., Talla, V., Kellogg, B., Gollakota, S., and Smith, J. (2016). Inter-technology backscatter: towards Internet connectivity for implanted devices. *Proc. ACM SIGCOMM Conf. (SIGCOMM '16)*, New York (22–26 August 2016), 356–369.

21 Yang, G., Liang, Y., Zhang, R., and Pei, Y. (2018). Modulation in the air: backscatter communication over ambient OFDM carrier. *IEEE Trans. Commun.* 66 (3): 1219–1233.

22 Correia, R. and Borges Carvalho, N. (2016). Design of high order modulation backscatter wireless sensor for passive IoT solutions. *Proc. IEEE Wireless Power Transf. Conf. (WPTC)*, Aveiro, Portugal (5–6 May 2016), 1–3.

23 Thomas, S.J. and Reynolds, M.S. (2012). A 96 Mbit/sec, 15.5 pJ/bit 16-QAM modulator for UHF backscatter communication. *Proc. IEEE Int. Conf. RFID*, Orlando, FL (3–5 April 2012), 185–190.

24 Thomas, S.J., Wheeler, E., Teizer, J., and Reynolds, M.S. (2012). Quadrature amplitude modulated backscatter in passive and semipassive UHF RFID systems. *IEEE Trans. Microwave Theory Tech.* 60 (4): 1175–1182.

25 Winkler, M., Faseth, T., Arthaber, H., and Magerl, G. (2010). An UHF RFID tag emulator for precise emulation of the physical layer. *Proc. Eur. Microw. Conf. (EuMC)*, Paris, France (28–30 September 2010), 273–276.

26 Correia, R., Boaventura, A., and Carvalho, N.B. (2017). Quadrature amplitude backscatter modulator for passive wireless sensors in IoT applications. *IEEE Trans. Microwave Theory Tech.* 65 (4): 1103–1110.

27 Agilent ATF-54143 low noise enhancement mode pseudomorphic HEMT in a surface mount plastic package. *Datasheet*. Avago Technologies. https://docs.broadcom.com/docs/AV02-0488EN (accessed 12 November 2020).

28 Peng, Y., Shangguan, L., Hu, Y. et al. (2018). PLoRa: a passive long-range data network from ambient LoRa transmissions. *Proc. ACM SIGCOMM Conf. (SIGCOMM'18)*, Budapest (20–25 August 2018), 147–160.

29 Talla, V., Hessar, M., Kellogg, B. et al. (2017). LoRa backscatter: enabling the vision of ubiquitous connectivity. *Proc. ACM Interact. Mobile Wearable Ubiquitous Technol.* 1 (3): 1–24.

30 Correia, R., Ding, Y., Daskalakis, S. et al. (2019). Chirp based backscatter modulation. *Proc. IEEE Int. Microw. Symp.*, Boston (2–7 June 2019).

**31** Belo, D., Correia, R., Ding, Y. et al. (2019). IQ impedance modulator front-end for low-power LoRa backscattering devices. *IEEE Trans. Microwave Theory Tech.* 67 (12): 5307–5314.

**32** Elshabrawy, T. and Robert, J. (2018). Closed-form approximation of LoRa modulation BER performance. *IEEE Commun. Lett.* 22 (9): 1778–1781.

**33** Croce, D., Gucciardo, M., Mangione, S. et al. (2018). Impact of LoRa imperfect orthogonality: analysis of link-level performance. *IEEE Commun. Lett.* 22 (4): 796–799.

**34** Kimionis, J. and Tentzeris, M.M. (2016). Pulse shaping: the missing piece of backscatter radio and RFID. *IEEE Trans. Microwave Theory Tech.* 64 (12): 4774–4788.

**35** Yang, S., Deng, Y., Tang, X. et al. (2019). Energy efficiency optimization for UAV-assisted backscatter communications. *IEEE Commun. Lett.* 23 (11): 2041–2045.

**36** Gong, S., Xu, J., Niyato, D. et al. (2019). Backscatter-aided cooperative relay communications in wireless-powered hybrid radio networks. *IEEE Netw.* 33 (5): 234–241.

# 3

# Electromagnetic Waves Scattering Characteristics of Metasurfaces

*Muhammad Ali Babar Abbasi*[1], *Dmitry E. Zelenchuk*[1], *and Abdul Quddious*[2]

[1] *The Centre for Wireless Innovation (CWI), Institute of Electronics, Communications and Information Technology (ECIT), Queen's University of Belfast, Belfast, United Kingdom*
[2] *KIOS Research Center, University of Cyprus, Nicosia, Cyprus*

## 3.1 Introduction

In recent years, it has been shown that almost extreme control of electromagnetic fields is achievable using metamaterial and metasurfaces. The benefits of this control can lead to applications from perfect lens designing to invisibility sheets/cloaks. The term "meta" means beyond, which accurately exhibits that it has properties not found in other naturally occurring materials. By proper definition, metamaterials can be defined as a subwavelength structure made of common objects such as glass, metal, or any other material. Its properties are not dependent on the composition of the material of which it is made but rather the geometrical arrangement of a periodic or aperiodic structure that gives it the ability to exhibit properties and behavior not found in any of the naturally occurring materials [1, 2]. Next evolution step for the metamaterials was the transition from volumetric to planar structures that would be more feasible for implementation and yet enable a similar degree of control over the electromagnetic waves. These structures are known as metasurfaces and find numerous applications by manipulating electromagnetic fields at chosen boundaries [3]. Metasurfaces work according to Schelkunoff's equivalence principle by manipulating electric and magnetic surface currents on a chosen planar boundary, and thus generating desired electromagnetic fields on either side of the surface, which for practical devices means an operation in either transmission or reflection mode [4, 5].

Controlling the properties of electromagnetic waves has long been an aspiration for scientists in optics and microwave physics. Although the work on the metasurface structures appeared recently, the pioneer work that brought it to the limelight was done by Shalaev and Capasso in the field of gradient metasurfaces [6]. Their work involved controlling the amplitude and phase of the incident wave using the artificially engineered structure. This initiated a revolution in the research community for controlling waves using the engineered structure for the applications that were not possible before. Keeping this concept in mind, many researchers attempted to experiment on synthesizing such material in the electromagnetic domain especially in the radio frequency (RF) and microwave region where the use of naturally occurring materials would be redundant for achieving targeted performances.

*Backscattering and RF Sensing for Future Wireless Communication,* First Edition.
Edited by Qammer H. Abbasi, Hasan T. Abbas, Akram Alomainy, and Muhammad Ali Imran.

Veselago, a Russian scientist, put forward the idea of such material back in the 1960s [7]. This work involved examining Maxwell's equation in a hypothetical medium exhibiting negative permittivity and permeability. His article theoretically described the existence of such material. The negative permeability came into the limelight again after the implementation of split-ring resonator (SRR) design as reported by Pendry in the late 1990s [8]. It was not until early 2000 when a group of researchers at the University of California at San Diego demonstrated a new material property called "negative refractive index" at microwave frequency, which till then have never been observed in any naturally occurring material. They demonstrated it by combining the SRR array [9] which many scientists used in the past and is very well known even today.

It has become apparent that many natural materials do not allow the realization of full control over various devices and active focus of the research has shifted toward artificial materials or metamaterials, where the bulk constitutive parameters were implemented with artificially designed scatterers or metamolecules [10, 11]. This research allowed the demonstration of novel physical phenomena such as negative refraction or compression of physical dimensions of the devices through transformation optics techniques [12]. The path for technology growth is always hampered by constraints which may exist at the operational level or the physical level. For making an optimized, efficient, and compact system, we are always limited by material properties. Since these properties are inflexible, therefore we are always at the crossroad with the same question: "What IF we can construct a material with complete control over its physical properties." By highlighting the term "physical," one does not infer to its external appearance but is focused on the material properties which pertain to its conductivity, permittivity, permeability, and so on. Since in our line of work we are focused on the use of such material in the electromagnetic spectrum, therefore it will be in our ample favor to have a complete control over how the waves would interact and scatter over the designed material. With the advent of latest nanoscale fabrication technology, it has become possible to synthesize such material with "Customized Properties."

### 3.1.1 General Classifications of Metasurfaces

Most commonly, the metasurfaces can be classified as periodic or aperiodic. The operation of periodic metasurfaces is simple, a standard unit cell (also called particle, unit, atom, etc.) is repeated periodically, forming a large surface. If boundary conditions are considered carefully, such metasurface can be made quasi-infinite for a specific application space. Periodic metasurfaces consisting of self-repeating structures have limited use in antennas, transmission lines, and absorbers. The other kind of metasurface is the aperiodic metasurface which does not follow a repetition of the unit cell. These surfaces are more general forms of metasurface and their analytical models are complex compared to periodic metasurfaces. Generally, aperiodic rigorous formulations of Huygens' principle is used to model metasurfaces operation. Huygens' principle states that each point of a wavefront acts as a source of the outgoing wave [13]. Thus, Huygens' metasurface in principle utilizes electric and magnetic polarizable elements to make arbitrary field pattern for a given wave source [14–16].

To understand a function of the general metasurface, consider the example provided in Figure 3.1a where a metasurface is placed in $xz$ plane. The metasurface is illuminated by a transverse magnetic plane wave of the following conditions: $E_z = 0$, $H_z = 0$, and $H_y = 0$.

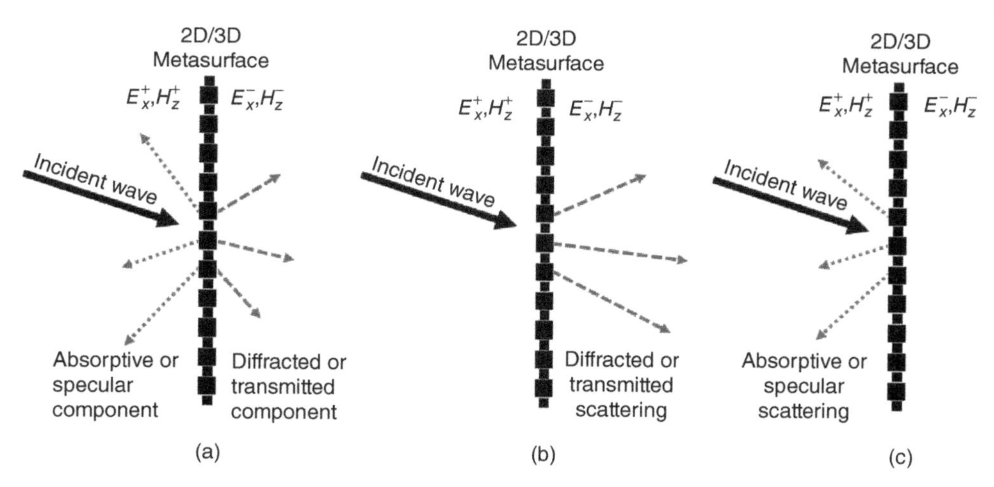


**Figure 3.1** (a) General simple illustration of metasurface with incident wave manipulation properties. (b) General form of transmission, refraction, diffraction, and scattering-type metasurface. (c) The general form of absorptive, specular, reflective, and diffusion-type metasurface.

The magnetic surface admittance and the electric surface impedance can be defined by

$$Y_{sm}(x) = \frac{[H_z^-(x) + H_z^+(x)]}{2[E_x^-(x) - E_x^+(x)]} \tag{3.1}$$

$$Z_{se}(x) = \frac{[E_x^-(x) + E_x^+(x)]}{2[H_z^-(x) - H_z^+(x)]} \tag{3.2}$$

The electric and magnetic field values on the two sides of the surface are $E_x^+$, $H_x^+$, $E_{x,}^-$ and $H_x^-$. The subscript represents the tangential components of the fields, while the superscript presents the side of the metasurface where incident wave hits (+) and the other side (−) [17]. For any incident electromagnetic (EM) wave hitting the metasurface, there are multiple possibilities. In general terms, the most common component division for a wave hitting metasurface is either reflection or transmission. In metasurfaces, the transmission can also involve scattering using the phenomenon of diffraction, which is directly defined by the properties of the metasurface. This is shown in Figure 3.1b. A class of metasurfaces operate only on the transmission-type scattering governed by wave transmission manipulation applications, e.g. invisibility surfaces and lenses. Ideally, $Z_{se}$ for the transmission-type metasurfaces should be 0.

The reflected component of the incident wave hitting metasurface depends on the application. Some metasurfaces try to absorb the incident wave and they are mostly for stealth technology. Some applications demand the reflected component of the metasurfaces to follow a certain reflective angle with a specialized phase and magnitude signature. This is achieved by shaping the metasurface using periodic or aperiodic structures such that $Z_{se}$ is close to infinity and all the energy is reflected. This is also shown in Figure 3.1c. The applications like formulation of perfect magnetic conducting surfaces, the electromagnetic bandgap (EBG) structures and perfect reflecting surfaces are the most common ones that use reflection-type metasurfaces [18–20]. A combination of transmission and reflection-type metasurface is the most common and the practical one. Mostly, all the metasurfaces have a diffracted, transmitted, specular, and absorptive components as shown in Figure 3.1a and one can control any of these components for the specific application.

Another classification, especially used for mmWave, sub-mmWave, and optical metasurfaces, can be based on the way its unit cell behaves. In literature, it is shown that specific forms of 2D metasurfaces can be subdivided into dielectric and plasmonic metasurfaces [21]. Dielectric metasurfaces unit cells have dielectric constant and behave as dielectric resonators, like that of silicon, tellurium, or germanium, etc., while the plasmonic metasurfaces are based on metallic structures with plasmon resonances that are formed by metallic particles. Dielectric metasurfaces generally support electric or magnetic dipole-like resonances. Applications of plasmonic metasurface range from orbital angular momentum (OAM) multicasting, polarization-sensitive mode generation, and terahertz coherent perfect absorption [21].

### 3.1.2 Characterization Approaches of Metasurfaces

Metasurface can be as simple as a passive/active circuit with capacitive and inductive component trying to achieve different than normal material behavior. These are also known as metamaterials or 0D metasurfaces. As compared to 0D metamaterials, 1D metamaterials are complex form of transmission line that support reverse wave propagation behavior in one dimension [22, 23]. 2D metasurfaces are generally comprised of two 1D metamaterial unit cells (or similar) and can support manipulation of electromagnetic waves. More complex form of 3D metasurfaces generally have a surface with a depth (adding the third dimension) and are developed using resonant/radiating 3D structures. Analytical modeling and characterization of 0D metamaterials are done using lumped parameters and can be analyzed via circuit theory. In other words, they can be represented in terms of resistance, inductance, and capacitance and solved via Kirchhoff's laws. 1D metamaterials can be solved using transmission line theory where the characteristic impedance of the line ($Z_0$) and the propagation constant of the propagating signal ($\beta$) are used as variables. 2D metasurfaces are mostly characterized using the equivalent surface susceptibility and solved using computational methods of electromagnetic theory. Lastly, the 3D metasurfaces with depth in it are generally solved by general electromagnetic theory where the material has permittivity ($\varepsilon$), permeability ($\mu$), and partial conductivity ($\sigma'$) [24]. This way, metasurface is decoupled to equivalent homogenous or heterogeneous materials resolved by mathematical formulations derived from fundamental Maxwell's equations.

Simple 2D metasurfaces can be approximated to have zero thickness (Figure 3.2a, b) with finite conductivity. This form can be analyzed using fundamental electromagnetic theory. The refraction and reflection coefficients through this kind of surfaces can be written in the form of standard Fresnel equations at the boundary of two mediums ($\varepsilon_1$ and $\varepsilon_2$) [25]:

$$r_{TE} = \frac{k_{x1} - k_{x2} - \sigma k_0}{k_{x1} + k_{x2} + \sigma k_0} \tag{3.3}$$

$$t_{TE} = \frac{2k_{x1}}{k_{x1} + k_{x2} + \sigma k_0} \tag{3.4}$$

$$r_{TM} = \frac{\sigma k_{x1} k_{x2} + \varepsilon_2 k_0 k_{x1} - \varepsilon_1 k_0 k_{x2}}{\sigma k_{x1} k_{x2} + \varepsilon_2 k_0 k_{x1} + \varepsilon_1 k_0 k_{x2}} \tag{3.5}$$

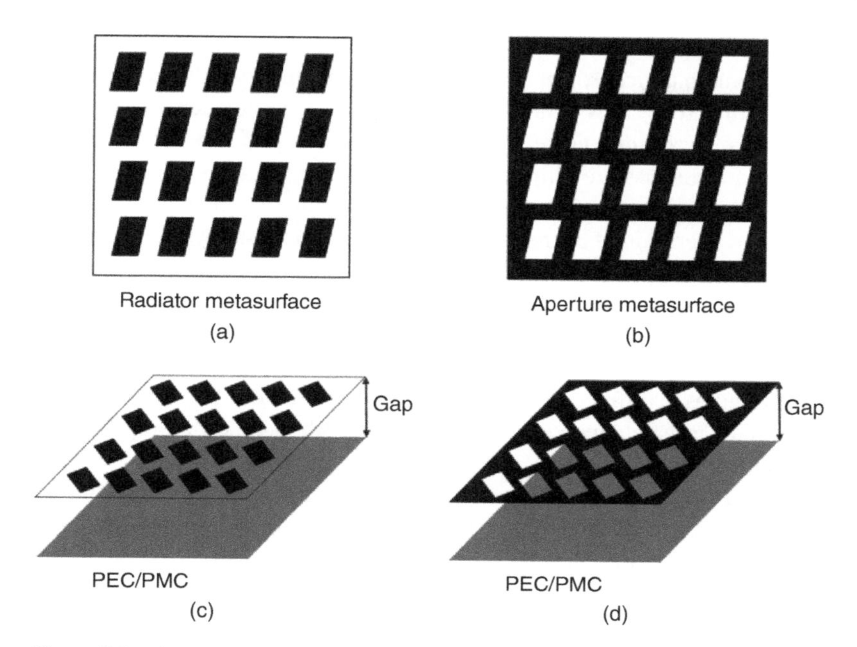


**Figure 3.2** A general simple form of 2D metasurface with (a) metallic, resonant, and/or quasi-resonant structures, (b) aperture, cavity, and/or plasmonic structures. General simplest forms of 3D metasurface with (c) metallic, resonant, or quasi-resonant structures backed by transmission-type, reflective, or absorptive surfaces, and (d) aperture, cavity, and plasmonic structures backed by transmission–type, reflective, or absorptive surfaces.

$$t_{TM} = \frac{2\varepsilon_2 k_0 k_{x1}}{\sigma k_{x1} k_{x2} + \varepsilon_2 k_0 k_{x1} + \varepsilon_1 k_0 k_{x2}} \tag{3.6}$$

where $k$ is the component of the wave vector perpendicular to the plane, and $x1$ and $x2$ represent corresponding mediums, $k_0 = 2\pi/\lambda_0$. TE and TM are transverse electric and transverse magnetic modes, respectively. While reflection and refraction can be mapped via discontinuities given by

$$\sigma(\omega) = \frac{e^2 \hbar v_f \tau \sqrt{\pi n_s}}{\varepsilon_0 c\pi\hbar^2 - i\tau\omega} \tag{3.7}$$

where $\tau$ represents the relaxation time and $v_f$ represents the Fermi velocity. Further details about the definitions can be seen in [25]. In addition to homogenous materials, the profile of a nonhomogenous metasurface consisting of continuously changing metallic strips is shown in [26]. The mathematical form of a metallic strip is defined by a typical sinusoidal function and can be added into an analytical model to estimate the transmission and reflection properties.

In some cases, 2D and 3D metasurfaces can be represented in terms of surface susceptibilities or surface impedances. Both modes can be used to find an analytical solution. For 3D metasurfaces, field properties like polarization and arbitrary incident angle are difficult to analyze using surface impedance. A combined approach where surface susceptibilities and impedance are used to solve complex multilayer problems is presented in [24]. A new kind of spectral numerical mode-matching method is shown in [27] to characterize

3D-layered multi-region complex metasurfaces as shown in Figure 3.2c and d. This method can be described as a semi-analytical solver that gets the dimensionality reduction to solve the multilayered structure. It is shown that even the simplified multilayered structures can provide high absorbance, anomalous reflection, and refraction responses. The method treats the reflection and transmission matrices via eigenmode expansions which help to reduce the original 3D problem to a simpler series of 2D eigenvalue problems for periodic structures. The optimum operation of 3D metamaterials requires a multilayered structure. This not only leads to high insertion losses but also complicates the fabrication process; hence, it is natural to ask if the same fascinating properties can be exhibited by a 2D arrangement of the same artificially engineered structure. The modus operandi of metasurfaces is based on the diffraction phenomenon which dictates that any flat array of periodic elements can be regarded as a diffraction grating, splitting the incident electromagnetic wave. This splitting characteristic depends upon the wavelength, geometric properties of the grating, and the angle of incidence.

The general form of an analytical model for any metasurface is almost the same. One of the most common ways to deal with the periodic structures is the formulation of transverse propagation-based problem. For aperiodic metasurfaces, this formulation does not apply, which increases the computational complexity. One of the investigations in [17] used an electrically thick Fabry–Pérot-based metasurface to mimic the desired anomalous component. The method is shown to produce closed-form solutions of the wave amplitudes and the results are shown to be comparable with the full-wave simulations. This work also defined a semi-analytical design procedure for field manipulations using metasurfaces. For metallic and dielectric structures used in metasurface, a new process of combining the impedance boundary condition and the electric field integral equations is demonstrated in [28]. One unique property of this process includes that only the electric current is used which is useful to approximate using standard current-based equations followed for purely metallic structures. A novel formulation for the lossy substrate is also introduced in this work which uses the impedance boundary conditions as well as the surface electric currents. This assists in formulating the problems related to a thick and thin substrate-coated conductors. The method is verified using numerical examples and is useful to estimate the performance of metal-cladded dielectric substrate used in metasurface design [28]. Field discontinuities across 2D metasurface are also generally investigated by applying appropriate boundary conditions at the interface. As mentioned before, metasurface features are analyzed based on boundary condition surface problems. These conditions are further evaluated in the form of surface reflection and transmission coefficients. At the end, numerical methods are used for validation. This process is used in [24] to characterize the isolated aperture and surface-wave mode analysis. Results are validated via comparison with full-wave electromagnetic simulations. On the other hand, the surface-wave characterization is done using transverse resonance method and verified by numerical examples as well.

## 3.2 Metasurface Applications and Practical Examples

In this section, design techniques, challenges, and possible solutions of diffraction type, transmitted, specular, and absorptive type of metasurfaces in an attempt to achieve extraordinary device performance for a particular application are detailed. A challenge for the design of metasurfaces is to control both polarizations of electromagnetic fields for devices that work with dual-polarized and circularly polarized excitation. The challenges are also studied for different applications of metasurfaces.

### 3.2.1 Absorptive, Reflective, and Diffusion–Type Metasurfaces

The absorptive metasurface can be viewed as a matching problem when looking from a perspective of transmission line theory. In the simplest form, metasurface behave as a single- or multilayered matched load where the impedance changes in a way to absorb all the energy. A standard resistive sheet, in this case, can only represent a real component of the required admittance. Imagine if we divide the standard resistive sheet into small periodic surface resistances and manipulate the spacing in a way to achieve the required imaginary component of the admittance, we can make an ideal absorptive metasurface. Heuristic techniques (like the one shown in [29]) generally uses metasurfaces when the properties are dynamically required to be changed from standard frequency selectivity to pure absorption. The discussed technique uses genetic algorithm-based optimization to do this job and computational electromagnetic simulations are used to verify the results. The unit cell discussed follows the logical structure of subwavelength radiators, while in the separate stacked layers, the absorption mechanism is introduced with a resistive sheet. Another layer of the perfect electric conductor is used beyond the stacked layers. The optimization algorithm (genetic algorithm) is shown to optimize the structure to alter the resistance value from 10 to 1000 Ω/square. The final goal of this absorptive metasurface is to get as low as −30 dB of the reflection amplitude in the Ku band. In another optimization case, another reflection amplitude goal of −15 dB is selected from 6 to 30 GHz band. Considering one example, it is obvious that there are countless possibilities to randomly determine the areas within the metasurface that can be optimized for a specific goal. Another way of absorption of wave incident to metasurface is to diffuse the energy away from the reflection direction. This is shown in [30] where optically transparent diffusion-type metasurface is presented. In this work, a nine-element structure cell is used to mimic the required boundary conditions. This structure is then extended to a $240 \times 280\ \text{mm}^2$ area. The coding scheme used in this work is 1 bit, while the full-wave electromagnetic simulations are used to verify the diffusion-type scattering. As a result, it is shown that the incoming wave is diffused in randomly formed polarizations at 9, 11, and 13 GHz frequencies. It is also argued that same working principle can be extended for oblique incident waves of 15°, 30°, and 45°, specifically at 13 GHz. The same diffusion-like scattering (mimicking absorption) can work for both polarizations of the incident waves [30]. Another absorptive metasurface of two EBGs lattices is shown in [31], with a goal of reduced radar cross section. The unit cell is arranged in a periodic and aperiodic arrangement such that the best reduction of monostatic and bistatic radar cross section can be achieved. Scattering characteristics of the metasurface layout are optimized through particle swarm optimization (PSO) such that high absorption with around 80% bandwidth is achieved. The normal incidence wave is shown to get −9.5 dB radar cross section (RCS) reduction, while the measured results are verified by simulations. A new metasurface composed of subwavelength metallic gratings with space-variant orientations is shown for the operation of microwave and infrared spectrums [32]. Equivalent transfer matrix is shown to characterize dispersion with high accuracy and broadband absorption response is observed.

### 3.2.2 Refractive and Transmission-Type Metasurfaces

Metasurfaces specialized for transmission are required to introduce as less loss as possible into the transmitted wave. Such metasurfaces are also referred to as selective surfaces which provide outstanding behavior for a wave that propagates through it. In [33], a similar surface is shown which shows a unique transmission method when the size of a periodic square metallic patch is increased. It is also shown that a similar metasurface on the infinite resistive sheet has a wideband frequency operation; however, due to evanescent diffraction, the metasurface of the rotated metallic patch do not behave as standard surfaces. Interestingly, these results are useful to realize evanescent-based transmission characteristic that is not predictable via general boundary conditions. The transmission characteristics depend upon the design frequency. In microwave frequencies, electrical continuity of a surface enables mutation that is realizable via Babinet's theory [33]. At a specific band, the signal transmission increases with the increase in the rotated metal patch size. On the other hand, on regular surfaces, the opposite phenomenon is obvious. Similar behavior is shown when a plane wave is transmitted through dielectric-loaded slots which are periodically formed on a finite-sized metallic plate [34]. When the dielectric slot has a depth, its surface can be considered as 3D metasurface. An analytical approach is used to identify the resonances due to higher-order modes generated inside the slots. The analytical model results match well with the full-wave electromagnetic simulations that are also presented in this work. The model used is also analog to Fabry–Perot resonances model in which dominant modes generated in the slot structures are considered. The Fano resonances in these kinds of structures arise from a phenomenon of extraordinary transmission as shown in [35], in which space between coupling slots is critical. The approximate formula for the plane-wave transmission can be considered general for any dielectric-loaded slot with depth and works well verified by high-frequency structure simulator (HFSS) simulations. Note that this response is extremely narrowband due to its resemblance to the modes of parallel plate waveguide and not the electromagnetic wave propagating through the resonating surface (general metasurface). Also, it is to be noted that that the loss within dielectric-loaded slots is high and can be seen in [34], and the broad transverse electric and transverse electromagnetic (TEM) mode resonances are not disturbed due to the addition of the high loss so the analytical model is still valid. It is hard to say whether similar models tend to work well for a high degree of obliging incident wave. Contrary to narrow bandwidth response, a broadband metasurface that also has a unique behavior is presented in [36]. The most important difference of this metasurface from the general form is that it does not work on local resonance prediction of unit cells, while it looks at the grid impedance when the unit cell is periodically placed. This way, it is easy to get to the global resonances of the metasurface. A higher-order metasurface mode is also shown with a comparatively wider band radiation and proof of concept is developed on 5 GHz Wi-Fi. The metasurface is tested using quasi-TM01, quasi-TM21, and dipole-based modes. The results indicated that the metasurface has transmission bandwidth of 45%.

At very high frequencies and optical spectrum, the propagation of surface plasmon polaritons from metasurfaces is an extremely complex phenomenon and the main reason is that analytical linkage between electronic properties (that are microscale) is required to be built to the optical properties. When the metasurface is composed of complex geometrical formation of the unit cell, transmission behavior and nature of the operation are hard to calculate. In studies like [25], this complex behavior is analyzed in the same way as in wave propagation through the homogeneous layer. Using this, fundamental laws of optics are used to easily predict the properties. A simple mathematical model is also shown which can be used to calculate the scattered and transmitted electromagnetic field from similar metasurfaces.

An active reflective metasurface in optical spectrum with tuneable phase is shown which works on electric and magnetic resonances [37]. The structure is designed on a silicon-based antenna on indium tin oxide (ITO) alumina which is controlled by multi-gate biasing. The reflection of the structure is controlled by resonance and as a result, good reflection amplitude in phase-change coverage of around 180° at the near-infrared spectrum is shown. In addition to this, electro-optical tuneability is added within the design with step-by-step increased complexity.

Ultrathin metasurface compromising of various subwavelength meta-particles offers promising advantages in controlling electromagnetic wave by spatially manipulating the wavefront characteristics across the interface. The recently proposed digital coding metasurface could even simplify the design and optimization procedures due to the digitalization of the meta-particle geometry. However, current attempts to implement the digital metasurface still utilize several structural meta-particles to obtain certain electromagnetic responses and requires a time-consuming optimization especially in multi-bit coding designs. In this regard, we present herein utilizing geometric phase-based, single-structured meta-particle with various orientations to achieve either 1-bit or multi-bit digital metasurface. Electromagnetic wave scattering patterns dependent on the incident polarization can be tailored by the encoded metasurfaces with regular sequences. On contrast, polarization-insensitive diffusion-like scattering can also be successfully achieved by digital metasurface encoded with randomly distributed coding sequences leading to substantial suppression of backward scattering in a broadband microwave frequency. The proposed digital metasurfaces provide simple designs and reveal new opportunities for controlling electromagnetic wave scattering with or without polarization dependence [38].

### 3.2.3 Digitally Encoded Metasurfaces

Digital coding of a metasurface can optimize the design and operation process of the surface due to geometry's digitalization. The collective response of each unit cell of the metasurface with arbitrarily digital response makes unique scattering field patterns, which depends on the coding sequence and vice versa. While the process is beneficial, the electromagnetic responses can be time-consuming to optimize large metasurfaces with multi-bit digitalization. The same structure can be 1-bit or multi-bit digital control of metasurface, like in [38], where the polarization of the scattering patterns is controlled using the metasurface. In addition to this, diffusion-like scattering is also shown to achieve polarization-insensitive randomization using encoding scheme. A simple 2-bit case of digitally encoded metasurface shows the manipulation of the polarization-dependent properties and is given in [38]. Another investigation reports a broadband backward scattering reduction digitally encoded metasurface that uses diffusion-like microwave reflection. The diffusion is caused via pseudorandom computer-generated coding sequence. The metasurface operates from 8 to 15 GHz with a 10-dB scattering reduction [30]. Moving higher in the frequency, a 2-bit random frequency-encoded metasurface for THz operation is shown in [39]. Another THz-encoded metasurface is shown to do a different job, i.e. focusing THz energy at a single point [40], while radar cross-section reduction in THz domain using coded metasurface is shown in [41].

### 3.2.4 Polarization-Sensitive Metasurface Spectral Filtering

The dual-polarized operation is of interest for satellite communication industry, and many quasi-optical systems for satellite antennas employ metasurfaces for spectral filtering as frequency-selective surfaces (FSSs). Whereas response of the metasurface to the dual-polarized excitation at normal incidence can be trivially dealt with symmetrical scatterers, such as square ring and crosses, making propagation through the metasurface identical for both polarizations, this is not the case for the oblique incidence. Incidentally, the oblique incidence at 45° is the main operational mode for the usage of the filtering metasurfaces for quasi-optical beam splitters in feeding systems of reflector antennas, employed in satellite communications and atmosphere sounding [42]. Polarization performance in the form of either dual polarization or circular polarization operation is being frequently required [43], as the former requires the amplitude response of both orthogonal polarizations to be identical, whereas the latter adds the condition of identical phase responses.

Operation under circular polarization excitation requires to concurrently preserve identical amplitude and phase response for both transverse electric (TE) and transverse magnetic (TM) polarization within a specified frequency band. The metasurfaces can be required to operate both in bandpass and bandstop mode. For spectral filtering, the metasurfaces designed as periodic structures with a specifically designed unit cell. Let us first consider a circuit model for a single-element metasurface, presented in Figure 3.3. In the simplest case, a bandpass metasurface can be modeled as parallel LC resonator and bandstop FSS as series LC resonator. The admittance $Y_0$ or impedance $Z_0 = 1/Y_0$ of the resulting Floquet waveguide is defined for both TM and TE polarizations as follows:

$$Z_0^{TE} = \frac{1}{Y_0^{TE}} = \frac{\eta_0}{\cos\theta} \tag{3.8a}$$

$$Z_0^{TM} = \frac{1}{Y_0^{TM}} = \eta_0 \cos\theta \tag{3.8b}$$

where $\eta_0$ is the plane-wave impedance in free space and $\theta$ is the incident angle [44]. The transmission coefficient for the bandpass circuit is

$$T_i = \frac{2}{2 + Y_i/Y_0} \tag{3.9}$$

where $Y_i = j\omega C_i + 1/j\omega L_i$, and $C_i$. and $L_i$ are capacitance and inductance of the circuit. For circular polarization excitation, the transmission should be constrained to be identical for

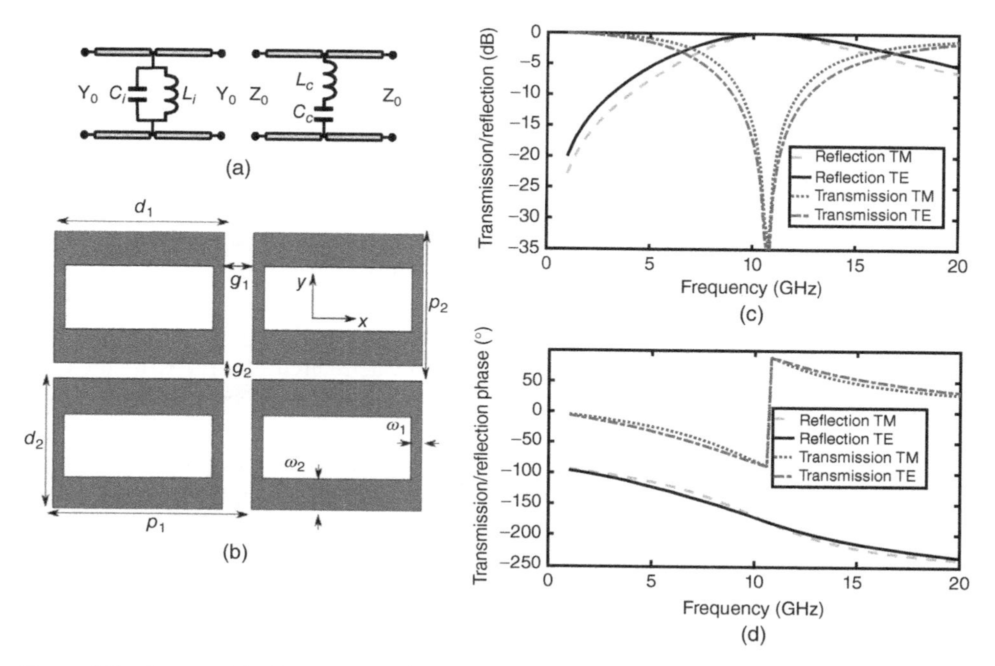


**Figure 3.3** Asymmetrical loop metasurface: (a) equivalent circuit model for bandpass and bandstop response, (b) unit cell geometry, simulated transmission, and reflection for both polarizations at the oblique incidence of 45°, (c) magnitude, and (d) phase.

both TE and TM components, thus,

$$T_i^{TE} = T_i^{TM} \tag{3.10}$$

It follows from (3.8a), (3.8b), and (3.10) that

$$Y_i^{TE} = Y_i^{TM}\cos^2\theta \tag{3.11}$$

for both inductance and capacitance, the condition (3.11) results in

$$C_i^{TE} = C_i^{TM}\cos^2\theta \tag{3.12a}$$

$$L_i^{TE} = \frac{L_i^{TM}}{\cos^2\theta} \tag{3.12b}$$

For the bandstop circuit, the reflection coefficient reads

$$\Gamma_i = -\frac{1}{1 + 2Z_c/Z_0} \tag{3.13}$$

where $Z_c = j\omega L_c + 1/j\omega C_c$. Now, for circular polarization, the reflection coefficient should be identical for both *TE* and *TM* components, thus

$$\Gamma_c^{TE} = \Gamma_c^{TM} \tag{3.14}$$

or substituting (3.8) into (3.14)

$$C_c^{TE} = C_c^{TM}\cos^2\theta \tag{3.15a}$$

$$L_c^{TE} = \frac{L_c^{TM}}{\cos^2\theta} \tag{3.15b}$$

It is worth noting that for the bandpass and bandstop scenarios, the conditions (3.12a), (3.12b), (3.15a), and (3.15b) are identical. In order to design the FSS with required circular polarization response, one may apply direct optimization of the structure with the aforementioned conditions comprising of the objective function. In some cases, closed-form expressions exist for the inductance and capacitance, thus making possible the direct synthesis with (3.12) or (3.15).

One of the structures with the known equivalent circuit is a strip grating and its derivatives such as grid FSS, square loop, and double square loop [45]. In [46], an asymmetric loop element metasurface was demonstrated. The equivalent circuit for the structure corresponds to the series LC resonator in order to synthesize a bandstop FSS for circular polarization excitation at oblique incidence.

The loop element has been made asymmetric, as shown in Figure 3.3, upon careful study of the expressions for the symmetrical square loop [45], it has been concluded in [47] that for oblique incidence, there is no solution satisfying conditions (3.15). The design requirements for a single-element FSS should specify a resonant frequency $f_{res}$, bandwidth $\Delta f$, which can be combined into the quality factor $Q = f_{res}/\Delta f$, and the angle of incidence, assuming single-mode operation at the maximum frequency, limits the period of the unit cell as $p(1 + \sin\theta) < \lambda$, where $\lambda$ is the free-space wavelength.

First, by multiplying together the normalized susceptances $\widetilde{B}^{TE} = (\omega_{res}^2 C_c^{TE})/Y_0$ and reactances $\widetilde{X}^{TE} = (\omega_{res}^2 L_c^{TM})/Z_0$ for both polarization at the resonant frequency and recalling that for an $LC$ resonator $\omega_{res}^2 = 1/LC$, one obtains

$$\widetilde{X}^{TE}\widetilde{B}^{TE} = \frac{\omega_{res}^2 L_c^{TE} C_c^{TE}}{Z_0^{TE} Y_0^{TE}} = 1 \tag{3.16a}$$

$$\widetilde{X}^{TM}\widetilde{B}^{TM} = \frac{\omega_{res}^2 L_c^{TM} C_c^{TM}}{Z_0^{TM} Y_0^{TM}} = 1 \tag{3.16b}$$

Then by applying the formulas for the quality factor as:

$$\frac{1}{Z_0^{TM}} \sqrt{\frac{L_c^{TM}}{C_c^{TM}}} = \frac{Q}{2} \tag{3.17}$$

two more equations are obtained. The normalized susceptance and reactances are functions of the incident angle and the geometrical parameters of the unit cell [47]. As the periods $p_1$ and $p_2$ can be determined from the grating lobe condition, the system of four implicit Equations (3.16a), (3.16b), and (3.17) contains four unknown geometrical parameters $g_1$, $g_2$, $w_1$, and $w_2$ (Figure 3.3) and can be solved numerically. The design procedure applied to find the example geometry in Figure 3.3b of a freestanding circular polarization FSS resonating at $f_{opt}$ = 12.4 GHz with quality factor $Q$ = 2.45. The angle of incidence and period were fixed as $\theta = 45°$, $p_1$ = 7.2 mm, and $p_2$ = 7.2 mm. Following the proposed design process, the rest of the geometrical parameters were found as $w_1 = w_2$ = 0.144 mm, $g_1$ = 0.8921 mm, and $g_2$ = 0.8383 mm. The resulting transmission and reflection amplitude and phase of both polarizations closely follow each other thus ensuring high-quality circular polarization bandstop filter, as shown in Figure 3.3.

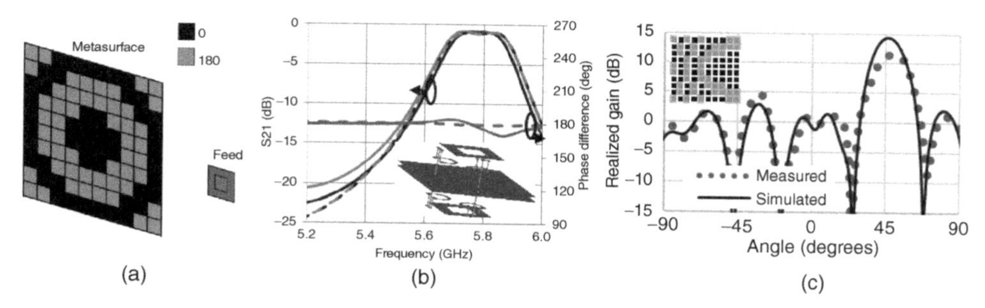


**Figure 3.4** Dual-polarized 1-bit transmitarray: (a) device architecture including individual unit cells (different colors indicate one of the two available phase states) mounted in a plastic grid frame and spatially fed by a focal source patch antenna, (b) the unit cell and simulated scattering parameters of the unit cell, the solid and dashed curves for the magnitude correspond to 0° and 180° phase states, and (c) metasurface phase distribution for 45° beam and corresponding simulated and measured radiation pattern.

### 3.2.5 Beamforming with Polarization-Controlling Metasurfaces

The metasurfaces can serve as beamforming structures that shape the spatial radiation properties of electromagnetic waves. To achieve this, quasi-periodical structures are employed with unit cells designed to superimpose necessary amplitude and phase profile onto an incident electromagnetic wave. The resultant distribution of electromagnetic field defines the spatial distribution of the electromagnetic waves emanating from the metasurface. A flat transmitarray antenna operates by adjusting the phase of the spherical wavefront emitted by the antenna feed positioned at the focal point, into the uniform plane wave propagating toward the intended recipient, as shown in Figure 3.4. Compared to the common dielectric lenses with a curved profile, the flat transmitarray structures can be manufactured in a low-cost printed circuit board (PCB) technology and subsequently can be made electrically reconfigurable. For each unit cell of a phase-modulated transmitarray [48], the received normalized wave amplitude reads

$$a_{mn} = \frac{\lambda e^{jkR_{mn}}}{4\pi R_{mn}} \boldsymbol{F}^{f}_{mn} \cdot \boldsymbol{F}^{u}_{mn} \tag{3.18}$$

where $n$ and $m$ are the row and column indexes within the array which determine the position of each unit cell, $k$ and $\lambda$ are free-space wavenumber and wavelength, respectively. $\boldsymbol{F}^{f}_{mn}$ is the vector complex field pattern of the focal plane source, $\boldsymbol{F}^{u}_{mn}$ is the field pattern of the unit cell on the receive side, and $R_{mn}$ is the distance between the focal plane source and the unit cell. Assuming that each unit cell has a known complex transmission coefficient $T_{mn}$, the complex amplitude of the waves radiated by unit cells reads

$$b_{mn} = T_{mn} a_{mn} \tag{3.19}$$

To find the resultant radiation pattern of the transmitarray, we assume for simplicity the case of linear polarization, where the unit cell pattern $F^u(\theta, \phi)$ is a complex scalar function. Therefore, the complex amplitude antenna pattern of the transmitarray, $F(\theta, \phi)$, reads

$$F(\theta, \phi) = |F^u(\theta, \phi)| \sum_{n=1}^{N} \sum_{n=1}^{N} b_{mn} e^{j\arg(F^u(\theta,\phi))} e^{jkd(n\sin\theta\cos\phi + m\sin\theta\sin\phi)} \tag{3.20}$$

where $d$ is the period of the array. This model is utilized for the design of a transmitarray by adjusting the unit cell transmission coefficients $T_{mn}$ for a given focal source and unit cell radiation patterns. The beam steering is achieved by modulating the phase distribution of the emitted wavefront across the array aperture. For symmetrical unit cells, the required local phase shift can be found as:

$$\arg(T_{mn}) = \arg(b_{mn}) - \arg(F^{f}_{mn}) - \arg(F^{u}_{mn}) + kR_{mn} \tag{3.21}$$

where the phase of the focal source pattern is calculated for the direction pointing at the unit cell. For any given technology, the metasurface is designed to satisfy these equations with a certain approximation. The simplest approximation is a discretization of the phase distribution with a 1-bit resolution as follows:

$$\arg(T_{mn}) = \begin{cases} 0°\forall \mid \arg(T_{mn}) \mid \leq 90° \\ 180° \text{otherwise} \end{cases} \tag{3.22}$$

However, one has to bear in mind that the simplicity of the 1-bit phase quantization comes at cost, e.g. in [49], it was shown that the 1-bit resolution results in the gain reduction of up to 4 dB, higher sidelobe level and appreciable beam squint.

Let us now look at an example of the dual-polarized transmittarray that was proposed in [48]. The transmitarray of a square array of dual-polarized multilayer unit cells with half-wavelength periodicity, e.g. in Figure 3.4, the receiving and transmitting sides of the unit cell are represented by the square-ring microstrip elements with electromagnetic feeds. The proximity-coupled feeds are implemented using the open-ended half-wavelength semi-annular microstrip loops in the layers beneath the square-ring antennas. A two-loop resonator is connected by a buried via hole that goes through a solid ground plane. Using feeds in the orthogonal sides of the square-ring patch allows a dual-polarized unit cell design. Each unit cell is 24 mm × 24 mm and manufactured on four dielectric substrates of 0.51-mm-thick Rogers RO4003 material ($\varepsilon_r$ = 3.5, $\tan\delta$ = 0.0018). The 1-bit phase shift is enabled by switching the feed point of the U-shaped resonator on the receiving side of the transmitarray [49]. This switching enables the currents in the patches at the receiving and transmitting sides to flow co-directionally, where this is referenced as the 0° state. In the 180° state, the resonators are connected at the opposite ends, so that the surface currents flow in the opposite directions. For this structure, the simulated results shown in Figure 3.4 demonstrate a 180° differential phase with the phase error $< \pm 3°$ for both polarizations in the frequency band 5.7–5.86 GHz (2.8%) corresponding to $>$ 10 dB return loss.

The beam-steering capabilities of the metasurface are demonstrated in a 10 × 10 transmitarray illuminated by a linearly polarized printed patch antenna, as shown in Figure 3.4. The array was assembled manually using the prefabricated 0° and 180° unit cells, arranging the unit cells within the grid frame in the prescribed phase distribution pattern to steer the beam in a given direction. The H-plane radiation patterns measured at 5.75 GHz for a 45° beam angle is shown in Figure 3.4. The corresponding phase distribution is shown in the inset. The measured results show that the beam-steering angle is in a very good agreement with the full-wave electromagnetic simulations.

### 3.2.6 Beamforming with Reflective-Type Metasurfaces

As mentioned above, the metasurfaces can be excited by a plane wave and the resultant radar cross section of the scattered fields can be shaped to the desired pattern. In particular, there is significant interest in the generation of helical beams with nonzero OAM [50], which presents a formidable task for beamforming. They find application in advanced optics for imaging [51] and may have significant future potential for enhanced spectrally efficient radio communications [52]. In [53], a metasurface is proposed with unit cell elements that are orientated to produce the same wavefront transformation as is made by the 3D spiral phase plate in [54] and thus to generate an OAM-scattered beam. A spiral phase plate adds an azimuthally progressive phase delay to an incoming wavefront. Once a Gaussian beam passes the spiral phase plate, it transforms into a Laguerre–Gaussian beam with a null in the center and helical exit wavefront [53]. Spiral phase distribution in the Cartesian $xy$ plane is given by

$$\psi(x,y) = \begin{cases} \tan^{-1}\frac{y}{x}, & x > 0, y > 0 \\ \tan^{-1}\frac{y}{x} + \pi, & x < 0 \\ \tan^{-1}\frac{y}{x} + 2\pi, & x > 0, y \leq 0 \end{cases} \tag{3.23}$$

To implement such phase distribution, a precise mechanism for phase variation is necessary. For circular polarization excitation, one can use a rotational phase shift which occurs once the unit cell is rotated by a chosen angle with respect to its center by a chosen angle. With reference to double split-ring slot element elevated by a solid ground plane by $h$ mm within a unit cell shown in Figure 3.5, let us assume that a right-hand circularly polarized (RHCP) wave of frequency $\omega$ propagating along negative z direction is incident on the reflective metasurface:

$$\boldsymbol{E}_{i,RHCP} = E_0(\boldsymbol{a}_x + j\,\boldsymbol{a}_y)e^{jkz} \tag{3.24}$$

$E_0$ is the amplitude of the electric field, $\boldsymbol{a}_x$ and $\boldsymbol{a}_y$ are the unit vectors of $x$ and $y$ axes correspondingly, $k$ is the wavenumber, and the time dependence $e^{j\omega t}$ is omitted here and in further considerations. As shown in [55], the reflected field is

$$\boldsymbol{E}_{r,RHCP} = 0.5(\Gamma_{x'} - \Gamma_{y'})\boldsymbol{E}_0(\boldsymbol{a}_x - j\,\boldsymbol{a}_y)e^{j2\theta}e^{-jkz} + 0.5(\Gamma_{x'} + \Gamma_{y'})\boldsymbol{E}_0(\boldsymbol{a}_x + j\,\boldsymbol{a}_y)e^{-jkz} \tag{3.25}$$

where $\Gamma_{x'}$ and $\Gamma_{y'}$ are the reflection coefficients for the waves linearly polarized along the axes $x'$ and $y'$, respectively. It is important to note that, since the reflected waves travel in the positive $z$ direction, the first term in (3.9) is RHCP wave and the second is left-hand circular polarization (LHCP) one. For the fully reflective structure without loss $|\Gamma_{x'}| = |\Gamma_{y'}| = 1$ and the phase difference of the reflection coefficients $\Delta = \angle\Gamma_{x'} - \angle\Gamma_{y'}$ the amplitudes of the reflected LHCP and RHCP waves were determined. It is clear that when $\Gamma_{x'} = -\Gamma_{y'}$, the wave is fully converted to RHCP, whereas in case of $\Gamma_{x'} = \Gamma_{y'}$, i.e. perfect electrically conducting mirror, the wave is fully converted to LHCP. The physical dimensions of the structure were optimized to obtain the 180° phase difference between the reflection coefficients at the specified operating frequency of 10 GHz. The optimal dimensions of the unit cell are given in Table 3.1. The unit cell is a square with a period of 19 mm.

The simulated amplitude and phase responses of the periodic structure when illuminated at normal incidence with a circular polarization wave are shown in Figure 3.5, respectively,

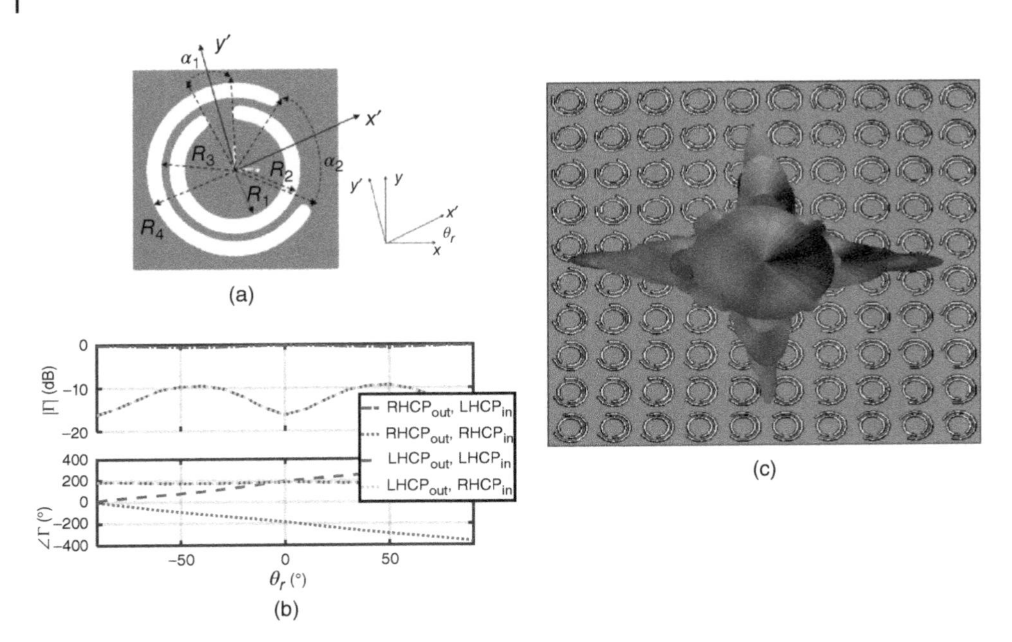


**Figure 3.5** Double split-ring reflective metasurface: (a) top layer unit cell of the metasurface, (b) S-parameters of the unit cell, and (c) scattered field from the metasurface.

**Table 3.1** Parameters of the unit cell.

| $R_1$ (mm) | $R_2$ (mm) | $R_3$ (mm) | $R_4$ (mm) | $\alpha_1$, ° | $\alpha_2$, ° | $h$ (mm) | $t$ (mm) |
|---|---|---|---|---|---|---|---|
| 4.7 | 5.9 | 6.8 | 8 | 21 | 95 | 7.5 | 1 |

as the unit cell is rotated. From the simulated results, one can conclude that the reflected wave preserves its hand opposite to the case of a perfectly conducting mirror. The study also shows that the circularly polarized wave upon reflection from the structure gains a phase shift twice that of the rotation angle. The cross-polarization conversion level is better than 10 dB.

Based on the results discussed above, the $10 \times 10$ structure with the dimensions given in Table 3.1 and the elements rotated according to the spiral phase shift was simulated. The scattered LHCP field upon excitation with normally incident LHCP plane wave at 10 GHz is shown in Figure 3.5. As one can observe, the main beam of the reflected field contains null in the center and azimuthal phase shift, which is the result of the spiral phase delay introduced by the screen.

### 3.2.7 Circular Polarization-Selective Metasurfaces

Stacked polarization-selective surfaces are used to conserve material on satellite antennas. Here, one surface acts to reflect one polarization while simultaneously transmitting the other, a second polarizing surface reflects the polarization transmitted by the first surface. Achieving this with circular polarization is a core critical path to enabling shared aperture multipurpose antennas [56]. Below, we discuss the operation of circular polarization-selective surfaces (CPSSs). The transmission coefficients for a structure subjected to *TM*/*TE* (*x*/*y*) or right/left circular polarization (*RHCP*/*LHCP*) excitation can be, respectively, expressed using the Jones matrices:

$$T^f_{lin} = \begin{pmatrix} S^{21}_{xx} & S^{21}_{xy} \\ S^{21}_{yx} & S^{21}_{yy} \end{pmatrix} = \begin{pmatrix} A & B \\ C & D \end{pmatrix} \tag{3.26}$$

or

$$T^f_{circ} = \begin{pmatrix} S^{21}_{RHCP,RHCP} & S^{21}_{RHCP,LHCP} \\ S^{21}_{LHCP,RHCP} & S^{21}_{LHCP,LHCP} \end{pmatrix} \tag{3.27}$$

where superscript $f$ denotes forward-directed ($+z$) incidence, and *lin* and *circ* denote linear and circular polarization base, respectively. The backward transmission coefficients (($-z$) incidence) $b$, are expressed similarly. We use IEEE notation for circular polarization waves and $e^{j\omega t}$ time dependence. Matrix $T^f_{circ}$ is obtained from $T^f_{lin}$ by the transformation of the basis as given in [57]:

$$T^f_{circ} = \widehat{\Lambda}^{t^{-1}} T^f_{lin} \widehat{\Lambda}^{t} \tag{3.28}$$

where the transformation matrix is written as:

$$\widehat{\Lambda}^{t} = \frac{1}{\sqrt{2}} \begin{pmatrix} 1 & 1 \\ -j & j \end{pmatrix} \tag{3.29}$$

in circular polarization basis:

$$T^f_{circ} = \frac{1}{2} \begin{pmatrix} A + D - j(B - C) & A - D + j(B + C) \\ A - D - j(B + C) & A + D + j(B - C) \end{pmatrix} \tag{3.30}$$

For backward transmission to preserve the hand of coordinate system, the axes change as $x^b = -x^f$ and $y^b = y^f$. With this change, the circular polarization eigenbasis is preserved. However, the linear transmission changes as follows:

$$T^b_{lin} = \begin{pmatrix} A & -C \\ -B & D \end{pmatrix} \tag{3.31}$$

And hence,

$$T^b_{circ} = \frac{1}{2} \begin{pmatrix} A + D - j(B - C) & A - D - j(B + C) \\ A - D + j(B + C) & A + D + j(B - C) \end{pmatrix} \tag{3.32}$$

Now, in order to realize an left-hand circular-polarization-selective surface (RHCPSS), structure that transmits *LHCP* and reflects *RHCP*, one needs

$$T^f_{circ} = \begin{pmatrix} 0 & 0 \\ 0 & S^{21}_{LHCP,LHCP} \end{pmatrix} \tag{3.33}$$

This is fulfilled when the following conditions hold

$$A = D = jB, B = -C \tag{3.34}$$

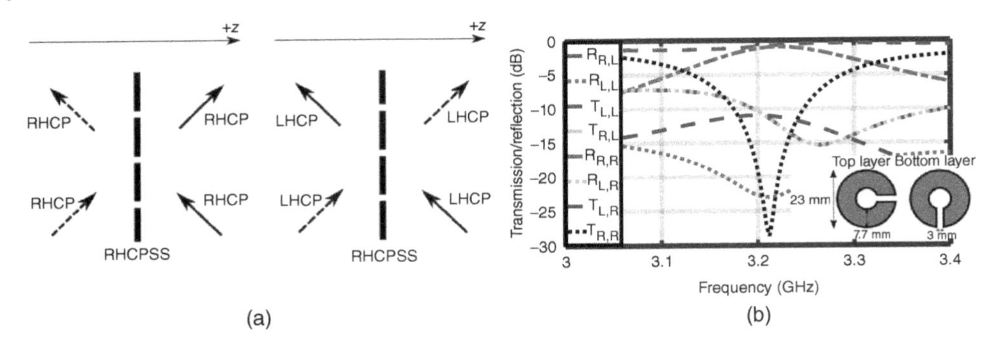


**Figure 3.6** Reciprocal symmetrical right-hand circular polarization-selective surface: (a) reflection and transmission operation of RHCPSS [58], (b) simulated performance of the metasurface, top and bottom rings are printed on 0.5-mm-thick RO4003, which are separated further by 5-mm-thick Rohacell. Source: (a) Modified from Roy and Shafai [58].

According to the analysis carried out above, this type of structure complies with the definition given in [58] for a symmetrical reciprocal CPSS whose properties are summarized in Figure 3.6. In order to realize a CPSS, a unit cell must respond to both TE and TM polarization while ensuring their cross-excitation. Originally, CPSSs were constructed with "single crank" structure called Pierrot element [59]. A coupled twisted SRR element with 90° turn is employed here to ensure the desired operation of the unit cell, similarly to [60]. The rings are coupled and aligned in the orthogonal planes. This coupling allows producing circular polarization scattered fields, provided that an appropriate phase and amplitude conditions are satisfied. A pair of SRRs separated by air substrate exhibits even and odd resonances with the currents in both SRR excited in phase or in antiphase. It has been shown in [61] that the even- and odd-mode resonant frequencies can be determined from a Lagrangian model as follows:

$$\omega_o = \omega_0 \sqrt{\frac{1 - \widetilde{k_e}}{1 - k_h}} \tag{3.35}$$

$$\omega_e = \omega_0 \sqrt{\frac{1 + \widetilde{k_e}}{1 + k_h}} \tag{3.36}$$

where $\omega_0$ is the resonant frequency of a single SRR, $\widetilde{k_e} = k_e(\cos\theta + a\cos^2\theta + b\cos^3\theta)$, $k_h$ is the magnetic coupling and $k_e$ is the electric coupling coefficient, $\theta$ is the twist angle, and $a$ and $b$ are fitting coefficients related to higher-order multipole expansion. It has to be noted that the magnetic coupling is isotropic and does not depend on the twist angle $\theta$ [61]:

$$k_m = \frac{\omega_e^2 - \omega_o^2}{\omega_e^2 + \omega_o^2} = \frac{k_h - \widetilde{k_e}}{1 - k_h\widetilde{k_e}} \tag{3.37}$$

An interest for circularly polarized selectivity presents the structure as shown in Figure 3.6 with a 90° twist. As one can see from the above formula for $\widetilde{k_e}$, for such a twist the electrical coupling vanishes and only magnetic coupling exists. Consider the dual-layer twisted ring structure unit cell consisting of an array of 3 × 3 twisted rings shown in Figure 3.6. Here, TE waves at normal incidence couple to the top split ring, while TM waves couple to the bottom split ring. If the rings are closely spaced, then mixed electric and magnetic coupling exists, but due to the lack of symmetry these will be different. The $3 \times 3$ array unit cell arrangement was optimized with an electromagnetic simulator to ensure (3.10) as:

$$S^{21}_{TE,TM} = -S^{21}_{TM,TE}, \tag{3.38}$$

$$S^{21}_{TE,TE} = S^{21}_{TM,TM}, \tag{3.39}$$

$$S^{21}_{TE,TE} = \pm jS^{21}_{TE,TM} \tag{3.40}$$

The results of the simulations are presented in Figure 3.6. It is shown that for $-z$ direction excitation, RHCP is being reflected while incident LHCP is being transmitted through the structure, i.e. the structure is LHCP transparent. The same holds true when the structure is excited from the $+z$ direction. This agrees with the general reciprocal symmetrical circular polarization-sensitive surface description in Figure 3.6.

### 3.2.8 Passive Lossless Huygens' Metasurfaces

In [62], a solution for passive lossless Huygens' metasurfaces is proposed for the purpose of wave manipulation based on the concept of a sheet of orthogonal electric and magnetic dipoles. The paper propagates the notion that a source wave/field can be converted into directive radiation using a passive and lossless Huygens' metasurface provided that two conditions are met, i.e. local impedance equalization and local power conservation. The design condition indicates that impedance equalization induces Fresnel reflections, while power conservation forms a radiating aperture which follows total excitation of the field magnitude.

For the Huygens' metasurface to become purely reactive, the real power must be conserved across the metasurface. For the purpose, it is required that the power (real) incident at region 1 on the metasurface is equal to the power (real) transmitted onto region 2 equally across the $z$ axis on the surface shown in Figure 3.7a:

$$R\{E_x H_y^*\}(y, 0^+) = R\{E_x H_y^*\}(y, 0^-) \tag{3.41}$$

where $E_x$ and $H_y$ are tangential electric and magnetic field components, respectively. The half-spaces below and above the metasurface are referred to as region 1 and region 2 of Figure 3.7a. The second condition implies the local equalization of the wave impedance of the fields on the two facets of the metasurface, i.e. for an arbitrary source, the reflected wave existence is necessary to satisfy this condition:

$$\frac{E_x^{trans}(y)}{H_y^{trans}(y)} = \frac{E_x^{inc}(y) + E_x^{ref}(y)}{H_y^{inc}(y) + H_y^{ref}(y)} \tag{3.42}$$

where $E_x^{trans}$ and $H_y^{trans}$ are transmitted electric and magnetic field components in region 2 along the surface of Figure 3.7a. $E_x^{inc}$ and $H_y^{inc}$ are incident electric and magnetic field components in region 1. $E_x^{ref}$ and $H_y^{ref}$ are reflected electric and magnetic field components in region 1. As referred in [13], transforming the complete incident field without inducing reflections results in the real part of admittances and surface impedance. This should be

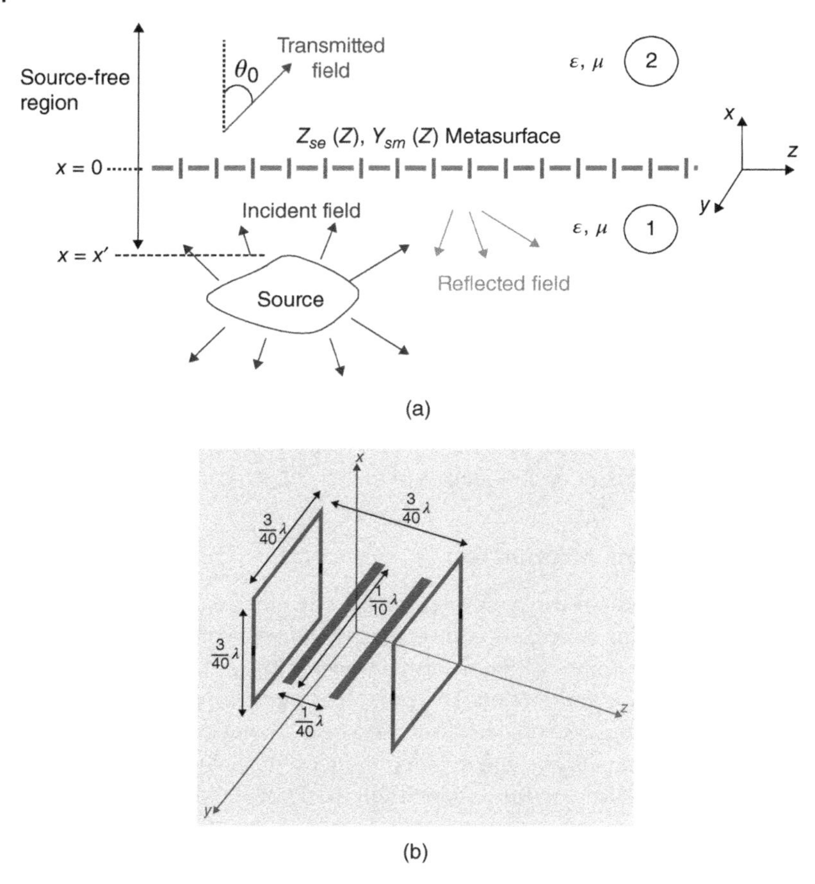


**Figure 3.7** (a) 2D configuration of the Huygens' metasurface positioned at $x = 0$ which excited by an arbitrary source placed at $x \leq x' < 0$ and (b) the implemented Huygens' metasurface unit cell geometric configuration in ANSYS HFSS. Source: Epstein and Eleftheriades [62]. © 2014, IEEE.

avoided at all cost if one has to realize a lossless and a passive metasurface. Thus, both local power conservation and local impedance equalization (matching) are the necessary prerequisites for achieving the goal.

The analytical design of the Huygens' metasurface in [62] is verified by the author via a finite element simulation of the design for dissimilar source configurations. The analytically designed Huygens' metasurface is implemented on a commercial solver. Each implemented unit cell is composed of two square of perfect electric conductor (PEC) closed-loop $3/40\lambda$ in length loaded with either lumped capacitor or lumped inductors symmetrically positioned to each other. Each Huygens' metasurface is implemented using $1/10\lambda$ long unit cell with a center frequency of $f = 1.5$ GHz. The design is shown in Figure 3.7b.

The comparison between the theoretical and simulated result of the metasurface showed a high level of correlation and similarity with the metasurface converting the field produced by electric line source at $x = -\lambda$ to directive radiation pattern at $\theta = 0°$ and $60°$ as shown in Figure 3.8.

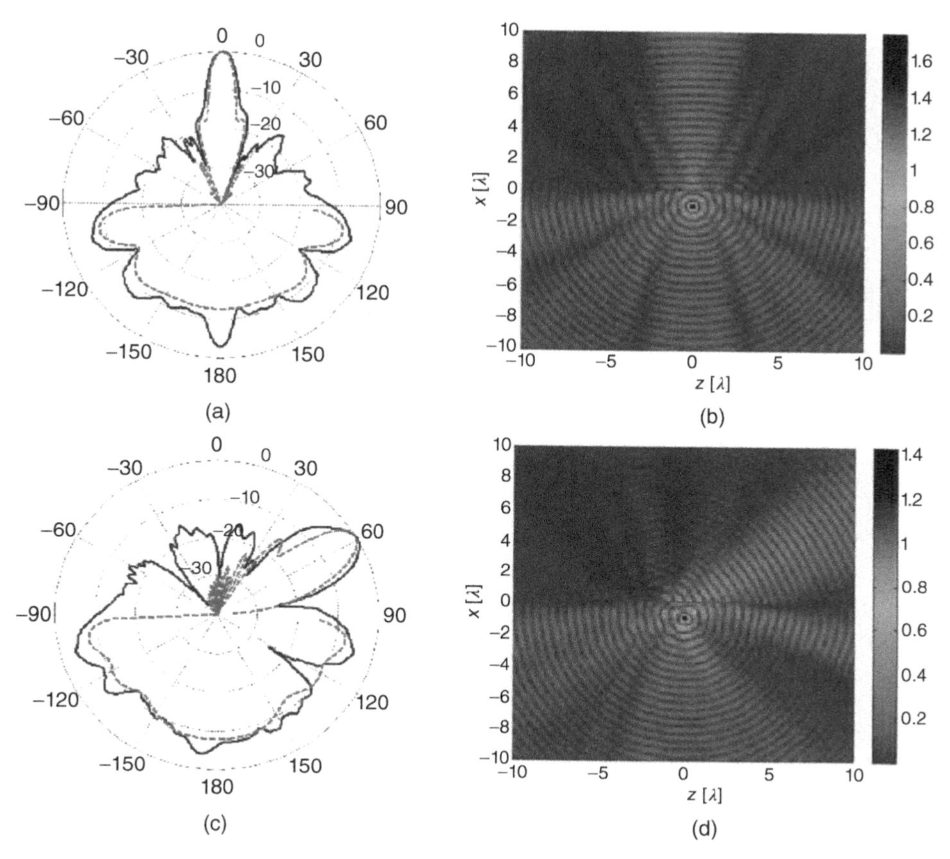


**Figure 3.8** Comparison between the full-wave simulated (solid lines) and theoretically calculated (dashed line), for $\theta = 0°$: (a) normalized radiation pattern and, (b) full-wave simulated real part of the electric field phasor and for $\theta = 60°$, (c) normalized radiation pattern, and (d) full-wave simulated real part of the electric field phasor. Source: Epstein and Eleftheriades [62]. © 2014, IEEE.

### 3.2.9 Cavity-Excited Huygens' Metasurface Antenna

Using a Huygens' metasurface, researchers resourcefully converted cavity fields into highly directive radiation cavities. No power is dissipated through edges with the use of three PEC walls that are surrounding the Fabry–Perot leaky-wave antennas structure, fed by an electric line source [63]. The reflecting surface is replaced by the metasurface as shown in Figure 3.9. With PEC walls, we overcome the efficiency problem in leaky-wave antenna (LWA). The higher-order mode is excited by adjusting the thickness of the cavity and the position of an incident field which results in favorable aperture illumination.

In [63], a common problem is addressed which is generating highly directed wave using low-profile devices by utilizing Huygens' metasurfaces. The concept is materialized by demonstrating a cavity-backed Huygens' metasurface in which a single source based cavity is designed for optimized aperture illumination as shown in Figure 3.9a. The Huygens' metasurface facilitates in disrupting the coupling between the source excitation and

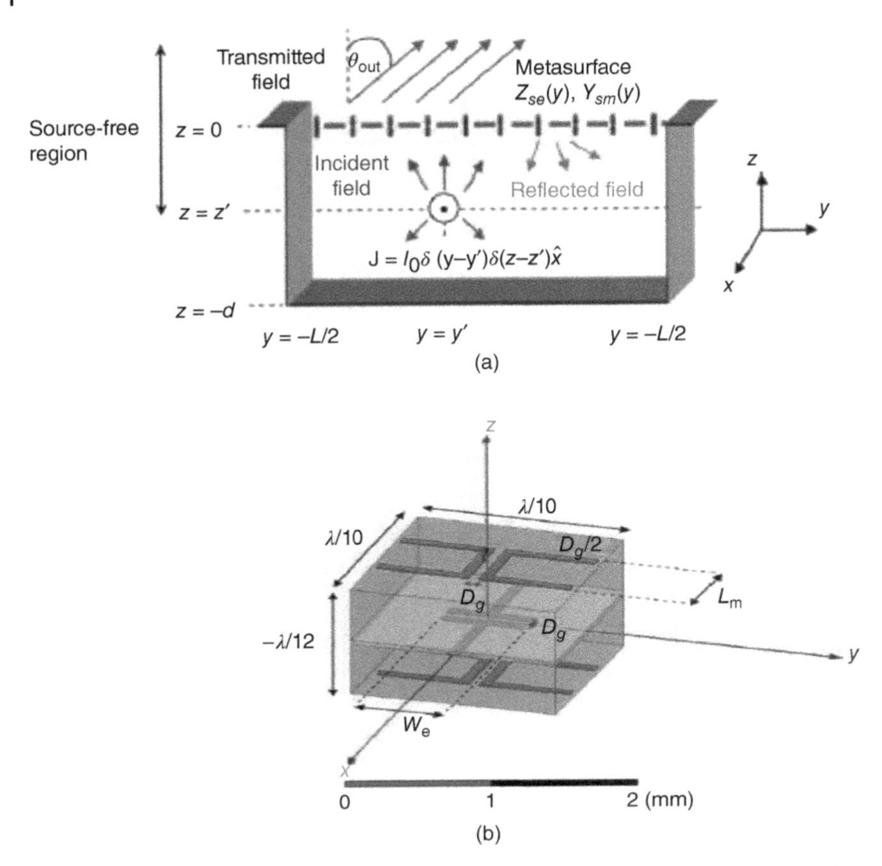


**Figure 3.9** (a) 2D configuration of the Huygens' metasurface positioned at $x = 0$ which excited by an arbitrary source placed at $x \leq x' < 0$ and (b) the implemented Huygens' metasurface unit cell geometric configuration in ANSYS HFSS. Source: Epstein et al. [63]. Licensed under CC BY 4.0.

radiation spectra resulting in tailored aperture properties that produce directive radiation pattern without inducing edge tapering losses.

The design consists of a spider shape unit cell shown in Figure 3.9b, which operates at 20 GHz [63]. The dimensions of the unit cell in the $x$, $y$, and $z$ axis are $\lambda/10$, $\lambda/10$, and $\lambda/12$, respectively. Rogers RT/duroid 6010LM and Rogers 2929 bondply are used between the three layers of metallic sheets for the unit cell. By changing the arm length of the top and bottom layer, one can tune the magnetic response of the unit cell, while the center sheet is responsible for the tuning of the electrical response. The modus operandi for the tuning is that the arm length affects the magnetic current caused by the tangential component of magnetic field, and the width of the center sheet plays a role in tuning the electrical current caused by the tangential electric fields. This metasurface is fed by a coaxial cable that is placed inside an aluminum cavity as shown in Figure 3.10a.

The novel design presented in [63] helped in solving an antenna design problem by generating highly directive waves/beams from a small low-profile structure fed through a single source without encountering edge taper losses. At $f = 20.04$ GHz, there is a good

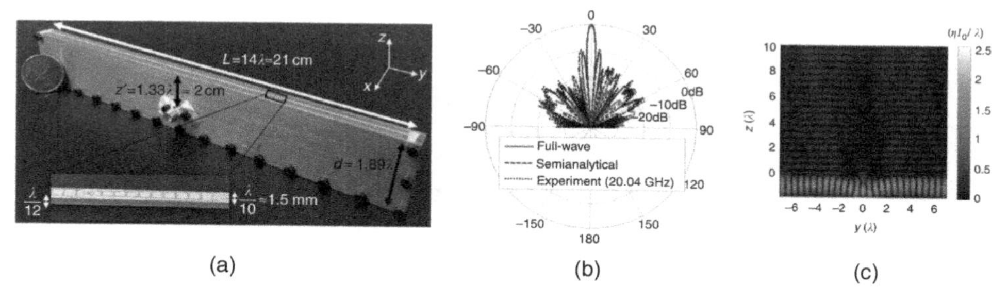


**Figure 3.10** (a) Fabricated prototype Huygens' metasurface antenna. (b) Comparison between the normalized radiation pattern of full-wave simulated (solid lines), theoretically calculated (dashed line), and measured (black dotted), for $L = 14\lambda$: (a) and (b) full-wave simulated electric field distribution. Source: Epstein et al. [63]. Licensed under CC BY 4.0.

agreement between the simulated and measured radiation pattern with the directivity, half power beamwidth (HPBW), and sidelobe level in complete coherence with each other, as shown in Figure 3.10b. This has been demonstrated through full-wave simulation and subsequent experimental validation of the result by fabricating a radiator with near to unity aperture illumination efficiency as shown in Figure 3.10c.

### 3.2.10 Passive Lossless Omega-Type Bianisotropic Metasurface

As evident, the Huygens' metasurfaces utilize a chain of orthogonal electric and magnetic elements for controlling directivity, angle, and polarization of the transmitted fields and for generating arbitrary field distribution for given incident field [13]. In comparison to Huygens' metasurface, omega-bianisotropic metasurface (O–BMS), along with electric impedance and magnetic admittance, also employs magnetoelectric coupling [64]. This property corresponds to an additional degree of freedom where one not only has control over the transmission wave properties of the metasurface but also the reflection wave properties as well.

A vast range of applications are explored with the systematic engineering of O-BMS-based devices [64, 65]. In contrast to Huygens' metasurfaces, O-BMS has control over both the magnitude and phase of not only the transmitted wave but also the reflected wave [66]. Also, on O-BMS, the electric field induces magnetic current and magnetic field induces electric current due to magnetoelectric coupling factor [67, 68]. The analytical theory for the design of an O-BMS for controlling the EM wave has been presented in [65]. They formulated an O–BMS structure contrary to Huygens' metasurface in which they not only controlled the phase of the transmitted wave but also achieved a high-level control over the amplitude and phase of the reflected wave. The demonstration of designing and simulating of the O-BMS structure is done via a full-wave simulation on the commercial solver. The physical realization of the O-BMS structure contains the spider unit cell consisting of three copper layers constructed on two 25 mil RT6010 laminates. At 20-GHz frequency, the unit cell has the dimension of 1.58 × 1.58 mm with a metasurface thickness of 1.32 mm as shown in Figure 3.11a. The three-layer design of the omega-shaped meta-atom is shown in Figure 3.11a that are equidistanced by placing substrate of thickness $t$ between them which acts as a transmission line for normal plane wave incident upon the structure. The

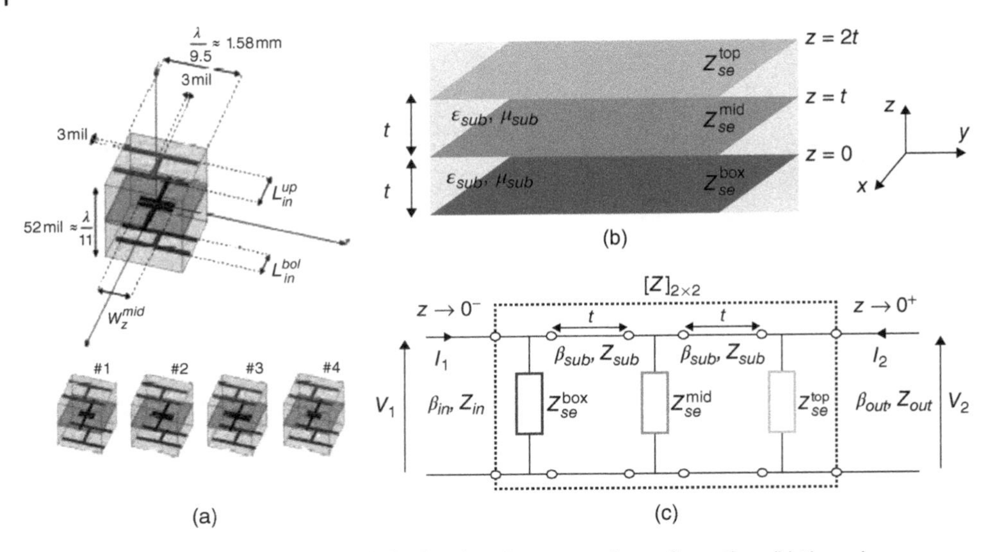


**Figure 3.11** (a) The implemented O-BMS unit cells geometric configuration, (b) three-layer impedance sheet configuration to represent O-BMS unit cell, and (c) the corresponding equivalent circuit model. Source: Epstein and Eleftheriades [65]. © 2016, IEEE.

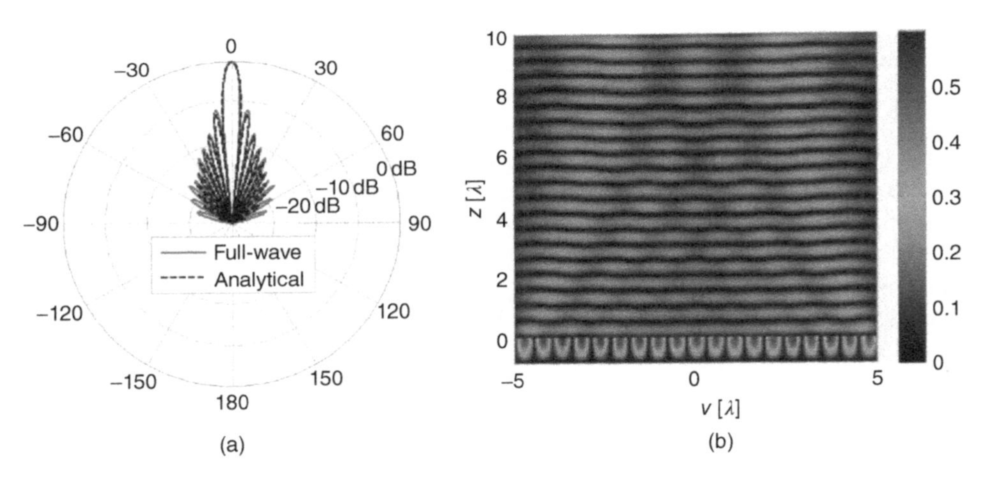


**Figure 3.12** Comparison between the full-wave simulated (solid lines) and theoretically calculated (dashed line), for $L = 10\lambda$: (a) normalized radiation pattern and (b) full-wave simulated electric field distribution. Source: Epstein and Eleftheriades [65]. © 2016, IEEE.

wave impedance $Z$ and wave number $\beta$ can be calculated using the model presented in Figure 3.11b.

The O-BMS-based structure is used to implement a simple antenna device [65]. The radiation patterns that are obtained from both the analytical calculation as well as from the full-wave simulations are in complete agreement with each other, as can be observed in Figure 3.12. The implemented design also shows a high aperture efficiency. The slight mismatch and discrepancy are attributed to the minute spatial dispersion and the finite length of the metasurface structure [65] demonstrated a generic theory for the design of a lossless and passive O-BMS structure. The result shows that such structures can be realized with standard available laminates and bonds thus making its fabrication possible via the common PCB fabrication technology. Thus, O-BMSs provides the independent control of both reflection and transmission magnitude and phase of normally incident plane waves, enabling a completely new line of research for the total control of electromagnetic waves.

## 3.3 Conclusion

It is believed that controlling the electromagnetic waves reflected from and transmitted through intelligently designed metasurfaces can shape the future of the communication industry. The ability to control surface waves by specifically designing the metasurface unit cell has opened a plethora of research areas. With varying geometric and periodic patterns, the metasurface can block, absorb, concentrate, guide, or disperse the incoming wave either in normal or oblique incidence angle. The electric and magnetic sources can be realized by use of inductance and capacitance elements within the unit cell of the metasurface for generating the required responses to the incident plane wave. From low radio frequencies to the ultraviolet spectrum, metasurface-based devices are showing extraordinary properties with potential applied in various fields that include but are not limited to spatial light and microwave signal modulation, beamforming, beam steering, beam shaping, quantum optic devices, low-, high-, and nanoscale-resolution imaging, microwave-, mmWave-, THz-, and bio–sensing, and even optical communication network formation.

## References


1 Lim, S., Caloz, C., and Itoh, T. (2004). Metamaterial–based electronically controlled transmission–line structure as a novel leaky–wave antenna with tunable radiation angle and beamwidth. *IEEE Transactions on Microwave Theory and Techniques* 52 (12): 2678–2690.

2 Erentok, A. and Ziolkowski, R.W. (2008). Metamaterial–inspired efficient electrically small antennas. *IEEE Transactions on Antennas and Propagation* 56 (3): 691–707.

3 Holloway, C.L., Kuester, E.F., Gordon, J.A. et al. (2012). An overview of the theory and applications of metasurfaces: the two–dimensional equivalents of metamaterials. *IEEE Antennas and Propagation Magazine* 54 (2): 10–35. https://doi.org/10.1109/MAP.2012.6230714.

4 Patel, A.M. and Grbic, A. (2013). Effective surface impedance of a printed–circuit tensor impedance surface (PCTIS). *IEEE Transactions on Microwave Theory and Techniques* 61 (4): 1403–1413. https://doi.org/10.1109/TMTT.2013.2252362.

5 Chen, M., Kim, M., Wong, A.M.H., and Eleftheriades, G.V. (2018). Huygens' metasurfaces from microwaves to optics: a review. *Nanophotonics* 7 (6): 1207–1231. https://doi.org/10.1515/nanoph-2017-0117.

6 Yu, N., Genevet, P., Kats, M.A. et al. (2011). Light propagation with phase discontinuities: generalized laws of reflection and refraction. *Science* 334 (6054): 333–337.

**7** Veselago, V.G. (1968). The electrodynamics of substances with simultaneously negative values of Img align = absmiddle alt = ε eps/Img and μ. *Physics-Uspekhi* 10 (4): 509–514.

**8** Pendry, J.B., Holden, A.J., Robbins, D.J., and Stewart, W.J. (1999). Magnetism from conductors and enhanced nonlinear phenomena. *IEEE Transactions on Microwave Theory and Techniques* 47 (11): 2075–2084.

**9** Shelby, R.A., Smith, D.R., and Schultz, S. (2001). Experimental verification of a negative index of refraction. *Science* 292 (5514): 77–79.

**10** Zheludev, N.I. and Kivshar, Y.S. (2012). From metamaterials to metadevices. *Nature Materials* 11 (11): 917–924. https://doi.org/10.1038/nmat3431.

**11** Smith, D.R. (2004). Metamaterials and negative refractive index. *Science* 305 (5685): 788–792. https://doi.org/10.1126/science.1096796.

**12** Su, Y. and Chen, Z.N. (2019). A radial transformation–optics mapping for flat ultra–wide–angle dual–polarized stacked GRIN MTM Luneburg lens antenna. *IEEE Transactions on Antennas and Propagation* 67 (5) https://doi.org/10.1109/TAP.2019.2900346.

**13** Pfeiffer, C. and Grbic, A. (2013). Metamaterial Huygens' surfaces: tailoring wave fronts with reflectionless sheets. *Physical Review Letters* 110 (19): 197401.

**14** Niemi, T., Karilainen, A.O., and Tretyakov, S.A. (2013). Synthesis of polarization transformers. *IEEE Transactions on Antennas and Propagation* 61 (6): 3102–3111.

**15** Rapoport, Y.G., Tretyakov, S., and Maslovski, S. (2013). Nonlinear active Huygens metasurfaces for reflectionless phase conjugation of electromagnetic waves in electrically thin layers. *Journal of Electromagnetic Waves and Applications* 27 (11): 1309–1328.

**16** Selvanayagam, M. and Eleftheriades, G.V. (2014). Polarization control using tensor Huygens surfaces. *IEEE Transactions on Antennas and Propagation* 62 (12): 6155–6168.

**17** Marcus, S.W. and Epstein, A. (2019). Fabry–Pérot Huygens' metasurfaces: on homogenization of electrically thick composites. *Physical Review B* 100 (11): 115144.

**18** Antoniades, M.A., Abbasi, M.A.B., Nikolic, M. et al. (2017). Conformai wearable monopole antenna backed by a compact EBG structure for body area networks. *2017 11th European Conference on Antennas and Propagation (EUCAP)*, Paris, France (19–24 March 2017), pp. 164–166, doi: https://doi.org/10.23919/EuCAP.2017.7928804.

**19** Quddious, A., Vryonides, P., Nikolaou, S. et al. (2019). Through–body communication measurements using wearable and implantable sensor antennas. *2019 Antennas Design and Measurement International Conference (ADMInC)*, St. Petersburg, Russia, Russia (16–18 October 2019), pp. 53–57, doi: https://doi.org/10.1109/ADMInC47948.2019.8969442.

**20** Abbasi, M.A.B., Quddious, A., Antoniades, M.A. et al. (2018). Wearable sensor for communication with implantable devices. *12th European Conference on Antennas and Propagation (EuCAP 2018)*, London, UK (9–13 April 2018), pp. 1–4, doi: https://doi.org/10.1049/cp.2018.0852.

**21** Wang, J. and Du, J. (2016). Plasmonic and dielectric metasurfaces: design, fabrication and applications. *Applied Sciences* 6 (9): 239.

**22** Abbasi, M.A.B., Antoniades, M.A., and Nikolaou, S. (2017). A compact reconfigurable NRI–TL metamaterial phase–shifter for antenna applications. *IEEE Transactions on Antennas and Propagation* 66 (2): 1025–1030.

**23** Abbasi, M.A.B., Antoniades, M.A., and Nikolaou, S. (2018). A compact microstrip crossover using NRI-TL metamaterial lines. *Microwave and Optical Technology Letters* 60 (11): 2839–2843.

**24** Liu, X., Yang, F., Li, M., and Xu, S. (2019). Generalized boundary conditions in surface electromagnetics: fundamental theorems and surface characterizations. *Applied Sciences* 9 (9): 1891.

**25** Inampudi, S. and Mosallaei, H. (2016). Fresnel refraction and diffraction of surface plasmon polaritons in two-dimensional conducting sheets. *ACS Omega* 1 (5): 843–853.

**26** Guo, Y., Yan, L., Pan, W., and Shao, L. (2016). Scattering engineering in continuously shaped metasurface: an approach for electromagnetic illusion. *Scientific Reports* 6: 30154.

**27** Liu, J., Liu, N., and Liu, Q.H. (2019). Spectral numerical mode matching method for 3D layered multi-region structures. *IEEE Transactions on Antennas and Propagation* 68 (2): 986–996.

**28** Wu, Q. (2019). Characteristic mode analysis of composite metallic–dielectric structures using impedance boundary condition. *IEEE Transactions on Antennas and Propagation* 67 (12): 7415–7424.

**29** Allen, K.W., Dykes, D.J., Reid, D.R. et al. (2017). Metasurface engineering via evolutionary processes. *2017 IEEE National Aerospace and Electronics Conference (NAECON)*, Dayton, OH, USA (27–30 June 2017), pp. 172–178.

**30** Chen, K., Cui, L., Feng, Y. et al. (2017). Coding metasurface for broadband microwave scattering reduction with optical transparency. *Optics Express* 25 (5): 5571–5579.

**31** Su, J., Lu, Y., Li, Z. et al. (2016). A wideband and polarization-independent metasurface based on phase optimization for monostatic and bistatic radar cross section reduction. *International Journal of Antennas and Propagation* 2016: 1–10.

**32** Xie, X., Pu, M., Huang, Y. et al. (2019). Heat resisting metallic meta-skin for simultaneous microwave broadband scattering and infrared invisibility based on catenary optical field. *Advanced Materials Technologies* 4 (2): 1800612.

**33** Chen, H.-Y., Guan, Y., Zhou, P.-H. et al. (2014). Microwave transmission of free-standing metasurfaces consisting of square metal patch. *Journal of Electromagnetic Waves and Applications* 28 (12): 1445–1454.

**34** Haddab, A.H. and Kuester, E.F. (2019). Plane-wave transmission through an array of dielectric-loaded slots in a thick metallic screen: normal incidence. *IEEE Transactions on Antennas and Propagation* 67 (7): 4668–4677.

**35** Yurduseven, O., Abbasi, M.A.B., Fromenteze, T., and Fusco, V. (2019). Frequency-diverse computational direction of arrival estimation technique. *Scientific Reports* 9 (1): 1–12.

**36** Lin, F.H. and Chen, Z.N. (2018). Truncated impedance sheet model for low-profile broadband nonresonant-cell metasurface antennas using characteristic mode analysis. *IEEE Transactions on Antennas and Propagation* 66 (10): 5043–5051.

**37** Forouzmand, A., Salary, M.M., Inampudi, S., and Mosallaei, H. (2018). A tunable multigate indium-tin-oxide-assisted all-dielectric metasurface. *Advanced Optical Materials* 6 (7): 1701275.

**38** Chen, K., Feng, Y., Yang, Z. et al. (2016). Geometric phase coded metasurface: from polarization dependent directive electromagnetic wave scattering to diffusion–like scattering. *Scientific Reports* 6 (1): 1–10.

**39** Li, S.-H. and Li, J.-S. (2019). Frequency coding metasurface for multiple directions manipulation of terahertz energy radiation. *AIP Advances* 9 (3): 35146.

**40** Hosseininejad, S.E., Rouhi, K., Neshat, M. et al. (2019). Reprogrammable graphene–based metasurface mirror with adaptive focal point for THz imaging. *Scientific Reports* 9 (1): 1–9.

**41** Yan, X., Liang, L.-J., Zhang, Y.-T. et al. (2015). A coding metasurfaces used for wideband radar cross section reduction in terahertz frequencies. *Acta Physica Sinica* 64 (15): 158101.

**42** Dickie, R., Cahill, R., Fusco, V. et al. (2011). THz frequency selective surface filters for earth observation remote sensing instruments. *IEEE Transactions on Terahertz Science and Technology* 1: 450–461. https://doi.org/10.1109/TTHZ.2011.2129470.

**43** Orr, R., Fusco, V., Zelenchuk, D. et al. (2015). Circular polarization frequency selective surface operating in Ku and Ka band. *IEEE Transactions on Antennas and Propagation* 63 (11): 5194–5197. https://doi.org/10.1109/TAP.2015.2477519.

**44** Luukkonen, O., Simovski, C., Granet, G. et al. (2008). Simple and accurate analytical model of planar grids and high–impedance surfaces comprising metal strips or patches. *IEEE Transactions on Antennas and Propagation* 56 (6): 1624–1632. https://doi.org/10.1109/TAP.2008.923327.

**45** Langley, R.J. and Parker, E.A. (1982). Equivalent circuit model for arrays of square loops. *Electronics Letters* 18 (7): 294. https://doi.org/10.1049/el:19820201.

**46** Zelenchuk, D.E. and Fusco, V.F. (2016). Design method for circularly polarized frequency selective surfaces. *2016 10th European Conference on Antennas and Propagation (EuCAP)*, Davos, Switzerland, (10–15 April 2016), pp. 1–5, doi: https://doi.org/10.1109/EuCAP.2016.7481777.

**47** Zelenchuk, D.E. and Fusco, V.F. (2016). Design method for circularly polarized frequency selective surfaces. *2016 10th European conference on Antennas and Propagation (EuCAP)*, Davos, Switzerland, (10–15 April 2016), pp. 3153–3158.

**48** Munina, I., Turalchuk, P., Kirillov, V. et al. (2019). A tiled C–band dual–polarized 1–bit transmitarray. *Proceedings of the 49th European Microwave Conference (EuMC)*, Paris, France (2–4 October 2019).

**49** Munina, I., Turalchuk, P., Kirillov, V. et al. (2019). A study of C–band 1–bit reconfigurable dual–polarized transmitarray. *Proceedings of the 13th European Conference on Antennas and Propagation (EuCAP)*, Krakow, Poland (31 March to 5 April 2018).

**50** Padgett, M. and Allen, L. (2000). Light with a twist in its tail. *Contemporary Physics* 41 (5): 275–285. https://doi.org/10.1080/001075100750012777.

**51** Jack, B., Leach, J., Romero, J. et al. (2009). Holographic ghost imaging and the violation of a bell inequality. *Physical Review Letters* 103 (8): 1–4. https://doi.org/10.1103/PhysRevLett.103.083602.

**52** Tamburini, F., Mari, E., Sponselli, A. et al. (2012). Encoding many channels on the same frequency through radio vorticity: first experimental test. *New Journal of Physics* 14 (3): 033001. https://doi.org/10.1088/1367-2630/14/3/033001.

53 Zelenchuk, D., Fusco, V., and Malyuskin, O. (2013). Split ring reflectarray FSS with spiral phase distribution. *2013 7th European Conference on Antennas and Propagation, EuCAP 2013*, Gothenburg, Sweden (8–12 April 2013), pp. 2709–2713.

54 Turnbull, G.A., Robertson, D.A., Smith, G.M. et al. (1996). The generation of free-space Laguerre–Gaussian modes at millimetre-wave frequencies by use of a spiral phaseplate. *Optics Communication* 127 (6): 183–188. https://doi.org/10.1016/0030--4018(96)00070–3.

55 Fox, A.G. (1947). An adjustable wave-guide phase changer. *Proceedings of the IRE* 35 (12): 1489–1498. https://doi.org/10.1109/JRPROC.1947.234574.

56 Cappellin, C., Sjöberg, D., Ericsson, A. et al. (2016). Design and analysis of a reflector antenna system based on doubly curved circular polarization selective surfaces. *2016 10th European Conference on Antennas and Propagation (EuCAP)*, Davos, Switzerland (10–15 April 2016), pp. 1–5, doi: https://doi.org/10.1109/EuCAP.2016.7481781.

57 Menzel, C., Rockstuhl, C., and Lederer, F. (2010). Advanced Jones calculus for the classification of periodic metamaterials. *Physical Review A* 82 (5): 053811. https://doi.org/10.1103/PhysRevA.82.053811.

58 Roy, J.E. and Shafai, L. (1996). Reciprocal circular-polarization-selective surface. *IEEE Antennas and Propagation Magazine* 38 (6): 18–32. https://doi.org/10.1109/74.556517.

59 Eal, D.E.M. and Lopez, H.I. (2013). New circular polarization selective surface concepts based on the pierrot cell using printed circuit technology. M.Sc.A. thesis. Ecole Polytechnique, Montreal, Canada.

60 Tang, W., Goussetis, G., Fonseca, N.J.G. et al. (2017). Coupled split-ring resonator circular polarization selective surface. *IEEE Transactions on Antennas and Propagation* 65 (9): 4664–4675. https://doi.org/10.1109/TAP.2017.2726678.

61 Liu, N., Liu, H., Zhu, S., and Giessen, H. (2009). Stereometamaterials. *Nature Photonics* 3: 157–162. https://doi.org/10.1038/nphoton.2009.4.

62 Epstein, A. and Eleftheriades, G.V. (2014). Passive lossless Huygens metasurfaces for conversion of arbitrary source field to directive radiation. *IEEE Transactions on Antennas and Propagation* 62 (11): 5680–5695.

63 Epstein, A., Wong, J.P.S., and Eleftheriades, G.V. (2016). Cavity-excited Huygens' metasurface antennas for near-unity aperture illumination efficiency from arbitrarily large apertures. *Nature Communications* 7 (1): 1–10.

64 Ra'di, Y., Asadchy, V.S., and Tretyakov, S.A. (2014). Tailoring reflections from thin composite metamirrors. *IEEE Transactions on Antennas and Propagation* 62 (7): 3749–3760.

65 Epstein, A. and Eleftheriades, G.V. (2016). Arbitrary power-conserving field transformations with passive lossless omega-type bianisotropic metasurfaces. *IEEE Transactions on Antennas and Propagation* 64 (9): 3880–3895.

66 Asadchy, V.S., Ra'Di, Y., Vehmas, J., and Tretyakov, S.A. (2015). Functional metamirrors using bianisotropic elements. *Physical Review Letters* 114 (9): 95503.

67 Ra'di, Y. and Tretyakov, S.A. (2013). Balanced and optimal bianisotropic particles: maximizing power extracted from electromagnetic fields. *New Journal of Physics* 15 (5): 53008.

68 Tretyakov, S.A. (2015). Metasurfaces for general transformations of electromagnetic fields. *Philosophical Transactions of the Royal Society A - Mathematical Physical and Engineering Sciences* 373 (2049): 20140362.

# 4

# Metasurfaces Based on Huygen's Wavefront Manipulation

A Review

*Abubakar Sharif[1,2], Jun Ouyang[1], Kamran Arshad[3], Muhammad Ali Imran[4], and Qammer H. Abbasi[4]*

[1] *School of Electronic Science and Engineering, University of Electronic Science and Technology of China, Chengdu, China*
[2] *Department of Electrical Engineering, Government College University, Faisalabad, Pakistan*
[3] *College of Engineering and IT, Ajman University, Ajman, United Arab Emirates*
[4] *James Watt Schoeedol of Engineering, University of Glasgow, Glasgow, United Kingdom*

## 4.1 Introduction

In microwave antenna engineering and optics, the electromagnetic waves shaping was carried out by employing well-established techniques such as dielectric lenses and metallic reflectors [1–3]. However, reflectors and dielectrics lenses are bulkier for many applications, especially at microwave and millimeter-wave applications. Moreover, the transmitarrays and reflectarrays were used to replace lenses and reflectors, respectively [4–8]. Transmitarrays and reflectarrays are usually fabricated by printing dipole or patch-type elementary antennas on flat dielectric substrates. The required phase characteristics to attain a certain wavefront transformation are achieved by spatially modifying the size or shape of elementary antennas. The required phase range is usually achieved by making these elementary antennas half-wavelength in size to work close to their resonance. Moreover, in the case of transmitarray, several layers are required to avoid reflections and to achieve good matching characteristics.

In this context, recent researches are focusing on developing new composite materials such as metamaterial and metasurfaces to achieve properties that are not existing in natural materials and further in order to achieve the same results as that of transmitarray and reflectarray with relatively low profile and high efficiency.

Metasurfaces or 2D metamaterials engineered to pose some electromagnetic properties by affecting the wavefronts through modified boundary conditions. Mostly, the metasurfaces composed of flat composite structures with subwavelength thickness. The difference between metamaterials and metasurfaces is the functionality of metamaterials totally realized by artificial $\mu$ and $\varepsilon$. In contrast, the functionality of metasurfaces is characterized by modifying wavefronts arbitrarily (as depicted in Figure 4.1). More precisely, metasurfaces can able to control the following parameters as a function of the position such as amplitude, phase, polarization, frequency (nonlinear), and reciprocity (nonlinear, anisotropic, etc.) [9, 10].

*Backscattering and RF Sensing for Future Wireless Communication,* First Edition.
Edited by Qammer H. Abbasi, Hasan T. Abbas, Akram Alomainy, and Muhammad Ali Imran.

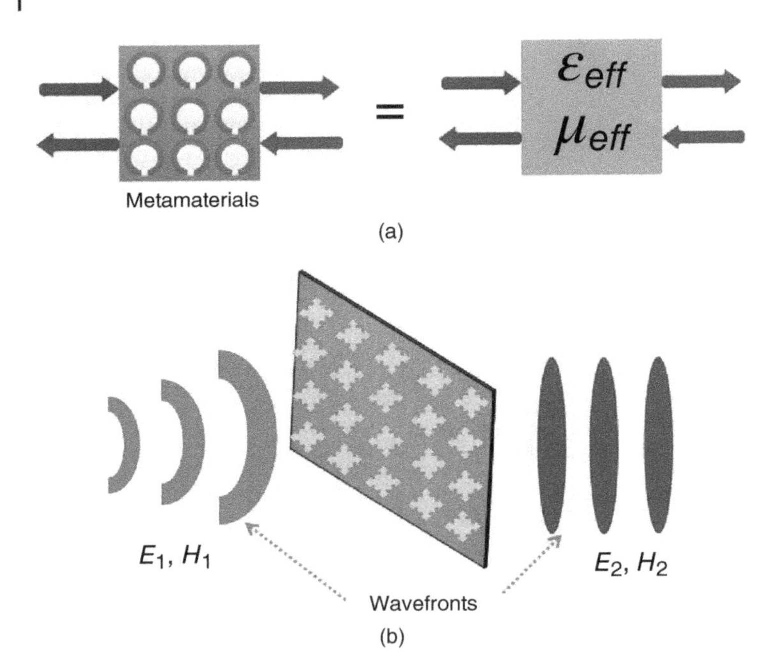


**Figure 4.1** (a) The functionality of metamaterials are realized by artificial $\mu$ and $\varepsilon$, and (b) the functionality of metasurfaces is characterized by modifying wavefronts arbitrarily.

Figure 4.2 shows some functionalities of metasurfaces, including focusing transmitarray, focusing reflectarray, linear to circular polarization converter, right-hand circular polarizer, polarization twister, and metasurface hologram [11]. Recently, Huygens' metasurfaces (HMSs) are gaining becoming prevalent in academia due to their manipulation feature of incident electromagnetic waves. These surfaces can behave similarly as transmitarrays and reflectarrays, but with some key differences [12, 13]. Figure 4.3 illustrates a future prospective application of metasurfaces decorated wall regarding vital sign detection for patients or normal persons both at the hospital and home environment. The idea is to collect the reflected signals from the human body and further apply signal processing and machine learning techniques to extract the vital signs information. In addition, metasurfaces can also be used for other healthcare applications such as magnetic resonance imaging (MRI) [14]. Artificial impedance boundaries, frequency-selective surfaces, and transmitting and reflective-phased metasurfaces were the most promising types of metasurfaces used for MRI applications.

## 4.2 Huygens' Metasurfaces (HMSs)

This section summarizes the fundamental principles HMSs and discusses different phenomenons, structures, and advances regarding HMSs.

HMSs typically consist of subwavelength unit cells with subwavelength thickness. Due to the involvement of subwavelength scales, these HMSs can be characterized and

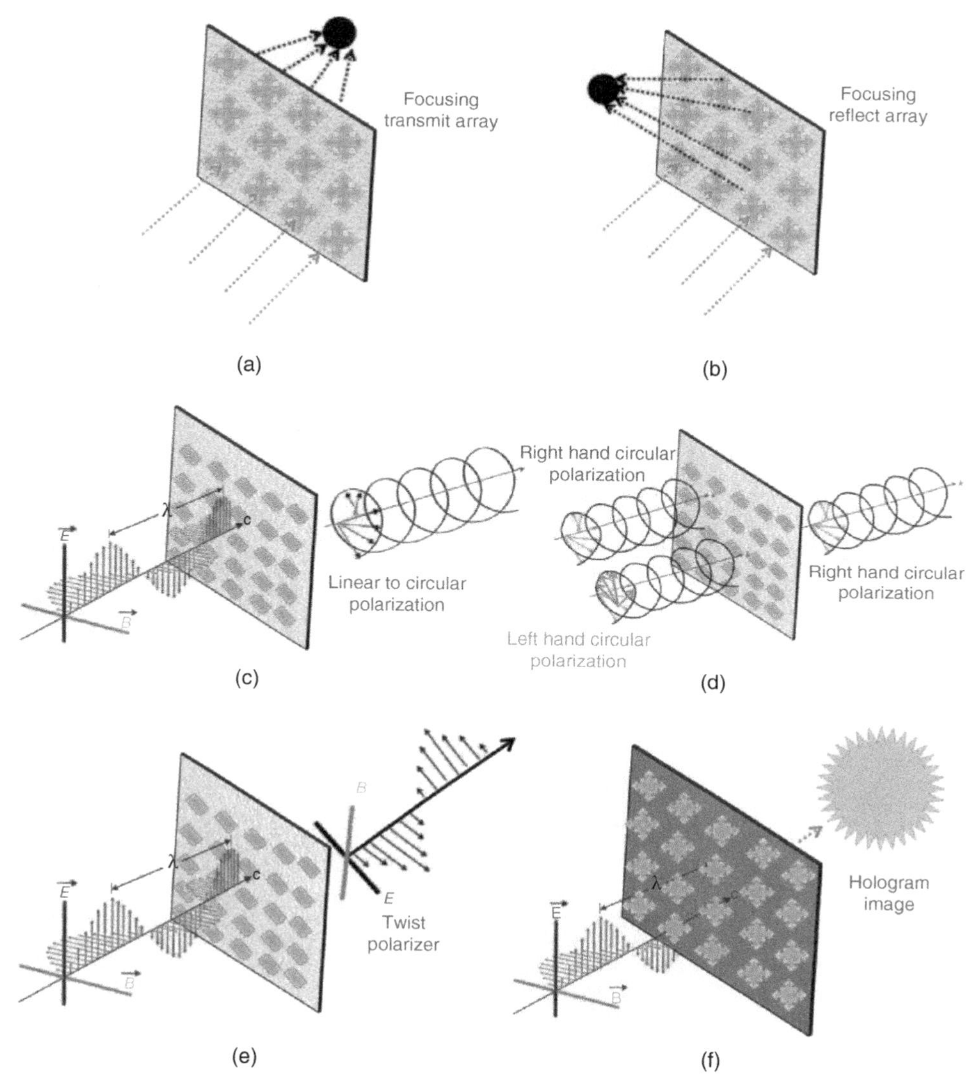


**Figure 4.2** Some functionalities of metasurfaces: (a) focusing transmitarray, (b) focusing reflect array, (c) linear to circular polarization converter, (d) right-hand circular polarizer, (e) polarization twister and (f) metasurface hologram. Source: Modified from Glybovski et al. [11].

homogenized in terms of spatially varying surface impedances and admittances. This fact justifies their name known as metasurfaces, which can be considered as 2D metamaterials and further can also be described regarding electric and magnetic susceptibilities [12].

The concept of metasurfaces can be realized using Schelkunoffs equivalence principle, which can be considered as a more general form of Huygens principle, which states that an incident electromagnetic wave to a desired one through field discontinuities resulted by induced electric and magnetic currents on the metasurface (interface) as illustrated

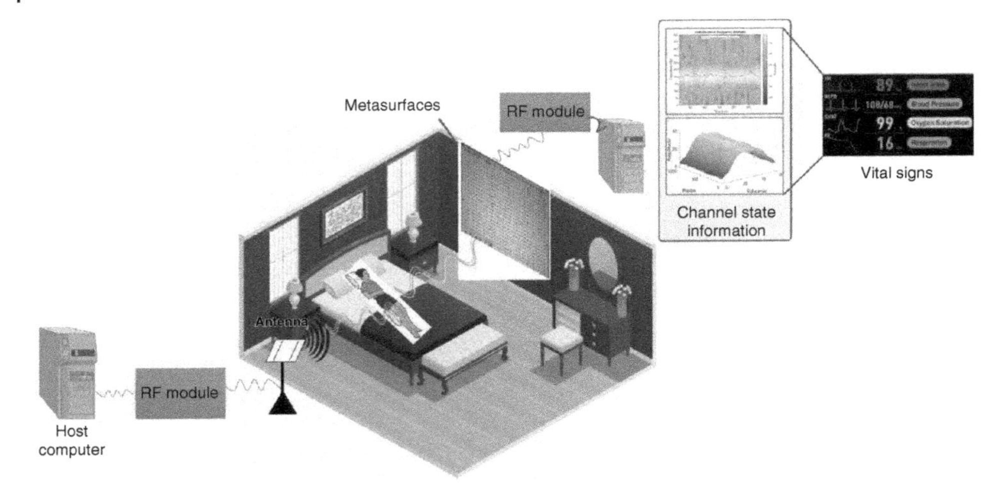


**Figure 4.3** A future prospective applications of metasurfaces decorated wall regarding vital sign detection for patients or normal persons both at the hospital and home environment.

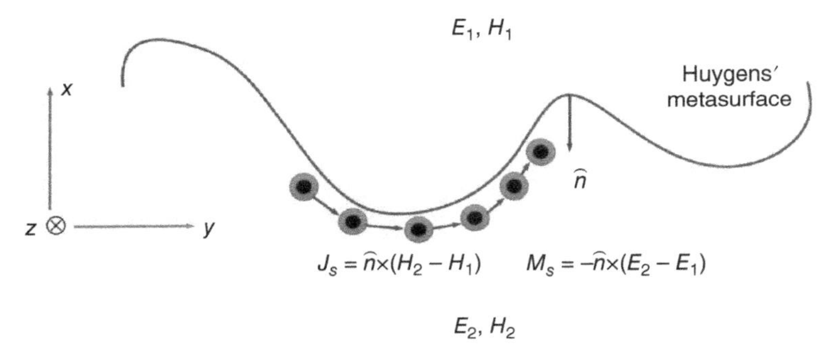


**Figure 4.4** A Huygens' metasurface transform an incident electromagnetic field $(E_1, H_1)$ to a transmitted field $(E_2, H_2)$ by utilizing a field discontinuity, sustained by induced orthogonal electric and magnetic currents $(J_s, M_s)$. Source: Modified from Selvanayagam and Eleftheriades [15].

in Figure 4.4 [15]. These magnetic and electric currents are physically obtained using orthogonal magnetic and electric dipoles that further corresponds to spatially varying impedance and admittance distributions [15]. Moreover, it is worth to mention here, the concept of active HMSs can be proposed by realizing impressed equivalent electric and magnetic currents instead of induced ones. Therefore, the active HMSs are preceded the passive ones.

The most basic boundary in electromagnetic theory is the interface between two media. When electromagnetic waves pass through the boundary of the medium, it produces reflections and refractions with different amplitudes and phases, which are closely related to the tangential electric field and magnetic field at the boundary of the medium. The electromagnetic theory and its boundary conditions indicate that the passive region satisfies the tangential field at the interface of the medium, which is known as continuity condition.

Moreover, this fact can be exploited by intentionally designing the boundary to introduce desired discontinuity on the boundary by synthesizing an electromagnetic flow surface to achieve the desired characteristics such as directional radiation or stealth characteristics. Particularly, an orthogonal current and a magnetic current can be applied to produces a field discontinuity at an interface in order to generate the electromagnetic current, which radiates the desired wavefront like the Huygens' source, as shown in Figure 4.1.

The HMSs pose very interesting features due to two main factors: (i) subwavelength nature of unit cells and (ii) rigorous boundary conditions, which can be considered as the induction of both electric and magnetic currents, according to Maxwell's equation. The interesting features of HMSs include (i) well-matching in spite of subwavelength nature and (ii) seizing the excitation of spurious radiation modes [16].

Therefore, the HMSs can be engineered to manipulate attributes of incident waves such as magnitude, phase and polarization, frequency (nonlinear), and reciprocity (nonlinear, anisotropic, etc.). Some negotiable achievements of HMSs include perfect refracting surfaces of subwavelength thickness that couples all incident power to the refracted beam and suffers no reflections. Similarly, perfect reflecting surfaces can be realized using HMSs that can transfer 100% of the incident power to an arbitrary reflection angle (working similar to reflectarrays).

Moreover, due to the field control property of metasurfaces, they can be used to design aperture antennas with required characteristics such as beamwidth, sidelobe-level distribution, beam-pointing direction, and as absorbing metasurfaces [17–20].

As mentioned before, HMSs can perform desired field transformations by utilization of the Schelkunoff's equivalence principle [12, 15, 21]. This phenomen is explained using Figure 4.1 and Equation (4.1) as, when fields $\vec{E}_1$,$\vec{H}_1$, $\vec{E}_2$, and $\vec{H}_2$ are electric and magnetic field densities specified in two half-spaces. To transform the fields from one domain to the other, the pivotal electric ($\vec{J}_s$) and magnetic ($\vec{M}_s$) current densities that are basically required [21], where $\vec{E}_1^{\,-}$, $\vec{H}_1^{\,-}$, $\vec{E}_2^{\,+}$, and $\vec{H}_2^{\,+}$ are the respective values of $\vec{E}_1^{\,-}$, $\vec{H}_1^{\,-}$, $\vec{E}_2^{\,+}$, and $\vec{H}_2^{\,+}$ at the boundary of the field discontinuity:

$$\vec{J}_s = \hat{n} \times (\vec{H}_2^{\,+} - \vec{H}_1^{\,-}),$$
$$\vec{M}_s = -\hat{n} \times (\vec{E}_2^{\,+} - \vec{E}_1^{\,-}) \tag{4.1}$$

There are two ways to achieve discontinuity, one is active and the other is passive. The active method is to place an electromagnetic flow antenna with orthogonal characteristics at the interface; the passive method can use similar equivalent surface impedance (see Equation (4.2), where $Z_e$ and $Z_m$ are equivalent electromagnetic surface impedances, respectively). Using the induced current to achieve this orthogonal current and magnetic current distribution, in particular, the second passive method has the potential of a single layer of thin surface (equivalent impedance surface) for arbitrary wavefront control:

$$\hat{n} \times \frac{E_1 + E_2}{2} = Z_e \hat{n} \times (H_1 - H_2)$$
$$\hat{n} \times \frac{H_2 + E_1}{2} = -(1/Z_m) \hat{n} \times (E_2 - E_1) \tag{4.2}$$

Based on the above theory, the ideal implementation of the design (orthogonal electromagnetic flow) is made as much as possible.

It conforms to the theoretical model and can be applied to the field of antenna radiation:

1. An ultrathin lens antenna that converts any source radiation into directional radiation;
2. To achieve an ideal uniform aperture, resulting in a highly directional antenna;
3. As a polarization converter to achieve the transformation between the circular antenna polarization and linear polarization;
4. An aperture field antenna that achieves a near-field focusing effect;
5. Control the propagation and leakage of surface waves to achieve any field distribution in the near-field;
6. Applied to a conformal antenna, designing a conformal antenna array having low cross-polarization and low sidelobes

In this approach, the HMS can be realized by modifying the desired structure to produce the required surface current densities conditions for transformation of any given field [12, 15]. In this technique, both magnetic and electric current densities are realized through external sources, which are actually act as impressed sources. The interaction of radiated fields originated from these impressed sources with the incident fields produce the desired output fields. However, the introduction of impressed current sources leads to some challenges including the power consumption, generation and distribution of microwave/optical power, and the precise realization of the required current weights [22]. Alternatively, these surface currents can be induced by synthesizing effective magnetic and electric properties of the incident fields. In this case, the metasurface is proposed by utilizing these intrinsic characteristics instead of using the arrays of impressed sources. After carefully synthesizing these electric and magnetic properties, the desired boundary surface current densities are excited by the incident fields, consequently leading towards lossless and passive metasurfaces [23]. Moreover, the interaction of scattered fields resulted from these exciting currents with incident wave further produces the output fields. Generally, the susceptibility [24–26], polarizability [26–28], and impedance/admittance approaches [12, 15] are three main perspectives for designing the metasurface properties. It is worth to mention that all these methods can achieve the same range of field manipulations irrespective of the chosen perspective [29, 30].

## 4.3 HMS Applications

### 4.3.1 HMS Refraction

The pioneer demonstrations of HMSs applications regarding low-reflection refraction were reported in [12, 31, 32]. The theoretical concepts, simulation results, and experimental verifications of low-reflection refraction at microwave frequencies were discussed in [12, 32]. In both cases, the wire/loop unit cells were used as Huygenssource. This Huygenssource was designed by realizing the magnetic response with loops and the electric response with metallic wires. However, the one main difference between these two designs is related to their orientations. In [12], the wire/loop unit cells were designed by placing the structures parallel to the direction of propagation or in other words we can say the unit cells were stacked in a transverse manner. In contrast, the unit cells in proposed [32] were designed by placing the unit cells perpendicular to the direction of propagation or we can say the unit cells are positioned in a single-plane configuration. Therefore, the physical orientation of unit cells in [12] and [32] were quite different. However, both of these structures successfully demonstrated the Huygens sources representation and good refractive properties. There are some other designs reported in literature to realize HMSs such as a multi-layer cascaded unit cell structure [33, 34]. The cascaded topology utilized inductive and capacitive properties of stacked planar metallic patterns to realize Huygens unit cell instead of unit cells consisting of wire/loop suggested in [12] and [32]. Since, the stacked planar structures realize Huygens sources without introduction of any physical wires and loops. Therefore, the cascading effect of these capacitive and inductive geometries is used to synthesize the required $Z_{se}$ and $Y_{sm}$ reactances to achieve the required boundary conditions. In [33], a three-layer structure was physically realized.

### 4.3.2 Antennas Beamforming

Other applications of metasurfaces include beamforming from arbitrary sources such as metasurfaces-based antennas and metasurfaces lenses. An example of beamforming using HMS-based cavity excited antenna was proposed in [35]. The cavity-excited HMS antennas used a single-source-fed cavity to optimize aperture illumination, and the main role associated with HMS was to provide the current distribution that guarantees the phase purity of aperture fields. The HMS also forced to reduce the edge-taper losses by breaking the coupling between the excitation and radiation spectra to achieve the desired radiation pattern. Therefore, the HMS-based low-profile antenna yielded near-unity aperture illumination efficiencies.

Also, another example of beamforming includes the full-wave simulations of cavity-excited bianisotropic HMS antenna for Taylor distribution showing controllable sidelobe levels [33].

### 4.3.3 Perfect Reflections and Focusing

It is very challenging to achieve an arbitrary reflection of incident waved with perfect efficiency. However, it can be achieved by designing multilayered metasurfaces structures, which can able to match input and output impedances of the EM wave. Figure 4.5 shows a schematic diagram depicting the concept of perfect reflection using metasurfaces [36].

In [37], a planar Huygens' meta-atoms with chiral coupling, balanced electric and magnetic responses was proposed. Three metasurfaces with different amplitude and polarization transformations were designed. These metasurfaces can be used for different applications such as passive antenna arrays, transparent lenses, and holograms.

### 4.3.4 Beam Scanning

Another vital application of metasurfaces is the achievement of beam scanning capabilities with low-cost and low-profile-phased arrays antennas. In [38], a phased array was proposed using metasurface having extended scanning range features. This metasurface utilized a bidirectional expansion scanning range. The proposed phase gradient metasurface consists

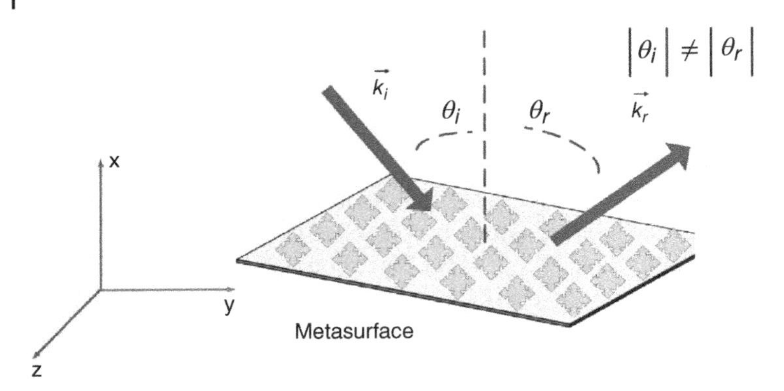


**Figure 4.5** A schematic diagram depicting the concept of perfect reflection using metasurfaces. Source: Based on Wong and Eleftheriades [36].

of multilayered periodic subwavelength elements (MPSEs). The MPSE contributes toward multiple incident angular stability and transmission phase shift for a wide range. The results demonstrated the scanning range of [−36°, 38°]–[−56°, 60°].

## 4.4 Conclusion and Key Scientific Issues to Be Addressed

This chapter has described some of the most recent results regarding HMSs in the context of working principle, design methodology of HMSs, and its applications in antennas and antenna array designs. Also, we described the use of metasurfaces for healthcare, a perfect reflection, beamforming, and beam scanning. The key scientific issues regarding HMSs was also elaborated as follows:

1. Equivalent analysis model of double anisotropic HMSs and its special unit design method. From the development history of HMSs, the traditional wire loop structure is adopted for the HMSs unit structure. The idea of this structure is the simplest form, that is, the ring structure simulating magnetic current and the linear structure of analog current. Later, an omega-bianisotropic implementation through a multilayered stacking structure was developed. Thus, it has evolved to realize its dual anisotropic properties based on asymmetric wire loop structures or multilayer cascade structures. In 2018, Ayman H. Dorrah et al. published a paper on IEEE TAP [39] to design a broadband double anisotropic HMSs unit. In 2017, Cui Tiejun et al. also published a paper on the IEEE TAP Anisotropic HMSs unit [40]. It can be seen that compared with other artificial electromagnetic materials, such unit structures have been relatively simple, and special requirements such as broadband, double multifrequency, polarization, and other HMSs have yet to be further studied. The influence of dielectric materials with different thicknesses on cell performance is also considered. In particular, the HMSs structure must comply with the principle of local energy conservation.
2. The energy distribution of the incident and reflected waves at the interface determines the energy distribution of the Huygens surface, which determines the aperture

distribution of the antenna surface. In other words, we need to design the aperture distribution by designing the reflection coefficient of the HMSs structure.
Therefore, in order to complete the requirements for designing arbitrary beamforming antennas mentioned earlier, an accurate mathematical modeling and electromagnetic analysis of the elements must be performed to achieve accurate reflection, transmission, and amplitude and phase control.

3. Virtual aperture field design and physical implementation based on double anisotropic HMSs structure. This technique is the key to achieving arbitrary beamforming, including near and far fields. It is necessary to first discuss "local energy conservation" and "and local impedance matching" from the two basic conditions of the HMSs surface. To achieve the goal of achieving arbitrary beams, the amplitude and phase modulation of the aperture field must be involved. However, the traditional passive lossless anisotropy HMSs must satisfy the following two basic conditions: "local energy conservation" and "match with local impedance." From the first condition, "conservation of local energy":

$$P_{\text{total power (incident field of source radiation+scattering field of other scatterers+reflection field of Huygens surface)}} = P_{\text{power at this point on the interface}} = P_{\text{antenna field power at the antenna}},$$

which determines the aperture of the antenna field amplitude distribution. It can be deduced from the second condition "local impedance matching" that it satisfies the Fresnel reflection principle. There is energy reflection with a large angle of incidence that is different from no reflection when the plane wave is transmitted. Due to the local impedance matching conditions, the size of the reflection field is challenging to design independently, which brings great difficulty to the amplitude distribution control in arbitrary beam control.

## 4.5 Future Trends

With the advent of 5G, 6G, and beyond, our living environments will become more connected sensible due to the emergence of new technological developments such as metasurfaces. Although this chapter provides an overview of the working principles, design methods, and application areas, however, there is still a vast topic regarding metasurface, especially HMSs, which are unexplored.

1. Design of unit structure of double anisotropic HMSs based on different application requirements. As a second-generation HMSs unit, the double anisotropic unit has more relative to the traditional anisotropic unit.
2. The degree of freedom is embodied in the electromagnetic coupling coefficient, which brings additional design space. Theoretical studies have proved that the double anisotropic HMSs electromagnetic structural unit can achieve a more refined design after having the degree of freedom of electromagnetic coupling coefficient, so that the structure has the characteristics of matching the wave impedance on both sides of the interface so that the design is flexible. Related literature has proved that such double anisotropic HMSs have the characteristics of being nonreflective at any angle for

plane-wave transmission and achieving arbitrary angular total reflection for plane-wave reflection. It has been found that the use of asymmetric wire loop structures or multilayer cascade structures can achieve double anisotropic properties, but specific performance comparisons and designs based on actual applications: bandwidth, dual multifrequency, polarization transformation, electronic control, size in terms of design, thickness, processing technology, precision requirements, etc. There are very few reported pieces of literature, which shows that the design space is still very large. This is also the premise and basis for subsequent antenna design, which deserves an in-depth study.

3. Research on antenna far-field arbitrary beamforming technology based on double anisotropic HMSs; antenna array design, especially beamforming, has always been the focus of antenna researchers, but traditional array feed networks suffer huge losses. These traditional methods (optical methods) provide low antenna efficiency with a large dimension. In 2016, the team of the University of Toronto in Canada published a design study on the back cavity high-efficiency pencil beam antenna based on anisotropic HMSs. The caliber efficiency is over 80%, which mostly proves the high-efficiency theory of HMSs [35]. In 2017, the team [40] theoretically demonstrated that higher anisotropy (up to 95%) and lower profiles could be achieved based on double anisotropic materials. (Anisotropic HMSs are half the thickness of the antenna.) The HMSs design based on the Huygens' equivalence principle brings an unprecedented perspective to antenna array synthesis. The HMSs can able to support the design of an antenna array with extremely high aperture efficiency, extremely lightweight, and extremely low profile. The amplitude/phase design control is more difficult due to the sharp amplitude-phase variation of the antenna aperture and the large-span amplitude-phase distribution. Therefore, the study of antenna far-field arbitrary beamforming technology based on the new mechanism of double anisotropic HMSs will be a significant research focus and difficulty of this topic.
4. Research on near-field energy distribution integrated design technology based on double anisotropic HMSs; "Near-field energy distribution integrated design based on double anisotropy HMSs" has a particular relationship with the previous research content "far-field arbitrary beamforming design." Because the far-field-based design method is more straightforward in describing the electromagnetic wave on the far-field radiation side of the interface, the structural design requirements for the double anisotropic HMSs are more complicated due to the tangential electromagnetic field distribution on the interface corresponding to the near-field energy distribution. In terms of applications, near-field focusing or shaping techniques are available in areas such as RFID identification and biomedical imaging. Large application prospects include careful research that will be extremely important for designing low-profile near-field antennas.
5. Design and implementation of active and controllable double anisotropic HMSs near- and far-field antennas. In the active control of artificial electromagnetic materials, Professor Feng Yijun of Nanjing University and Professor Cui Tiejun of Southeast University have many research results in this respect. The publication also confirms its value in real-world applications. From the literature data, the active loading and reconfigurable techniques of anisotropic HMSs were reported in the literature in 2016–2018. However, the active loading studies of double anisotropic HMSs have not

been reported in detail. The advantage of the double anisotropic unit and the study of its active loading technology will be deployed to focus on wide application for more flexible design ideas. In general, based on the above-mentioned near- and far-field antenna design, the performance variation of the double anisotropic HMSs can be realized by active loading to provide more design freedom in the antenna design, which is in line with the current demand for antennas.

6. Research on integrated technology of active Huygens source ideal antenna array based on real orthogonal electromagnetic flow. All of the above studies are antenna designs that use passive, lossless structures to meet equivalent surface impedance boundary conditions. As described in the previous background, active structures can be used in addition to passive structures (which can be achieved using conventional loop antenna plus dipole antennas).
7. From the literature, the active Huygens' source array has been used in stealth cloak design to eliminate the target's reflection field and achieve active stealth. At the same time, the use of similar techniques has also achieved the false target scattering to confuse the reporter and other literature reports. It has been partially verified by experiments that the principle behind stealth cloak design is the Huygens source antenna radiation synthesis.
8. To design an ideal conformal antenna, it is expected to solve the problems of cross-polarization and aperture utilization of traditional conformal antennas using ideal Huygens' source radiation.

## References


**1** Rudge, A.W. and Adatia, N.A. (1978). Offset-parabolic-reflector antennas: a review. *Proceedings of the IEEE* 66 (12): 1592–1618.

**2** Jones, E. (1954). Paraboloid reflector and hyperboloid lens antennas. *Transactions of the IRE Professional Group on Antennas and Propagation* 2 (3): 119–127.

**3** Wu, X., Eleftheriades, G.V., and van Deventer-Perkins, T.E. (2001). Design and characterization of single- and multiple-beam mm-wave circularly polarized substrate lens antennas for wireless communications. *IEEE Transactions on Microwave Theory and Techniques* 49 (3): 431–441.

**4** Huang, J. Capabilities of printed reflectarray antennas. *Proceedings of International Symposium on Phased Array Systems and Technology*, Boston, MA, USA, pp. 131–134.

**5** Pozar, D.M. (1996). Flat lens antenna concept using aperture coupled microstrip patches. *Electronics Letters* 32 (23): 2109.

**6** Ryan, C.G.M., Chaharmir, M.R., Shaker, J. et al. (2010). A wideband transmitarray using dual-resonant double square rings. *IEEE Transactions on Antennas and Propagation* 58 (5): 1486–1493.

**7** Li, M. and Behdad, N. (2013). Wideband true-time-delay microwave lenses based on metallo-dielectric and all-dielectric lowpass frequency selective surfaces. *IEEE Transactions on Antennas and Propagation* 61 (8): 4109–4119.

8 Hum, S.V. and Perruisseau-Carrier, J. (2014). Reconfigurable reflectarrays and array lenses for dynamic antenna beam control: a review. *IEEE Transactions on Antennas and Propagation* 62 (1): 183–198.
9 Hsiao, H.-H., Chu, C.H., and Tsai, D.P. (2017). Fundamentals and applications of metasurfaces. *Small Methods* 1 (4): 1600064.
10 Bukhari, S.S., Vardaxoglou, J., and Whittow, W. (2019). A metasurfaces review: definitions and applications. *Applied Sciences (Switzerland)* 9 (13): 2727.
11 Glybovski, S.B., Tretyakov, S.A., Belov, P.A. et al. (2016). Metasurfaces: from microwaves to visible. *Physics Reports* 634: 1–72.
12 Pfeiffer, C. and Grbic, A. (2013). Metamaterial Huygens' surfaces: tailoring wave fronts with reflectionless sheets. *Physical Review Letters* 110 (19): 197401.
13 Yu, N., Genevet, P., Kats, M.A. et al. (2010). Light propagation with phase discontinuities: Generalized laws of reflection and refraction. *Science* 334 (6054): 333–337.
14 Craeye, C., Glybovski, S., and Simovski, C. (2017). Comparative Study of Metasurfaces Stressed on MRI Application WP4 Report: Deliverable 4. *Report No. 1*.
15 Selvanayagam, M. and Eleftheriades, G. (2013). Discontinuous electromagnetic fields using orthogonal electric and magnetic currents for wavefront manipulation. *Optics Express* 21: 14409–14429.
16 Epstein, A. and Eleftheriades, G.V. (2014). Floquet-Bloch analysis of refracting Huygens metasurfaces. *Physical Review B* 90: 235127.
17 Asadchy, V.S., Faniayeu, I.A., Ra'di, Y. et al. (2015). Broadband reflectionless metasheets: frequency-selective transmission and perfect absorption. *Physical Review X* 5: 031005.
18 Landy, N.I., Sajuyigbe, S., Mock, J.J. et al. (2008). Perfect metamaterial absorber. *Physical Review Letters* 100: 207402.
19 Liu, N., Mesch, M., Weiss, T. et al. (2010). Infrared perfect absorber and its application as plasmonic sensor. *Nano Letters* 10 (7): 2342–2348.
20 Diem, M., Koschny, T., and Soukoulis, C.M. (2009). Wide-angle perfect absorber/thermal emitter in the THz regime. *Physical Review B* 79: 033101.
21 Schelkunoff, S.A. (1936). Some equivalence theorems of electromagnetics and their application to radiation problems. *Bell System Technical Journal* 15 (1): 92–112.
22 Selvanayagam, M. and Eleftheriades, G.V. (2013). Experimental demonstration of active electromagnetic cloaking. *Physical Review X* 3: 041011.
23 Kuester, E.F., Mohamed, M.A., Piket-May, M., and Holloway, C.L. (2003). Averaged transition conditions for electromagnetic fields at a metafilm. *IEEE Transactions on Antennas and Propagation* 51 (10): 2641–2651.
24 Holloway, C.L., Kuester, E.F., Gordon, J.A. et al. (2012). An overview of the theory and applications of metasurfaces: the two-dimensional equivalents of metamaterials. *IEEE Antennas and Propagation Magazine* 54 (2): 10–35.
25 Achouri, K., Salem, M.A., and Caloz, C. (2015). General metasurface synthesis based on susceptibility tensors. *IEEE Transactions on Antennas and Propagation* 63 (7): 2977–2991.
26 Albooyeh, M., Tretyakov, S., and Simovski, C. (2016). Electromagnetic characterization of bianisotropic metasurfaces on refractive substrates: general theoretical framework. *Annalen der Physik* 528 (9): 721–737.

27 Niemi, T., Karilainen, A.O., and Tretyakov, S.A. (2013). Synthesis of polarization transformers. *IEEE Transactions on Antennas and Propagation* 61 (6): 3102–3111.

28 Holloway, C.L., Mohamed, M.A., Kuester, E.F., and Dienstfrey, A. (2005). Reflection and transmission properties of a metafilm: with an application to a controllable surface composed of resonant particles. *IEEE Transactions on Electromagnetic Compatibility* 47 (4): 853–865.

29 Epstein, A., Wong, J.P.S., and Eleftheriades, G.V. (2015). Design and applications of Huygens metasurfaces. *2015 9th International Congress on Advanced Electromagnetic Materials in Microwaves and Optics (METAMATERIALS)*, Oxford, UK. pp. 67–69.

30 Epstein, A. and Eleftheriades, G.V. (2016). Huygens' metasurfaces via the equivalence principle: design and applications. *Journal of the Optical Society of America B: Optical Physics* 33 (2): A31.

31 Wong, J.P.S., Selvanayagam, M., and Eleftheriades, G.V. (2014). Design of unit cells and demonstration of methods for synthesizing Huygens metasurfaces. *Photonics and Nanostructures - Fundamentals and Applications* 12: 360–375.

32 Wong, J.P.S., Selvanayagam, M., and Eleftheriades, G.V. (2015). Polarization considerations for scalar Huygens metasurfaces and characterization for 2-D refraction. *IEEE Transactions on Microwave Theory and Techniques* 63 (3): 913–924.

33 Chen, M., Abdo-Sanchez, E., Epstein, A., and Eleftheriades, G.V. (2017). Experimental verification of reflectionless wide-angle refraction via a bianisotropic Huygens' metasurface. *2017 XXXIInd General Assembly and Scientific Symposium of the International Union of Radio Science (URSI GASS)*, Montreal, QC, Canada, pp. 1–4.

34 Wong, J.P.S., Epstein, A., and Eleftheriades, G.V. (2016). Reflectionless wide-angle refracting metasurfaces. *IEEE Antennas and Wireless Propagation Letters* 15: 1293–1296.

35 Epstein, A., Wong, J.P.S., and Eleftheriades, G.V. (2016). Cavity-excited Huygens' metasurface antennas for near-unity aperture illumination efficiency from arbitrarily large apertures. *Nature Communications* 7: 10360.

36 Wong, A.M.H. and Eleftheriades, G.V. (2018). Perfect anomalous reflection with a bipartite Huygens' metasurface. *Physical Review X* 8: 011036.

37 Cuesta, F.S., Faniayeu, I.A., Asadchy, V.S., and Tretyakov, S.A. (2018). Planar broadband Huygens' metasurfaces for wave manipulations. *IEEE Transactions on Antennas and Propagation* 66 (12): 7117–7127.

38 Lv, Y.H., Ding, X., Wang, B.Z., and Anagnostou, D.E. (2020). Scanning range expansion of planar phased arrays using metasurfaces. *IEEE Transactions on Antennas and Propagation* 68 (3): 1402–1410.

39 Dorrah, A.H., Chen, M., and Eleftheriades, G.V. (2018). Bianisotropic Huygens' metasurface for wideband impedance matching between two dielectric media. *IEEE Transactions on Antennas and Propagation* 66 (9): 4729–4742.

40 Wan, X., Zhang, L., Jia, S.L. et al. (2017). Horn antenna with reconfigurable beam-refraction and polarization based on anisotropic Huygens metasurface. *IEEE Transactions on Antennas and Propagation* 65 (9): 4427–4434.

# 5

# Metasurface: An Insight into Its Applications

*Fahad Ahmed and Nosherwan Shoaib*

*Research Institute for Microwave and Millimeter-Wave Studies (RIMMS), National University of Sciences and Technology (NUST), Islamabad, Pakistan*

A metamaterial is an artificially engineered material that has properties which are typically not found in natural materials. The "metamaterial" word is composed of two parts, i.e., meta + material, where its significance comes from the Greek word "meta" which refers to characteristics of the material that are typically not found in natural materials. The metamaterial is a three-dimensional (3D) structure, composed of special arrangement of unit cells, of precise shape and size, which can be used to tune the refractive index values to near zero, positive, or negative values. The metasurfaces are basically two-dimensional (2D) analogous of metamaterials which offer unique properties to control and manipulate the magnitude, polarization, and phase of the electromagnetic (EM) waves [1]. Traditionally, the polarization manipulation can be attained through conventional methods, for instance, by impinging a wave through an anisotropic material where the phase difference is provided as the wave propagates through the material or the Faraday effect. Nevertheless, these methods required bulky size and have narrow bandwidth when the operating wavelength is much larger. Similarly, the absorption could only be achieved through conventional absorbers. The natural materials, e.g. wedge [2] and ferrite [3], have been used for absorption purposes. However, their performance is incident angle-dependent. Moreover, their bulky size makes these types of absorbers unsuitable for many practical applications. To attain absorption and polarization manipulation through compact design, scholars have stimulated their motivation toward metasurfaces. They received special attention due to their low profile, ease of fabrication, and most important on-chip realization. Since the sub-wavelength unit cell of these metasurfaces is designed to acquire the desired functionality, thus, metasurfaces find numerous usage in several applications, including, but not limited to, the beam splitting [4], antenna gain enhancement [5], radar cross-sectional reduction [6], real-time hologram [7], flat lensing [8], sensors [9], emitters [10], planar optics [11], vortex generation [12], microscopy [13], contrast imaging [14], and polarization conversion [15–17]. In this concern, the absorbers, polarization rotators, and beam splitters have been proposed in the literature for microwave regime in reflection and transmission modes.

*Backscattering and RF Sensing for Future Wireless Communication,* First Edition.
Edited by Qammer H. Abbasi, Hasan T. Abbas, Akram Alomainy, and Muhammad Ali Imran.

This chapter presents an insight to the readers on several concepts related to metasurface-based polarizers, beam splitters, and absorbers. A clear distinction between the absorbers and polarizers is presented. To achieve these objectives, the basic requirements and unit cell configurations will be discussed to understand the basic concept behind polarizer, splitter, and absorber.

## 5.1 Polarization

EM waves are composed of linear, circular, and elliptical polarized waves. As the name suggests, the electric field of linearly polarized wave lies in one dimension, whereas the circularly polarized waves are composed of two components equal in magnitude with a phase difference of odd multiple of 90°. The elliptical polarization has two components with unequal magnitudes and arbitrary phase difference. Likewise, each polarization offers advantages over other polarization. Keeping in view this background, researchers have developed interest in manipulating the polarization of EM waves. Methods and techniques were developed to convert one polarization into another form (i.e., linear to circular or vice versa). The manipulation of EM waves has enormous advantages that drove the researchers to propose feasible techniques to convert one polarization into another polarization. Initial techniques used were complex and bulky as they used Faraday's principle to manipulate EM waves. Currently, the 3D artificial surfaces known as metamaterials are designed to achieve EM wave's polarization conversion, beam splitting, and other manipulations. Metamaterial is not only effective, but it is also compact, broadband, and angular stable. The subsequent section provides insight into the importance of metasurfaces in detail.

## 5.2 Polarizers

The polarizer is defined as a device which can transform the polarization of incident EM wave. Different conventional techniques, i.e., Brewster effect, optical activity, and Faraday effect, were employed in the past to control the polarization state of an EM wave. However, these techniques have several disadvantages, i.e., higher cost, bulky size, sensitivity for oblique angle, and narrow bandwidth. To overcome the aforementioned limitations, the artificially engineered EM structures, i.e., metamaterials are introduced [5, 6, 12, 15]. In the subsequent sections, the basic principle and the geometric configurations for the polarizers will be discussed.

### 5.2.1 Basic Principle

To understand the polarization phenomena, it is important to understand the underlying concepts of scalar and vector fields. Let us assume a scalar field "$S$" that, at each point, has an amplitude that changes in time as shown in Figure 5.1. The dark colors in Figure 5.1 represent higher intensity, and light color represents low intensity. The field "$S$" is a function of space and time as represented by Equation (5.1), and the intensity "$I$" is given by the

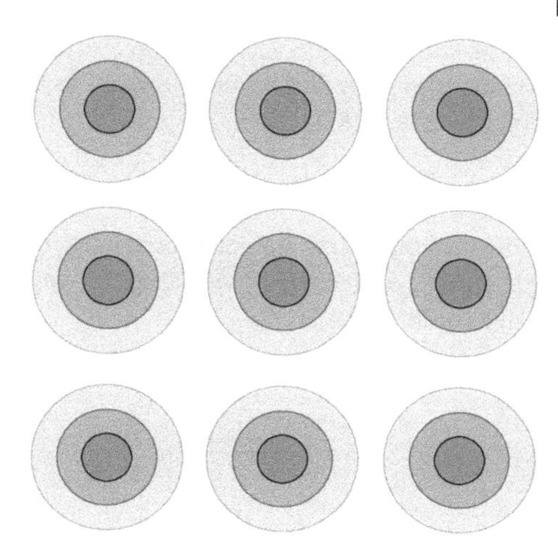

**Figure 5.1** Amplitude change of scalar field with time.

squared amplitude as mentioned in Equation (5.2).

$$S = U(x, t) \tag{5.1}$$

$$I(x) =< |U(x, t)|^2 > \tag{5.2}$$

In reality, one can use Maxwell's equations and observe that the field consists of an oscillating electric and magnetic components, which are not scalar but vectorial fields (see Equations 5.3 and 5.4). This means that in each point, the amplitude oscillates in a certain direction as shown in Figure 5.2, and the intensity is given by the squared length of the field vector as mentioned in Equation (5.4).

$$E(x, t) = \begin{bmatrix} E_x\,(x, t) \\ E_y\,(y, t) \end{bmatrix} \tag{5.3}$$

$$I(x) =< |E(x, t)|^2 >=< |E_x(x, t)|^2 > + < |E_y(x, t)|^2 > \tag{5.4}$$

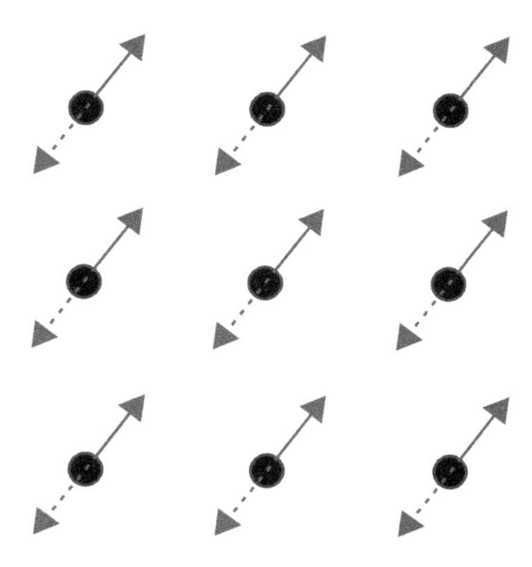

**Figure 5.2** Field comprising electric and magnetic components oscillating in certain direction.

Let us start from the vectorial wave equation and see how different types of polarization follow from it. In diffraction optics, let us assume scalar fields, which obey the scalar wave equation, i.e., [18].

$$\nabla^2 U(x,t) - \frac{1}{c^2}\frac{\partial^2 U(x,t)}{\partial t^2} = 0 \tag{5.5}$$

However, the Maxwell's equations (in vacuum, i.e. no current or charges) involve the vectorial electric and magnetic fields, so the wave equation that follows from them is a vectorial wave equation, i.e., [19].

$$\nabla \cdot E(x,t) = 0 \tag{5.6}$$

$$\nabla \cdot E(x,t) = -\frac{\partial^2 B(x,t)}{\partial t} \tag{5.7}$$

$$\nabla \cdot B(x,t) = 0 \tag{5.8}$$

$$\nabla \cdot B(x,t) = \frac{1}{c^2}\frac{\partial^2 E(x,t)}{\partial t} \tag{5.9}$$

Let us consider the propagation through vacuum, the vectorial wave equations tell us that each component of the electric field satisfies the scalar wave equation, i.e.

$$\nabla E_x(x,t) - \frac{1}{c^2}\frac{\partial^2 E_x(x,t)}{\partial t^2} = 0 \tag{5.10}$$

$$\nabla E_y(x,t) - \frac{1}{c^2}\frac{\partial^2 E_y(x,t)}{\partial t^2} = 0 \tag{5.11}$$

$$\nabla E_z(x,t) - \frac{1}{c^2}\frac{\partial^2 E_z(x,t)}{\partial t^2} = 0 \tag{5.12}$$

$$\nabla^2 E(x,t) - \frac{1}{c^2}\frac{\partial^2 E(x,t)}{\partial t^2} = 0 \tag{5.13}$$

From the plane-wave solution, it can infer that each component of the field is a plane wave with the same vector $\boldsymbol{K}$ that indicates the wavelength and the direction of propagation, and each field component has a different complex amplitude $A_x$, $A_y$, and $A_z$, which are contained in the vector $\boldsymbol{A}$, i.e.

$$E(x,t) = Ae^{i(k\cdot x - \omega t)} \tag{5.14}$$

$$\begin{bmatrix} E_x \\ E_y \\ E_z \end{bmatrix} = \begin{bmatrix} A_x e^{i(k\cdot x-\omega t)} \\ A_y e^{i(k\cdot x-\omega t)} \\ A_z e^{i(k\cdot x-\omega t)} \end{bmatrix} \tag{5.15}$$

Using the first Maxwell's equation, it can be written as:

$$\nabla \cdot E(x,t) = ik \cdot Ae^{i(k\cdot x-\omega t)} = 0 \tag{5.16}$$

$$k \cdot A = 0 \tag{5.17}$$

which implies that the amplitude vector and the wave vector must be perpendicular to each other as shown in Figure 5.3.

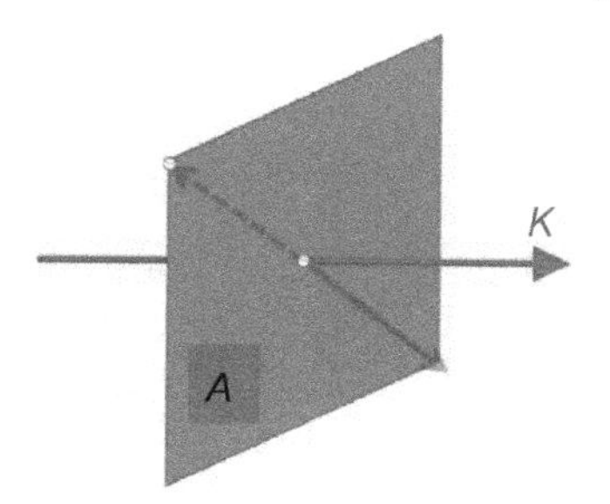


**Figure 5.3** Field oscillations are perpendicular to the direction of propagation.

It means that oscillating field vector should always be perpendicular to the direction of propagation if the field is propagating in the vacuum. Let us assume that the field in propagating in the $z$ direction as shown in Figure 5.4.

This means that the field should be oscillating in the $x$- and $y$-plane and the $z$-component of the field is zero, i.e. [20].

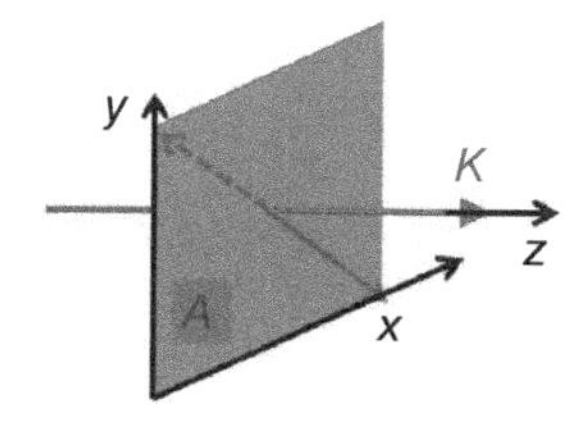


**Figure 5.4** Field propagation in $z$-direction.

$$E(x,t) = \begin{bmatrix} E_x(x,t) \\ E_y(x,t) \\ E_z(x,t) \end{bmatrix} = \begin{bmatrix} A_x e^{-i\omega t} \\ A_y e^{-i\omega t} \\ 0 \end{bmatrix} \tag{5.18}$$

Therefore, the electric field can be described using only complex amplitude $A_x$, $A_y$, and $A_z$. In vector representation, these are known as a Jones vector, i.e.

$$J = \begin{bmatrix} A_x \\ A_y \end{bmatrix} \tag{5.19}$$

Jones vectors are commonly used to describe the polarization states. Consider the Jones vector as [1, 0], i.e.

$$J = \begin{bmatrix} A_x \\ A_y \end{bmatrix} = \begin{bmatrix} 1 \\ 0 \end{bmatrix} \tag{5.20}$$

To understand its significance, recall that the physical real-valued electric field is given by the real part of the complex amplitude multiply by a time-dependent phase factor $e^{-i}\omega t$. In this case, the $x$-component of the field oscillates as $\cos(\omega t))$, while $y$-component of the field is zero. Therefore, it can be concluded that the light is linearly polarized, or more specifically, that is horizontally polarized as shown in Figure 5.5 and supported by the following mathematical formulations [21].

$$\begin{bmatrix} \varepsilon_x(x,t) \\ \varepsilon_y(x,t) \end{bmatrix} = \begin{bmatrix} Re\{A_x(x,t)e^{-i\omega t}\} \\ Re\{A_y(x,t)e^{-i\omega t}\} \end{bmatrix} = \begin{bmatrix} Re\{e^{-i\omega t}\} \\ 0 \end{bmatrix} = \begin{bmatrix} \cos\omega t \\ 0 \end{bmatrix} \tag{5.21}$$

Now, let us consider the Jones vector as [0, 1], i.e.

$$J = \begin{bmatrix} A_x \\ A_y \end{bmatrix} = \begin{bmatrix} 0 \\ 1 \end{bmatrix} \tag{5.22}$$

If $x$-component is zero, while $y$-component of the field oscillates as $\cos(\omega t)$. Therefore, the light is still linearly polarized but now vertically as shown in Figure 5.6 and supported by

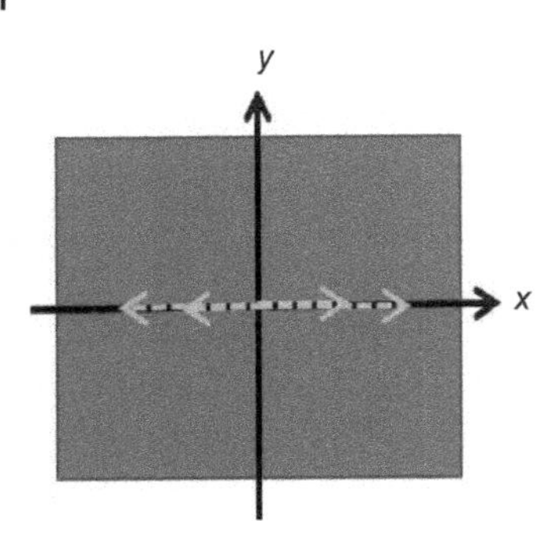


Figure 5.5 Demonstration of linearly polarized wave in x-direction.

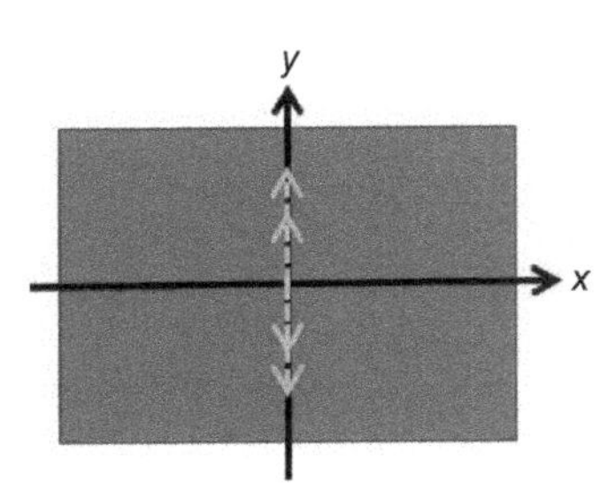


Figure 5.6 Demonstration of linearly polarized wave in y-direction.

the following mathematical expressions [4].

$$\begin{bmatrix} \varepsilon_x(x,t) \\ \varepsilon_y(x,t) \end{bmatrix} = \begin{bmatrix} Re\{A_x(x,t)e^{-i\omega t}\} \\ Re\{A_y(x,t)e^{-i\omega t}\} \end{bmatrix} = \begin{bmatrix} 0 \\ Re\{e^{-i\omega t}\} \end{bmatrix} = \begin{bmatrix} 0 \\ \cos\omega t \end{bmatrix} \tag{5.23}$$

Now, consider the third case, when the Jones vector is [1, 1], i.e.

$$J = \begin{bmatrix} A_x \\ A_y \end{bmatrix} = \begin{bmatrix} 1 \\ 1 \end{bmatrix} \tag{5.24}$$

In this case, both the $x$- and $y$-components oscillate as $\cos(\omega t)$, which means that light is linearly polarized at an angle of 45° as shown in Figure 5.7 and supported by the following expressions:

$$\begin{bmatrix} \varepsilon_x(x,t) \\ \varepsilon_y(x,t) \end{bmatrix} = \begin{bmatrix} Re\{A_x(x,t)e^{-i\omega t}\} \\ Re\{A_y(x,t)e^{-i\omega t}\} \end{bmatrix} = \begin{bmatrix} Re\{e^{-i\omega t}\} \\ Re\{e^{-i\omega t}\} \end{bmatrix} = \begin{bmatrix} \cos\omega t \\ \cos\omega t \end{bmatrix} \tag{5.25}$$

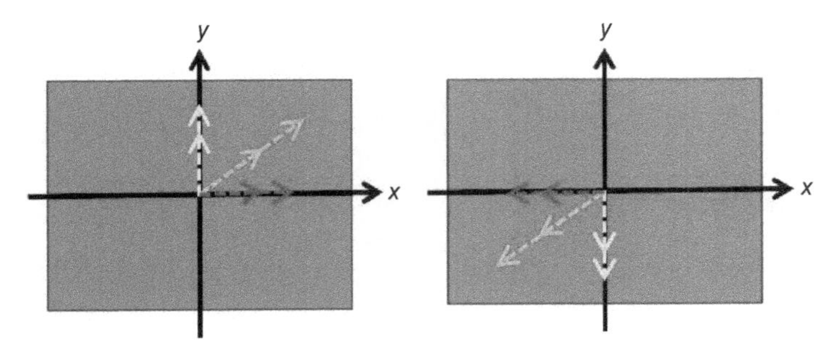


Figure 5.7 Demonstration of linearly polarized wave at 45°.

One interesting case will be to consider Jones vector having complex values i.e. [1, $i$] i.e.

$$J = \begin{bmatrix} A_x \\ A_y \end{bmatrix} = \begin{bmatrix} 1 \\ i \end{bmatrix} \tag{5.26}$$

In this case, the mathematical expressions will be as follows:

$$\begin{bmatrix} \varepsilon_x(x,t) \\ \varepsilon_y(x,t) \end{bmatrix} = \begin{bmatrix} Re\{A_x(x,t)e^{-i\omega t}\} \\ Re\{A_y(x,t)e^{-i\omega t}\} \end{bmatrix} = \begin{bmatrix} Re\{e^{-i\omega t}\} \\ Re\{i\cos\omega t + \sin\omega t\} \end{bmatrix} = \begin{bmatrix} \cos\omega t \\ \sin\omega t \end{bmatrix} \tag{5.27}$$

$$\begin{bmatrix} \varepsilon_x(x,t) \\ \varepsilon_y(x,t) \end{bmatrix} = \begin{bmatrix} Re\{A_x(x,t)e^{-i\omega t}\} \\ Re\{A_y(x,t)e^{-i\omega t}\} \end{bmatrix} = \begin{bmatrix} Re\{e^{-i\omega t}\} \\ Re\{ie^{-i\omega t}\} \end{bmatrix} \tag{5.28}$$

$$\begin{bmatrix} \varepsilon_x(x,t) \\ \varepsilon_y(x,t) \end{bmatrix} = \begin{bmatrix} Re\{A_x(x,t)e^{-i\omega t}\} \\ Re\{A_y(x,t)e^{-i\omega t}\} \end{bmatrix} = \begin{bmatrix} Re\{e^{-i\omega t}\} \\ Re\left\{e^{i\frac{\pi}{2}}e^{-i\omega t}\right\} \end{bmatrix} \tag{5.29}$$

$$\begin{bmatrix} \varepsilon_x(x,t) \\ \varepsilon_y(x,t) \end{bmatrix} = \begin{bmatrix} Re\{A_x(x,t)e^{-i\omega t}\} \\ Re\{A_y(x,t)e^{-i\omega t}\} \end{bmatrix} = \begin{bmatrix} Re\{e^{-i\omega t}\} \\ Re\left\{e^{-i\left(\omega t-\frac{\pi}{2}\right)}\right\} \end{bmatrix} \tag{5.30}$$

$$\begin{bmatrix} \varepsilon_x(x,t) \\ \varepsilon_y(x,t) \end{bmatrix} = \begin{bmatrix} Re\{A_x(x,t)e^{-i\omega t}\} \\ Re\{A_y(x,t)e^{-i\omega t}\} \end{bmatrix} = \begin{bmatrix} Re\{e^{-i\omega t}\} \\ Re\left\{e^{-i\left(\omega t-\frac{\pi}{2}\right)}\right\} \end{bmatrix} = \begin{bmatrix} \cos\omega t \\ \cos\left(\omega t - \frac{\pi}{2}\right) \end{bmatrix} \tag{5.31}$$

The field in the $x$-direction is given by the real part of $e^{-i\omega t}$, while the field in the $y$-direction is given by the real part of $ie^{-i\omega t}$ (see Equations 5.27, 5.28). By rewriting $i$ as $e^{-i\frac{\pi}{2}}$, and expressing the product as one complex exponential, it is observed that the field is oscillating in a cosine, but with a $\frac{\pi}{2}$ phase delay compared to the $x$-component. It can also be written in a sine form instead of a cosine. Another way to achieve these results is to consider the expression $ie^{-i}\omega t$ where the complex exponential can be written as $\cos\omega t - i\sin\omega t$, and after multiplying this with $i$, the real part becomes $\sin(\omega t)$, i.e.,

$$\begin{bmatrix} \varepsilon_x(x,t) \\ \varepsilon_y(x,t) \end{bmatrix} = \begin{bmatrix} Re\{A_x(x,t)e^{-i\omega t}\} \\ Re\{A_y(x,t)e^{-i\omega t}\} \end{bmatrix} = \begin{bmatrix} Re\{e^{-i\omega t}\} \\ Re\{ie^{-i\omega t}\} \end{bmatrix} = \begin{bmatrix} \cos\omega t \\ \sin\omega t \end{bmatrix} \tag{5.32}$$

$$\begin{bmatrix} \varepsilon_x(x,t) \\ \varepsilon_y(x,t) \end{bmatrix} = \begin{bmatrix} Re\{A_x(x,t)e^{-i\omega t}\} \\ Re\{A_y(x,t)e^{-i\omega t}\} \end{bmatrix} = \begin{bmatrix} Re\{e^{-i\omega t}\} \\ Re\{i(\cos\omega t - i\sin\omega t)\} \end{bmatrix} = \begin{bmatrix} \cos\omega t \\ \sin\omega t \end{bmatrix} \tag{5.33}$$

It can be seen at time $t = 0$, the $x$-component is 1, and the $y$-component is 0. As the time increases, the $x$-component decreases, while the $y$-component becomes larger. When the $y$-component is maximal, the $x$-component is zero. It means that electric field vector is rotating in circles; therefore, this state of polarization may be called as circular polarization. Equation (5.34) defines the left-hand circularly polarized wave, while Figure 5.8 provides its demonstration.

$$\begin{bmatrix} \varepsilon_x(x,t) \\ \varepsilon_y(x,t) \end{bmatrix} = \begin{bmatrix} Re\{A_x(x,t)e^{-i\omega t}\} \\ Re\{A_y(x,t)e^{-i\omega t}\} \end{bmatrix} = \begin{bmatrix} Re\{e^{-i\omega t}\} \\ Re\{i\cos\omega t + \sin\omega t\} \end{bmatrix} = \begin{bmatrix} \cos\omega t \\ \sin\omega t \end{bmatrix} \tag{5.34}$$

For right-hand circular polarization, Jones vector and field equations can be written as follows:

$$J = \begin{bmatrix} A_x \\ A_y \end{bmatrix} = \begin{bmatrix} 1 \\ -i \end{bmatrix} \tag{5.35}$$

$$\begin{bmatrix} \varepsilon_x(x,t) \\ \varepsilon_y(x,t) \end{bmatrix} = \begin{bmatrix} Re\{A_x(x,t)e^{-i\omega t}\} \\ Re\{A_y(x,t)e^{-i\omega t}\} \end{bmatrix} = \begin{bmatrix} Re\{e^{-i\omega t}\} \\ Re\{-i\cos\omega t - \sin\omega t\} \end{bmatrix} = \begin{bmatrix} \cos\omega t \\ -\sin\omega t \end{bmatrix} \tag{5.36}$$

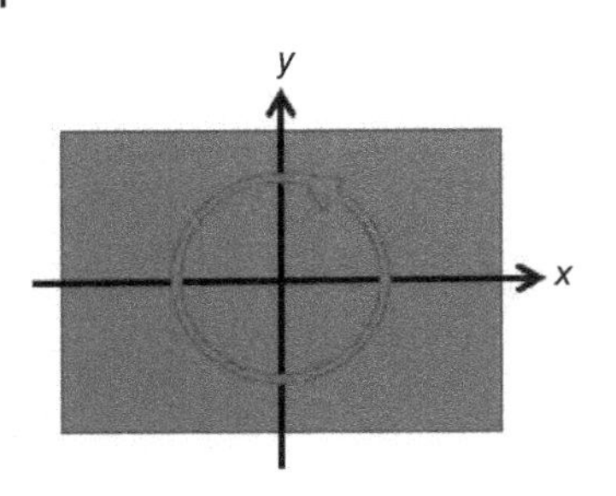


**Figure 5.8** Demonstration of left-hand circularly polarized wave.

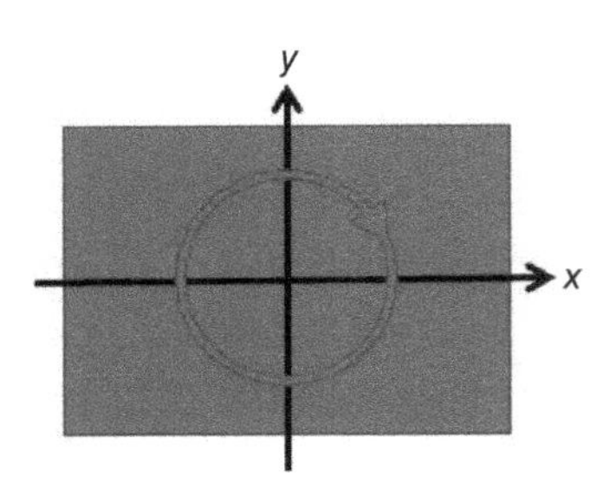


**Figure 5.9** Demonstration of right-hand circularly polarized wave.

The demonstration of right-hand circularly polarized wave is shown in Figure 5.9. After defining and understanding the polarization concepts, the next step is to create and manipulate them. A simple optical component, i.e., the linear polarizer, may be considered that only transmits the horizontal component of the field. Mathematically, it sets *the* $y$-component equal to zero, while the $x$-component is left unchanged. This operation can be described by a matrix [1, 0; 0, 0] (see Equation 5.37), which can be called as the Jones matric for a horizontal polarizer.

$$\begin{bmatrix} A_x \\ A_y \end{bmatrix} \rightarrow \begin{bmatrix} A_x \\ 0 \end{bmatrix} = \begin{bmatrix} 1 & 0 \\ 0 & 0 \end{bmatrix} \begin{bmatrix} A_x \\ A_y \end{bmatrix} \tag{5.37}$$

If there is a vertical polarizer, the $x$-component is a set of 0 and the $y$-component is unchanged and it is described by the matrix [0, 0; 0, 1] (see Equation 5.38). The illustrations for horizontal and vertical polarizers are shown in Figure 5.10.

$$\begin{bmatrix} A_x \\ A_y \end{bmatrix} \rightarrow \begin{bmatrix} 0 \\ A_y \end{bmatrix} = \begin{bmatrix} 0 & 0 \\ 0 & 1 \end{bmatrix} \begin{bmatrix} A_x \\ A_y \end{bmatrix} \tag{5.38}$$

Now suppose there is linearly diagonal polarized light and it has to be converted to circularly polarized light. Mathematically, the Jones matrix for this operation is given by [1, 0; 0, i], i.e.

$$\begin{bmatrix} 1 \\ 1 \end{bmatrix} \rightarrow \begin{bmatrix} 1 \\ i \end{bmatrix} = \begin{bmatrix} 1 & 0 \\ 0 & \pm i \end{bmatrix} \begin{bmatrix} 1 \\ 1 \end{bmatrix} \tag{5.39}$$

The illustration for circular polarizers is shown in Figure 5.11. To understand its physical realization, let us recall the Jones vector interpretation. It represents the complex amplitude of the plane-wave solution for the $x$- and $y$-component of the field. Let us assume that the plane waves are propagating in the $z$-direction, so, one can write the complex-valued field components as the Jones vector multiplied by the phase factor $e^{-i(kZ - \omega t)}$. The factor $e^{i\omega t}$ may

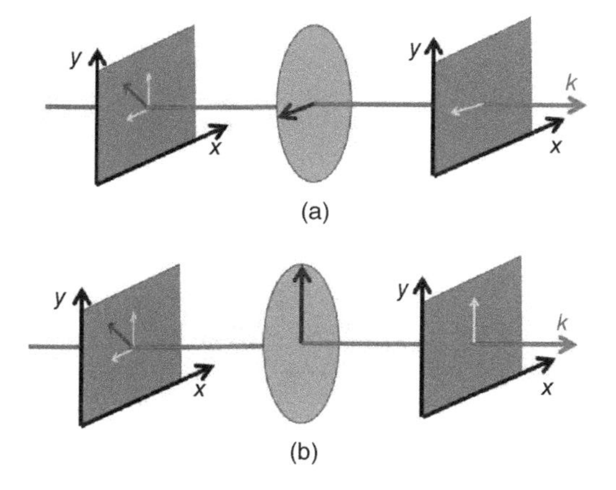


**Figure 5.10** (a) Horizontal polarizer. (b) Vertical polarizer.

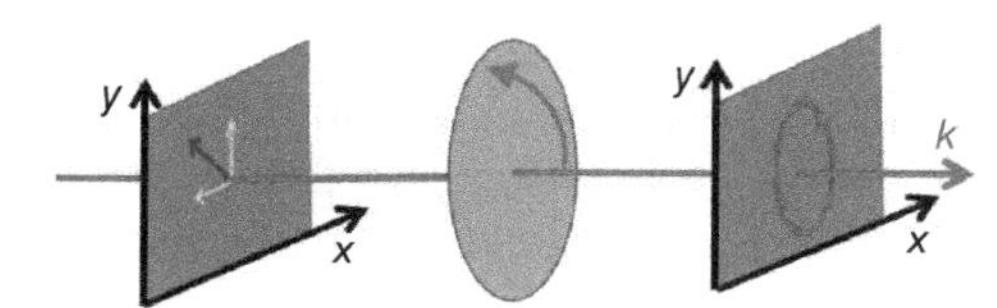


**Figure 5.11** Circular polarizer.

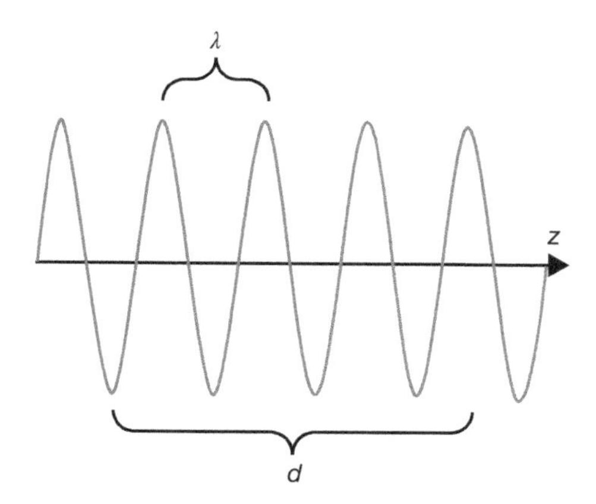


**Figure 5.12** Field propagation by a distance "*d*."

be ignored due to that fact that it is a global phase factor and it does not affect the phase difference between the *x*- and *y*-components; therefore, it does not affect the polarization state. The trimmed form of the phase factor $e^{-ikZ}$ depicts that as the field propagates by a distance "*d*," it undergoes a phase shift of $e^{-ikd}$. Intuitively, this can be understood by realizing that if there is a field with wavelength $\lambda$, the phase shift due to propagation can be found by counting number of $\lambda$ fit in the propagation distance "*d*," where each $\lambda$ yields a phase shift of $2\pi$ as shown in Figure 5.12.

The wave number $k$ is defined as $\frac{2\pi}{\lambda}$; therefore, phase shift can be written as, $kd$. In this case, the wavelength is reduced by a factor $n$ (see Figure 5.13); therefore, the phase shift is multiplied by a factor $n$, which implies that the field components will become:

$$\begin{bmatrix} A_x \\ A_y \end{bmatrix} \rightarrow \begin{bmatrix} A_x \\ A_y \end{bmatrix} e^{iknd} \tag{5.40}$$

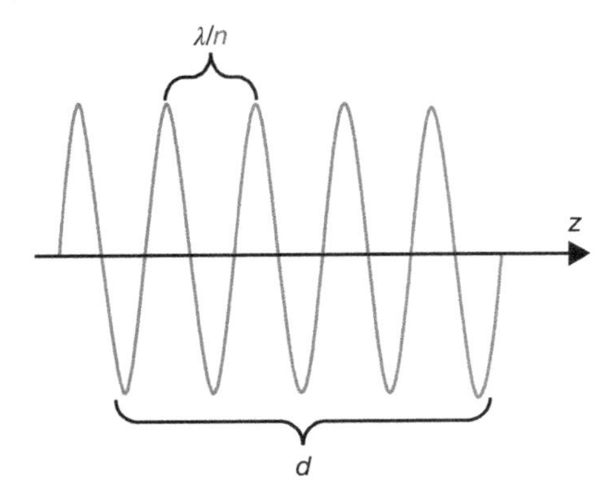


**Figure 5.13** Propagation by distance "*d*" through medium with refractive index "*n*."

The aforementioned phenomena happen only if there is a material which has different refraction indices for the $x$-component and the $y$-component of the field. Such a material is called a birefringent medium, and an example of birefringent media would be crystalled such as calcite. In this case, each component is multiplied by a different phase factor and this operation is represented by a diagonal matrix with $e^{ikn_xd}$ and $e^{ikn_yd}$ are placed on diagonal.

$$\begin{bmatrix} A_x \\ A_y \end{bmatrix} \rightarrow \begin{bmatrix} e^{ikn_xd}A_x \\ e^{ikn_yd}A_y \end{bmatrix} = \begin{bmatrix} e^{ikn_xd} & 0 \\ 0 & e^{ikn_yd} \end{bmatrix} \begin{bmatrix} A_x \\ A_y \end{bmatrix} \tag{5.41}$$

$$\begin{bmatrix} A_x \\ A_y \end{bmatrix} \rightarrow \begin{bmatrix} e^{ikn_xd}A_x \\ e^{ikn_yd}A_y \end{bmatrix} = e^{ikn_xd} \begin{bmatrix} 1 & 0 \\ 0 & e^{ik(n_y-n_x)d} \end{bmatrix} \begin{bmatrix} A_x \\ A_y \end{bmatrix} \tag{5.42}$$

The global phase factor $e^{ikn_xd}$ is irrelevant to the polarization state and can be factor out as follows:

$$\begin{bmatrix} A_x \\ A_y \end{bmatrix} \rightarrow \begin{bmatrix} e^{ikn_xd}A_x \\ e^{ikn_yd}A_y \end{bmatrix} \propto \begin{bmatrix} 1 & 0 \\ 0 & e^{ik(n_y-n_x)d} \end{bmatrix} \begin{bmatrix} A_x \\ A_y \end{bmatrix} \tag{5.43}$$

The above equation left with a complex exponential $e^{ik(n_y-n_x)d}$ containing the term $n_y - n_x$, which is referred to the birefringence of a material. It can be seen that there is a need to write the correct form for an optical component that transform linearly polarized light to circularly polarized light as follows:

$$\begin{bmatrix} A_x \\ A_y \end{bmatrix} \rightarrow \begin{bmatrix} e^{ikn_xd}A_x \\ e^{ikn_yd}A_y \end{bmatrix} \propto \begin{bmatrix} 1 & 0 \\ 0 & e^{ik\Delta nd} \end{bmatrix} \begin{bmatrix} A_x \\ A_y \end{bmatrix} \tag{5.44}$$

Here, the complex exponential $e^{ik\Delta nd}$ should be equal to i. By performing the following simplifications, i.e.,

$$e^{ik\Delta nd} = i \text{ or } e^{ik\Delta nd} = e^{\frac{\pi}{2}i+m2\pi i} \text{ or } ik\Delta nd = \frac{\pi}{2}i + m2\pi i \text{ or } k\Delta nd = \frac{\pi}{2} + m2\pi \text{ or}$$
$$\frac{2\pi}{\lambda}\Delta nd = \frac{\pi}{2} + m2\pi \text{ or } \frac{\Delta nd}{\lambda} = \frac{1}{4} + m \text{ or } \Delta nd = \frac{1}{4}\lambda + m\lambda \tag{5.45}$$

It is concluded that $\Delta nd$ should be equal to a quarter wavelength plus an integer multiply of the wavelength. Therefore, this optical component is called a quarter-wave plate (QWP). In principle, it does not matter much whether it contains an $i$ or $-i$. Similarly, by changing

**Figure 5.14** Half-wave plate (HWP) operation.

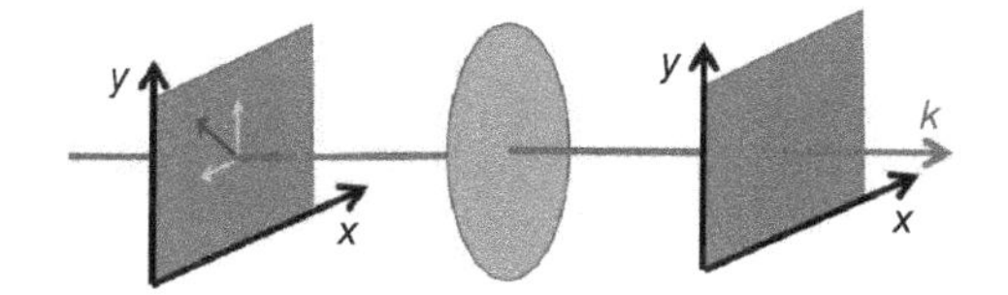


the requirement for $\Delta nd$, it can also be defined as a half-wave plate (HWP). The illustration for HWP operation is shown in Figure 5.14.

The HWP operation can be represented by Equation (5.46) which shows that the sign of the $y$-component is flipped. If a vector is drawn with a flipped $y$-component, then the vector is mirrored in the $x$-axis. If it is written in the wave plate, the mirror axis is rotated; therefore, the resulting polarization vector is also rotated. This show the usefulness of a HWP that it can be used to rotate the polarization direction of linearly polarized light without any loss of intensity.

$$\begin{bmatrix} A_x \\ A_y \end{bmatrix} \rightarrow \begin{bmatrix} A_x \\ -A_y \end{bmatrix} = \begin{bmatrix} 1 & 0 \\ 0 & -1 \end{bmatrix} \begin{bmatrix} A_x \\ A_y \end{bmatrix} \tag{5.46}$$

### 5.2.2 Geometrical Configuration

A polarizer, designed either in reflection and transmission mode, finds many applications in the optical or microwave regime. Due to the several potential applications, this concept is extensively studied. Although the conventional techniques used to achieve polarization conversion phenomena were narrowband, angularly unstable, complex, and bulky as they were based on Faraday's principle. However, metasurfaces provided a solution to these problems with added advantages. The following sections will be the design approach of the polarizers.

To design a polarizer, the geometry of the metasurface plays an important role. The basic principle of polarization conversion is anisotropy in the unit cell design. However, the anisotropy should be introduced in the unit cell in such a way that mirror symmetry is created which is also known as C2-symmetry. The mirror geometry can be obtained if unit cell is designed based on anisotropy concept. The mirror symmetry phenomenon is demonstrated in Figure 5.15.

Let us assume a matrix $M$ that reflects any vector in the $xy$-plane about $v$-axis shown in Figure 5.15. Consider the vector $Y = \begin{bmatrix} 0 \\ 1 \end{bmatrix}$. When the matrix $M$ operates on it, it is rotated to "$-X$", i.e.

$$MY = \begin{bmatrix} 0 & -1 \\ -1 & 0 \end{bmatrix} \begin{bmatrix} 0 \\ 1 \end{bmatrix} \tag{5.47}$$

$$=> \begin{bmatrix} -1 \\ 0 \end{bmatrix} = -X \tag{5.48}$$

Similarly, applying $M$ to "$-X$" gives $Y$, i.e.

$$M(-X) = \begin{bmatrix} 0 & -1 \\ -1 & 0 \end{bmatrix} \begin{bmatrix} -1 \\ 0 \end{bmatrix} \tag{5.49}$$

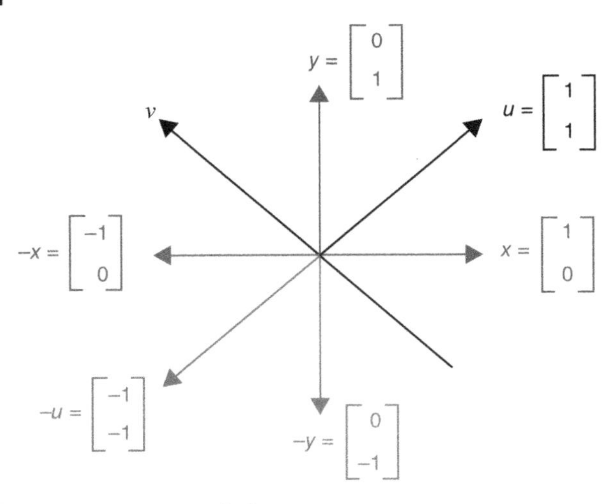


**Figure 5.15** Demonstration of mirror symmetry phenomenon.

$$=> \begin{bmatrix} 0 \\ 1 \end{bmatrix} = Y \tag{5.50}$$

When the vector $M$ operates on $X$, it is rotated to "$-Y$":

$$MX = \begin{bmatrix} 0 & -1 \\ -1 & 0 \end{bmatrix} \begin{bmatrix} 1 \\ 0 \end{bmatrix} \tag{5.51}$$

$$=> \begin{bmatrix} 0 \\ -1 \end{bmatrix} = -Y \tag{5.52}$$

When the vector $M$ operates on $U$, it is rotated to "$-U$":

$$MU = \begin{bmatrix} 0 & -1 \\ -1 & 0 \end{bmatrix} \begin{bmatrix} 1 \\ 1 \end{bmatrix} = \begin{bmatrix} -1 \\ -1 \end{bmatrix} = -U \tag{5.53}$$

In the similar manner, when $M$ operates on an image, the image is reflected about $v$-axis in the same manner as discussed above for the vector. The matrix $MTM^{-1} = \overline{T}$ represents the transmission coefficient matrix of the structure which is obtained after the application of the matrix $M$. In simple words, if $T$ is the transmission coefficient matrix for a unit cell, then the transmission coefficient matrix of the reflected image of the unit cell along $v$-axis will be $\overline{T} = MTM^{-1}$. For example, if the transmission coefficient matrix for the unit cell is represented by the following matrix:

$$T = \begin{pmatrix} T_{xx} & T_{xy} \\ T_{yx} & T_{yy} \end{pmatrix} \tag{5.54}$$

Then, the transmission coefficient matrix for the image will be:

$$MTM^{-1} = \begin{bmatrix} 0 & -1 \\ -1 & 0 \end{bmatrix} \begin{bmatrix} T_{xx} & T_{xy} \\ T_{yx} & T_{yy} \end{bmatrix} \begin{bmatrix} 0 & -1 \\ -1 & 0 \end{bmatrix} \tag{5.55}$$

$$=> \overline{T} = \begin{bmatrix} T_{yy} & T_{yx} \\ T_{xy} & T_{xx} \end{bmatrix} \tag{5.56}$$

The $T_{xx}$ is replaced by $T_{yy}$ because after the application of matrix $M$ on the unit cell, the part of the unit cell, (such as split) which was previously along $y$-axis, is now shifted toward

$x$-axis. Similarly, the part of the unit cell, which was previously along $x$-axis, is now shifted toward $y$-axis. $T_{xx}$ and $T_{yy}$ are interchanged in transmission matrix for the reflected image because the parts of the unit cell along $x$- and $y$-axis are interchanged by applying matrix $M$.

If the unit cell and its reflected image are same, then the transmission matrix for both must also be same, i.e., $T = \overline{T}$. Hence, it can be expressed as:

$$MTM^{-1} = T \tag{5.57}$$

and

$$T_{xx} = T_{yy} \text{ and } T_{yx} = T_{xy}. \tag{5.58}$$

The same is the case with the reflection matrix. Keeping this in view, the cross-polarizer in reflection mode may be designed as followed in the upcoming section. It should be noted that the terms "half-wave plate" and "QWP" are also used to represent cross-polarizer and circular polarizer, respectively.

#### 5.2.2.1 Cross-Polarizer – HWP

This first step to design a cross-polarizer is to choose the dimensions of the metasurface to achieve the resonances at desired frequency band. The rough estimation of the resonances can be calculated using the following equation:

$$C = n\lambda_{eff} = \frac{nc}{\lambda\sqrt{\varepsilon_{reff}}} \tag{5.59}$$

where $C$ is the circumference of the circle or the largest dimension of the patch and $\varepsilon_{reff}$ is the effective permittivity. To achieve magnetic resonance, "$n$" should be considered unity. On the other hand, the electric resonances can be achieved for $n = 2$ [22].

The second important step is to design a surface which acts as a double refractive material. In double refraction phenomenon, two waves are generated after passing the medium. The one which follows the Snell's law is known as ordinary wave (O-ray), while the other one is known as extraordinary wave (E-ray) which does not obey the Snell's law.

In order to achieve the HWP operation at obtained resonances, the thickness of the substrate shall be chosen such as to produce a path difference of $\frac{\lambda}{2}$ or a phase difference $\pi$ between the ordinary wave and extraordinary wave, when monochromatic plane polarized light of wavelength $\lambda$ is incident normal to the surface. The device for producing and detecting the linearly polarized light is known as HWP. The conventional plates were made up of thin sheets of quarts or calcite placed parallel to the optical axis. To calculate thickness "$t$" of the HWP, let us assume the negative crystal, which is also known as calcite, in which velocity of E-ray is greater than the O-ray. In this case, the path difference may be defined as follows:

$$\text{Path difference} = t\,(\mu_0 - \mu_e) \tag{5.60}$$

The phase difference can be defined as follows:

$$\text{Phase difference } (\theta) = \frac{2\pi}{\lambda} \text{ path difference}$$

or

$$\theta = \frac{2\pi}{\lambda} = t(\mu_0 - \mu_e) \tag{5.61}$$

For HWP, the phase difference will become:

$$\theta = \pi = \frac{2\pi}{\lambda} = t(\mu_0 - \mu_e) \tag{5.62}$$

or

$$t = \frac{\lambda}{2(\mu_0 - \mu_e)} \tag{5.63}$$

In case of positive crystal, the thickness will become:

$$t = \frac{\lambda}{2(\mu_e - \mu_o)} \tag{5.64}$$

From aforementioned mathematical formulations, it can be seen that Equation (5.63) or (5.64) may be used to calculate the thickness of the HWP. To compute the thickness for the QWP, Equations (5.60)–(5.63) may be used; however, one must use $\theta = \pi/2$.

After defining the mathematical formulations for the polarizers, it is now important to discuss the polarization conversion criterion either in transmission or reflection mode. For better understanding cross-polarization, let us assume that $E_{yi}$ and $E_{xi}$ are the incident electric field components in $y$- and $x$-directions. Likewise, let us consider $E_{yr}$ and $E_{xr}$ as reflected $E$-field components in $y$- and $x$-direction. The Jones matrix, highlighting the conversion of incident EM wave into its cross- and co-components reflection coefficients, can be written as follows:

$$\begin{pmatrix} E_{xr} \\ E_{yr} \end{pmatrix} = \begin{pmatrix} R_{xx} & R_{xy} \\ R_{yx} & R_{yy} \end{pmatrix} \begin{pmatrix} E_{xi} \\ E_{yi} \end{pmatrix} \tag{5.65}$$

where, $R_{xy} = \left|\frac{E_{xr}}{E_{yi}}\right|$ and $R_{yx} = \left|\frac{E_{yr}}{E_{xi}}\right|$ are cross-polarized coefficients, while the $R_{yy} = \left|\frac{E_{yr}}{E_{yi}}\right|$ and $R_{xx} = \left|\frac{E_{xr}}{E_{xi}}\right|$ are co-polarized coefficients. To achieve the highly efficient cross-polarization conversion, the co-components of both TE and TM incident waves remain $<-10$ dB, while cross-component should be above $-3$ dB over required frequency band. In this context, a metasurface containing the unit cells shown in Figure 5.16 is designed. The co- and cross-reflection coefficients, i.e., $R_{yy}$ and $R_{xy}$, respectively, are shown in Figure 5.17 which are in accordance with the aforementioned criteria within the frequency range of 8.2–15 and 19.3–20.8 GHz for highly efficient cross-polarization conversion [15]. The reflection coefficient values in tabular form are presented in Table 5.1.

Let us discuss the unit cell configuration, shown in Figure 5.16b. The unit cell should be symmetric along both $u$- and $v$-axes. To explain the incident and reflected electric fields in terms of $u$- and $v$-axes, Equations (5.66) and (5.67) are helpful.

$$E_i = uE_{ui} + vE_{vi} \tag{5.66}$$

$$E_r = uR_{uu}E_{ur} + vR_{vv}E_{vr} \tag{5.67}$$

$E_i$ is incident $E$-field which is decomposed into $u$- and $v$-components, i.e. $\pm 45°$ along $y$-axis. On the other hand, the $E_r$ is the reflected electric field which consists of co-polarized reflection coefficients $R_{uu}$ and $R_{vv}$. The directions along $u$- and $v$-axes of incident electric field are $E_{ui}$ and $E_{vi}$, and reflected electric field is $E_{ur}$ and $E_{vr}$, respectively. The response of the structure can be observed from reflection coefficients and their phase difference

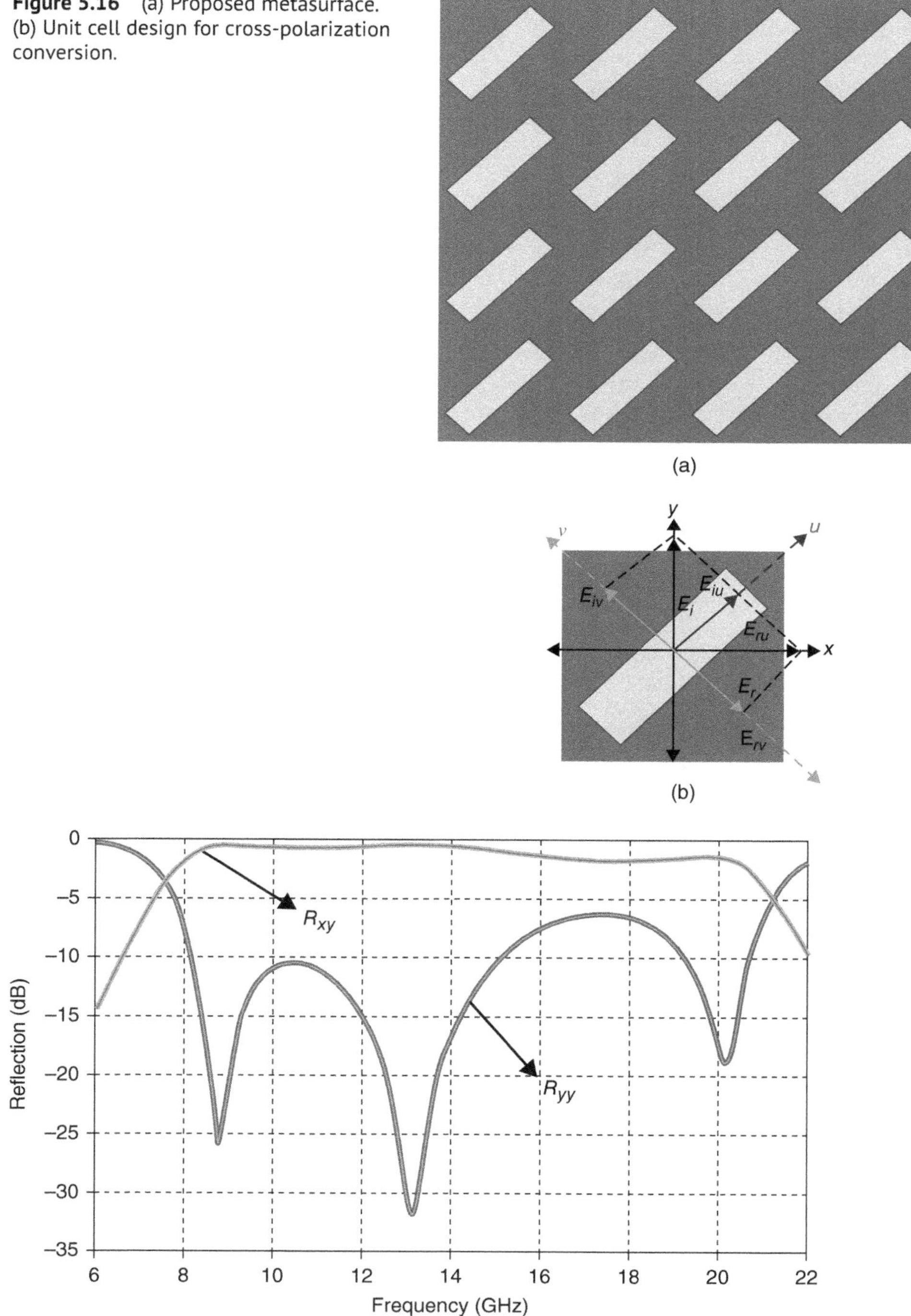


**Figure 5.16** (a) Proposed metasurface. (b) Unit cell design for cross-polarization conversion.

**Figure 5.17** The co- and cross-reflection coefficients of metasurface.

**Table 5.1** Co- and cross-reflection coefficients of metasurface.

| Frequency (GHz) | $R_{yy}$ (dB) | $R_{xy}$ (dB) |
|---|---|---|
| 6 | −0.28 | −14.34 |
| 8 | −7.50 | −1.00 |
| 10 | −10.92 | −0.78 |
| 12 | −15.00 | −0.56 |
| 14 | −16.39 | −0.58 |
| 16 | −8.00 | −1.42 |
| 18 | −6.52 | −1.80 |
| 20 | −17.00 | −1.76 |
| 22 | −1.94 | −9.54 |

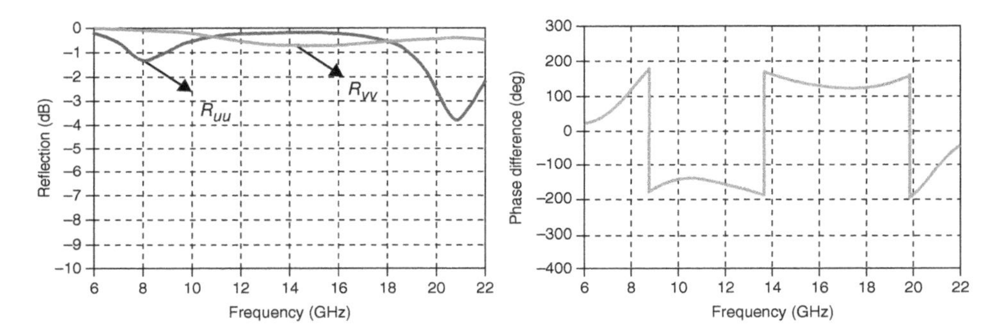


**Figure 5.18** Magnitude and phase difference of *u*- and *v*-reflected wave.

shown in Figure 5.18. It can be observed that the magnitude of the reflected components is approximately equal, i.e., $|R_{uu}| \approx |R_{vv}|$ and phase difference is $\Delta\phi = \pm 180°$, which rotates the reflected EM waves 90° with respect to incident EM waves. It can be noted that the anisotropic behavior of proposed multifunctional metasurface converts the *y*-polarized incident EM wave into *x*-polarized EM wave upon reflection from the structure.

The aforementioned phenomenon can be explained by observing the current distribution. If the currents on the top and bottom layers of the metasurface are in the same direction, then it causes electric resonances. However, if the currents on top and bottom of the metasurface flow in the opposite direction, then it causes magnetic resonance. The three resonances have been chosen at frequency points of 8.7, 13.1, and 20.14 GHz and the respective current distributions are shown in Figure 5.19.

It can be observed from Figure 5.19 that the first two resonances are magnetic, while the third one is electric resonance. At these resonances, one component acts as high impedance surface (HIS), while the other component acts as a perfect electric conductor. At 8.7 and 20.14 GHz, the *v*-component is reflecting with 0° phase (HIS), while *u*-component is reflecting with 180° phase (PEC). Similarly, at 13.1 GHz, the *u*-component is reflecting

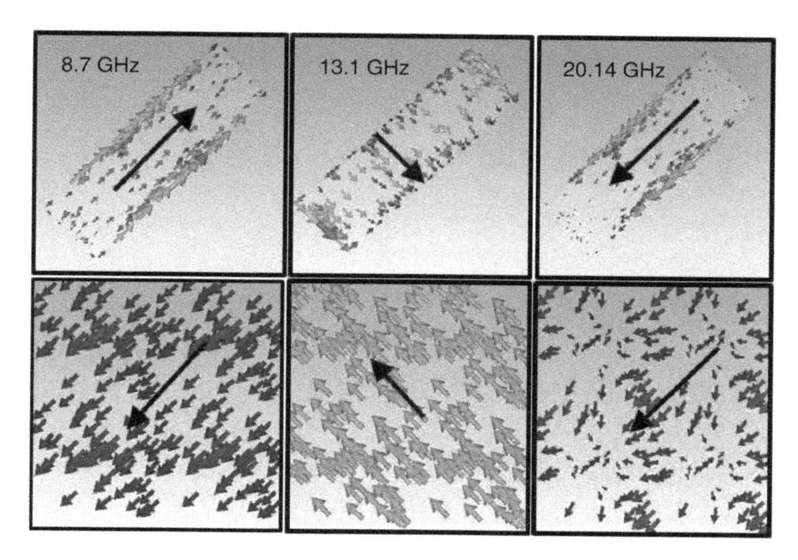


**Figure 5.19** Current distribution at frequency points 8.7, 13.1, and 20.14 GHz.

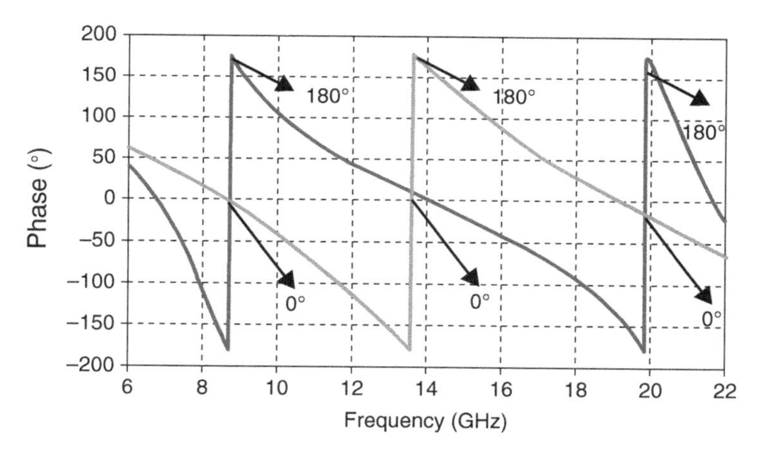


**Figure 5.20** Phase of reflection coefficient. (a) Red for $R_{uu.}$ (b) Brown for $R_{vv.}$

with 0° phase (HIS), while $v$-component is reflecting with 180° phase (PEC). To achieve cross-polarization conversion, the symmetry along $u$- and $v$-axes is necessary due to aforementioned reason since if the wave comes across an open circuit, it reflects back with 0° (i.e., act as HIS), while if it comes across short circuit, it reflects back with 180° phase (i.e., act as PEC). Such behavior can be observed in Figure 5.20 which shows the achieved phase of the reflection coefficient.

#### 5.2.2.2 Circular Polarizer – QWP

To achieve circular polarization, a suitable geometry shall be selected in which one component acts as PEC or HIS, while other component acts as a capacitor or inductor. It is pertinent to mention that the circular-polarization conversion cannot be achieved without

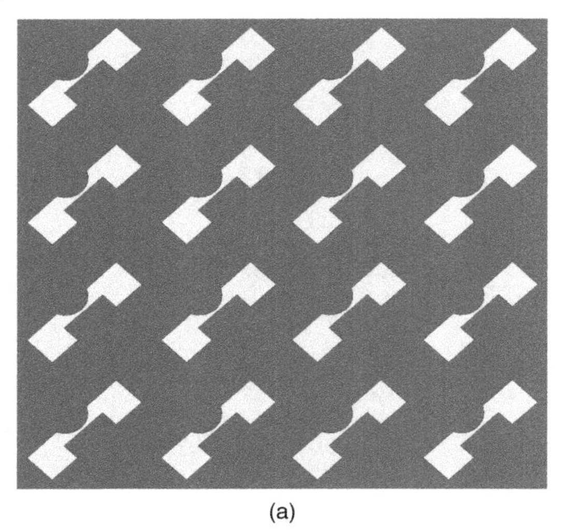

(a)

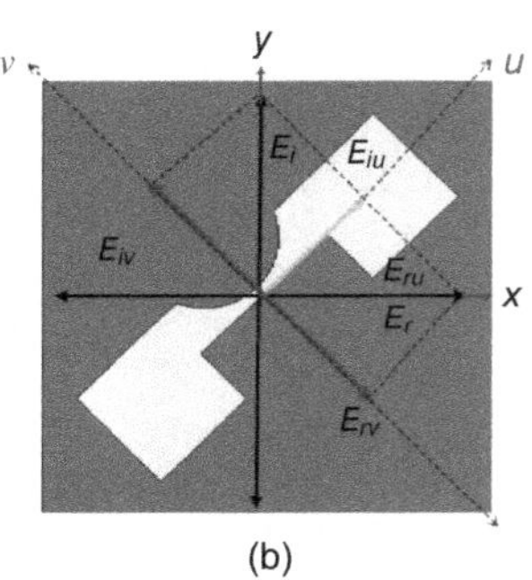


(b)

**Figure 5.21** (a) Proposed metasurface. (b) Unit cell with a slit at 45° for circular polarization conversion.

a resonance, i.e., it starts right before or after the resonance. At resonance, it is quite possible that the structure acts as a cross-polarizer or if the resonance is weak, it does not perform any conversion.

A metasurface with a unit cell having slit at 45° is designed to achieve the circular-polarization conversion as shown in Figure 5.21. The gap in the unit cell acts as a capacitor, while the solid part acts as an inductor. A slit is introduced in the structure at the angle of 45° to maintain the anisotropy and symmetry. It can be seen in Figure 5.21b that there exists a mirror symmetry along $v$-axis; however, no mirror symmetry exists along $u$-axis.

The reflection coefficient magnitude and phase difference response are shown in Figure 5.22. It can be observed from Figure 5.22 that circular-polarization conversion is achieved at X-band in addition to cross-polarization conversion at 6.9–7.4 GHz. The ratio of magnitude of both reflection coefficients (i.e. $R_{xy}$ and $R_{yy}$) is almost unity, and the phase difference is almost −90° over the entire X-band [15]. Table 5.2 shows the co- and cross-reflection coefficient values of the proposed metasurface.

As shown in Figure 5.21b, the $y$-axis may be decomposed into $u$- and $v$-axes at ±45°. The equations in term of $u$- and $v$-axes for circular-polarization conversion may be written as follows:

$$\begin{pmatrix} E_u^r \\ \overline{E_v^r} \end{pmatrix} = R_{uv} \begin{pmatrix} E_u^i \\ \overline{E_v^i} \end{pmatrix} = \begin{pmatrix} r_{uu} & 0 \\ 0 & r_{vv} \end{pmatrix} \begin{pmatrix} E_u^i \\ \overline{E_v^i} \end{pmatrix} \tag{5.68}$$

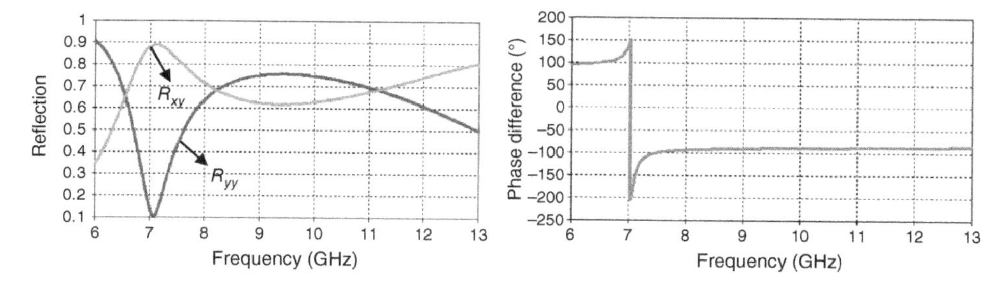


**Figure 5.22** Reflection coefficient magnitude and phase difference response.

**Table 5.2** Co- and cross-reflection coefficients of metasurface.

| Frequency (GHz) | $R_{yy}$ (dB) | $R_{xy}$ (dB) |
|---|---|---|
| 6 | −0.70 | −10.12 |
| 7 | −19.17 | −1.11 |
| 8 | −5.30 | −2.19 |
| 9 | −2.50 | −4.29 |
| 10 | −2.50 | −4.30 |
| 11 | −3.09 | −3.22 |
| 12 | −4.44 | −2.62 |
| 13 | −6.29 | −1.72 |

$$\vec{E}_i = E_o\hat{e}_y = E_{ui}\hat{e}_u + E_{vi}\hat{e}_v = E_o\cos(45°)(\hat{e}_u + \hat{e}_v) \tag{5.69}$$

$$\vec{E}_r = E_{ur}\hat{e}_u + E_{vr}\hat{e}_v = r_{uu}E_{ui}\hat{e}_u + r_{vv}E_{vi}\hat{e}_v = E_o\cos(45°)(r_{uu}\hat{e}_u + r_{vv}\hat{e}_v) \tag{5.70}$$

The $u$-component of the surface behaves almost as HIS at 9.5 GHz, while the $v$-component is acting as a low impedance surface (LIS) and thus creating phase difference of 90° as shown in Figure 5.23. The 90° phase difference may be attributed to the anisotropy and mirror symmetry along with only one axis as shown in Figure 5.21b.

The above explanation for cross- and circular-polarization conversion can also be further elaborated via polarization conversion ratio (PCR) and polarization extinction ratio (PER). The PCR is defined as the ratio of power in reflected cross-polarized component to the power in both co- and cross-polarized components, while PER defines the efficiency and handedness of circularly polarized wave. The mathematical formulations for the PCR and PER are given below [22–24].

$$\mathrm{PCR} = \frac{|R_{xy}|^2}{|R_{xy}|^2 + |R_{yy}|^2} \tag{5.71}$$

$$\mathrm{PER} = 20\log_{10}\frac{|R_{+y}|}{|R_{-y}|} \tag{5.72}$$

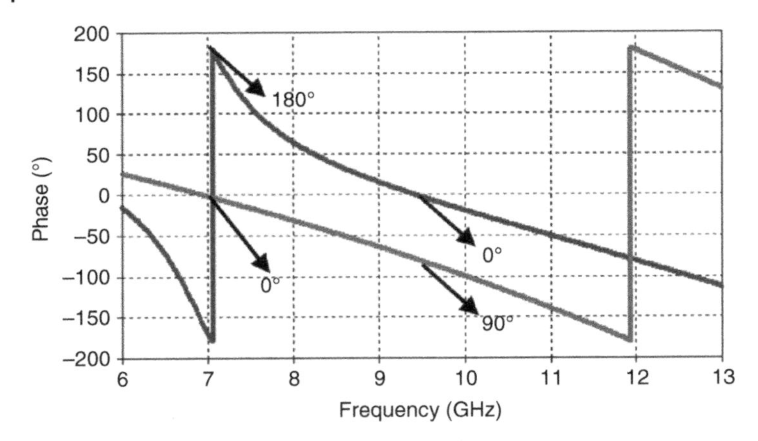


**Figure 5.23** Phase of reflection coefficient. (a) Red for $R_{uu}$. (b) Brown for $R_{vv}$.

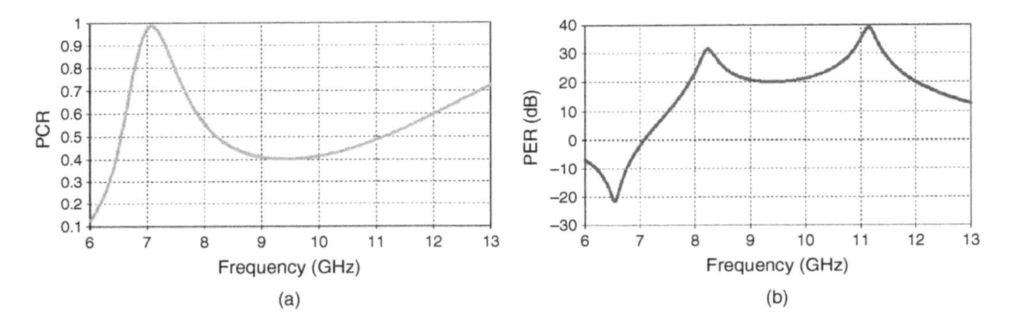


**Figure 5.24** (a) PCR response. (b) PER response.

Using Equations (5.71) and (5.72), the PCR and PER responses are plotted in Figure 5.24 for the proposed metasurface. It can be seen from Figure 5.24a that the PCR is >0.9 (i.e., 90%) in frequency band of 6.9–7.4 GHz which means cross-polarization conversion is achieved over this frequency band. On the other hand, it can be observed from Figure 5.24b that the PER is above +20 dB over entire X-band which means that the circular polarization conversion, in particular right-hand circular polarization (RHCP), is achieved over this frequency band.

#### 5.2.2.3 HWP and QWP in Transmission Mode

In the previous sections, the HWP and QWP in reflection mode are presented. In this section, the HWP and QWP in transmission mode will be presented. The multilayered and bi-anisotropic metasurfaces are required to achieve symmetric and asymmetric transmission. In addition, for asymmetric transmission, there should exists chirality in the unit cell design. In this section, the symmetric transmission will be discussed.

The symmetric transmission can be achieved if there is a mirror symmetry along *u*-, *v*-, or both *u*- and *v*-axes. If the unit cell is printed on only one side of the metasurface, only beam splitting or 1 : 1 half-mirror operation can be achieved. For polarization conversion in transmission mode, metasurface should be multilayered and bi-anisotropic. Here, a

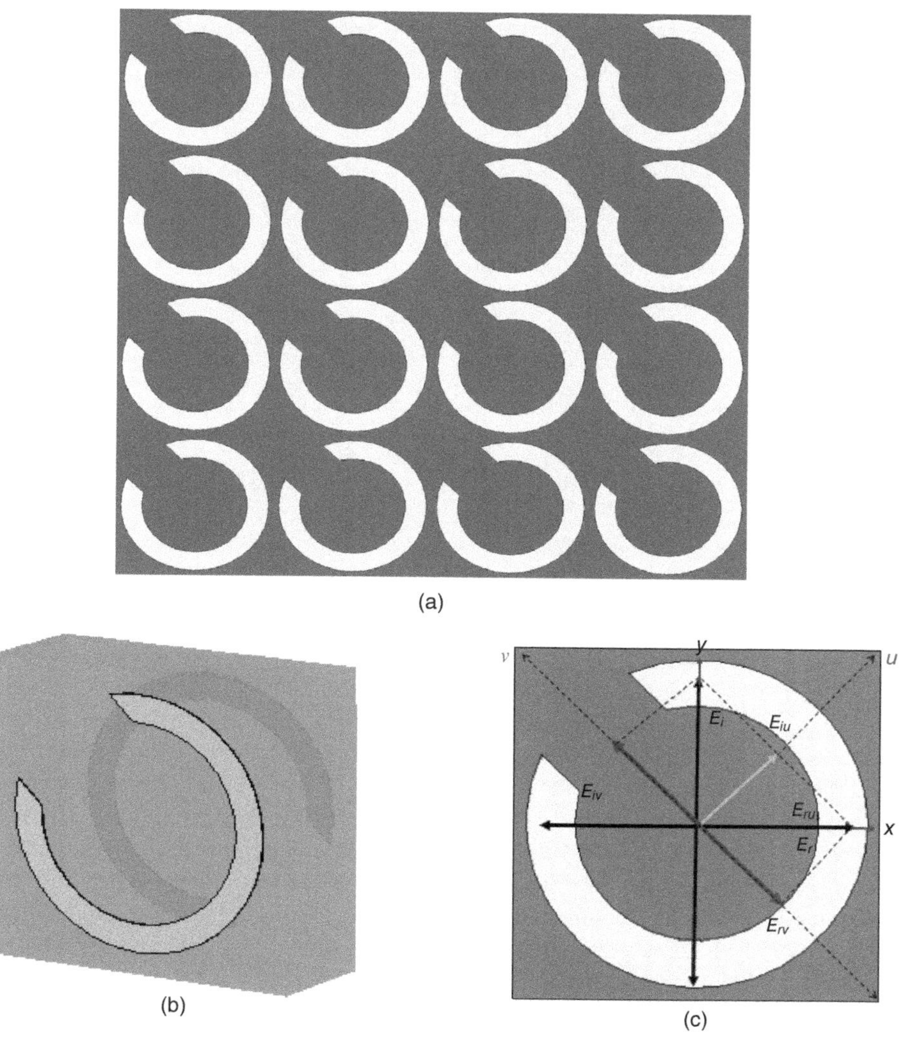


**Figure 5.25** (a) Proposed metasurface (front-side) for polarization conversion in transmission mode. (b) Unit cell configuration (front- and back-side). (c) Unit cell configuration (front-side highlighting axes).

bi-anisotropic metasurface for polarization conversion (linear to linear and linear to circular) is chosen as shown in Figure 5.25. FR-4 substrate with 3.2 mm thickness is selected. The C-shaped unit cell is designed on the upper layer of the substrate. The unit cell is rotated at the angle of 180° and then mirrored on the bottom layer (see Figure 5.25b). From both sides, the unit cell is anisotropic and symmetric along $v$-axis as shown in Figure 5.25c.

The transmission coefficient response is shown in Figure 5.26. It can be seen that the cross-polarization conversion (HWP) criteria are fulfilled at 12, 20, and 22 GHz since the cross-component $T_{xy}$ is above −3 dB and the co-component $T_{yy}$ remains below −10 dB at all the three resonances. On the other hand, the circular-polarization conversion criteria are

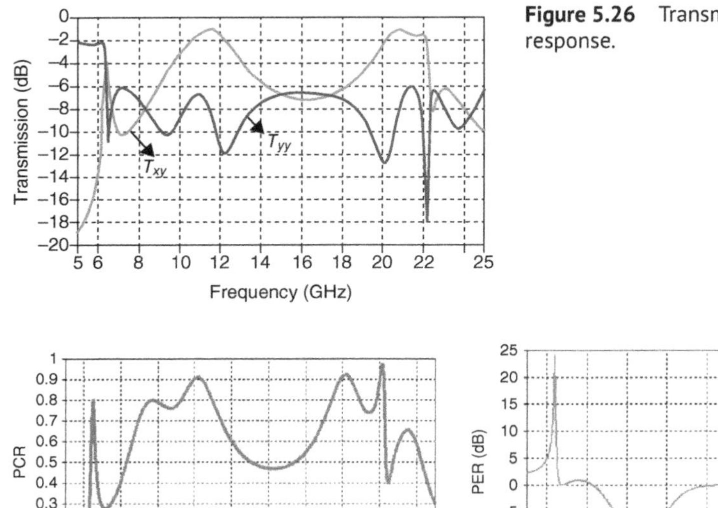


**Figure 5.26** Transmission coefficient response.

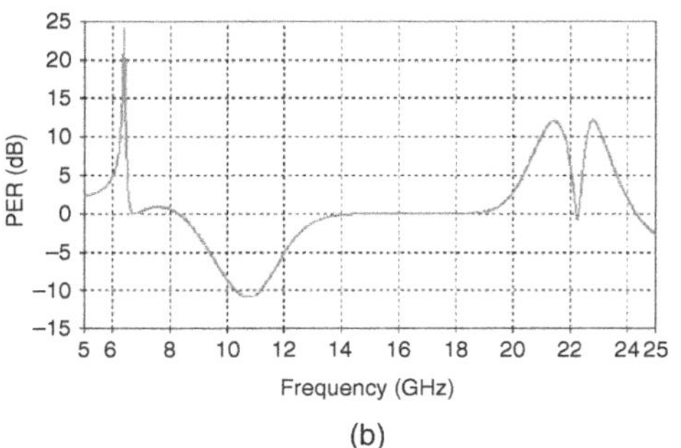


**Figure 5.27** (a) PCR response. (b) PER response.

fulfilled at 6.5 GHz where the ratio of magnitudes of transmission coefficients remains near unity and the phase difference (not shown here) is around 90° at 6.5 GHz. For further elaboration, the PCR and PER are plotted in Figure 5.27. It can be seen from Figure 5.27a that the PCR remains above 0.9 (i.e., 90%) at 12, 20, and 22 GHz which means cross-polarization conversion is achieved. On the other hand, it can be observed from Figure 5.27b that the PER is above +20 dB at 6.5 GHz which means that the circular-polarization conversion, in particular, the RHCP, is achieved over aforementioned frequency. The PCR and PER values are presented in tabular form in Table 5.3.

It can be concluded from last few sections that achieving HWP and QWP in transmission mode is quite challenging as compared to HWP and QWP in reflection mode since multi-layered and bi-anisotropic metasurfaces are required to achieve symmetric and asymmetric transmission.

## 5.3 Beam Splitter

A beam splitter is an optical device which plays an important role in various applications, i.e., optical routing, interferometry, and optical switching. Although different conventional techniques, i.e., cube-shaped beam splitter, trigonal prism, Wollaston prisms, and Rochon prisms, employed for beam splitting, have been of great interest to the scientific community. However, beam splitters based on those techniques are obsoleted by metasurface-based beam splitters because of their narrowband operation, bulky size, expensiveness, and power

**Table 5.3** PCR and PER.

| Frequency (GHz) | PCR | PER (dB) |
|---|---|---|
| 5 | 0.03 | 3 |
| 6.5 | 0.6 | 24 |
| 7 | 0.3 | 2 |
| 8 | 0.4 | 1 |
| 9 | 0.7 | −3 |
| 10 | 0.79 | −9 |
| 11 | 0.77 | −11 |
| 12 | 0.91 | −5 |
| 13 | 0.71 | −2.5 |
| 14 | 0.6 | 0 |
| 15 | 0.5 | 0 |
| 16 | 0.46 | 0 |
| 17 | 0.485 | 0 |
| 18 | 0.55 | 0 |
| 19 | 0.6 | 0.5 |
| 20 | 0.92 | 0.25 |
| 21 | 0.8 | 9.5 |
| 22 | 0.98 | 5 |
| 23 | 0.62 | 12 |
| 24 | 0.61 | 3 |
| 25 | 0.4 | −3 |

loss. In the subsequent sections, a design example for metasurface-based beam splitter is presented followed by mathematical analysis of the matter.

### 5.3.1 Design Example

The beam splitter can be designed by introducing the anisotropy and mirror symmetry in the unit cell design. One of the best ways to design the beam splitter is to print a unit cell only single side of the substrate, and bottom side should not be grounded to ensure transmission [4, 25, 26]. To demonstrate the beam splitting operation, a metasurface and the unit cell configuration are proposed as shown in Figure 5.28. The ultrathin flexible substrate (polyimide) is chosen to design a beam splitter. The transmission and reflection coefficients response are shown in Figure 5.29 which depicts beam splitting operation at 22.5 GHz. The cross- and co-components of both reflected and transmitted waves combine at 22.5 GHz. Each component has 1/4th of the total incident power, i.e., $R_{yy}^2 + R_{yy}^2=0.5$ and $T_{yy}^2 + T_{xy}^2 = 0.5$. It means half of the power is reflected (i.e., $R_{yy}^2 + R_{xy}^2 = 0.5$), and the other half is transmitted (i.e.,

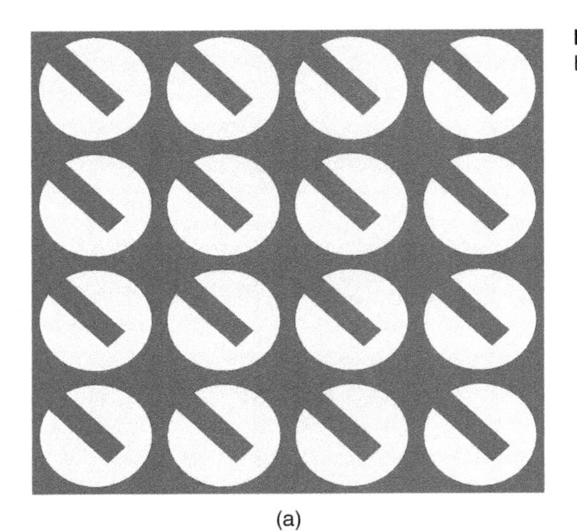

(a)

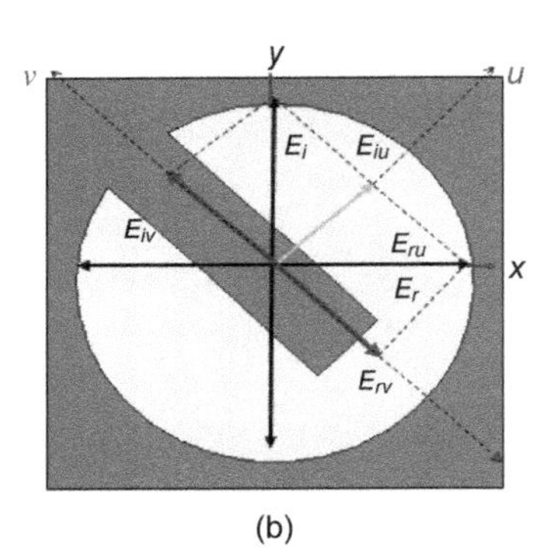


(b)

**Figure 5.28** (a) Proposed metasurface for beam splitting. (b) Unit cell configuration.

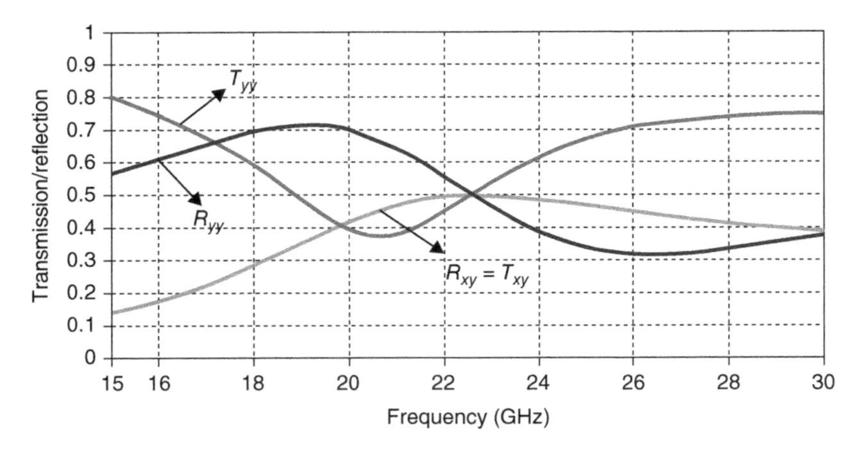


**Figure 5.29** Transmission and reflection coefficients response.

$T_{yy}^2 + T_{xy}^2 = 0.5$) at 22.5 GHz. The metasurface absorbed no power and performed a beam splitting operation at 22.5 GHz as desired.

### 5.3.2 Mathematical Background

The beam splitting operation may also be explained via mathematical formulations for *uv*-analysis. The eigenvalues and eigenvectors are obtained from the transmission matrix which is given by the following matrix:

$$T = 0.5\begin{bmatrix} e^{-i\frac{\pi}{4}} & e^{i\frac{\pi}{4}} \\ e^{i\frac{\pi}{4}} & e^{-i\frac{\pi}{4}} \end{bmatrix} \tag{5.73}$$

An eigenvector of the above matrix will be a vector $X$ which will satisfy:

$$TX = \lambda X \tag{5.74}$$

where $X$ is the eigenvector and $\lambda$ is the eigenvalue of matrix $T$. Let us consider the vector $u = \begin{bmatrix} 1 \\ 1 \end{bmatrix}$ and check whether $u$ is an eigenvector of $T$ or not, using Equation (5.74) as follows:

$$Tu = 0.5\begin{bmatrix} e^{-i\frac{\pi}{4}} & e^{i\frac{\pi}{4}} \\ e^{i\frac{\pi}{4}} & e^{-i\frac{\pi}{4}} \end{bmatrix}\begin{bmatrix} 1 \\ 1 \end{bmatrix} \tag{5.75}$$

$$=> \begin{bmatrix} \cos\left(\frac{\pi}{4}\right) \\ \cos\left(\frac{\pi}{4}\right) \end{bmatrix} \tag{5.76}$$

$$=> \frac{\sqrt{2}}{2}\begin{bmatrix} 1 \\ 1 \end{bmatrix} = \frac{\sqrt{2}}{2}u \tag{5.77}$$

It can be observed from Equation (5.77) that $u$ is an eigenvector of matrix $T$ with eigenvalue of $\frac{\sqrt{2}}{2}$.

Now, to check whether $v = \begin{bmatrix} -1 \\ 1 \end{bmatrix}$ is an eigenvector of $T$, one may use Equation (5.74) as follows:

$$Tv = 0.5\begin{bmatrix} e^{-i\frac{\pi}{4}} & e^{i\frac{\pi}{4}} \\ e^{i\frac{\pi}{4}} & e^{-i\frac{\pi}{4}} \end{bmatrix}\begin{bmatrix} -1 \\ 1 \end{bmatrix} \tag{5.78}$$

$$=> \begin{bmatrix} i\left(\sin\left(\frac{\pi}{4}\right)\right) \\ -i\left(\sin\left(\frac{\pi}{4}\right)\right) \end{bmatrix} \tag{5.79}$$

$$=> e^{-i\frac{\pi}{2}}\begin{bmatrix} -\left(\sin\left(\frac{\pi}{4}\right)\right) \\ \left(\sin\left(\frac{\pi}{4}\right)\right) \end{bmatrix} \tag{5.80}$$

$$=> \frac{\sqrt{2}}{2}e^{-i\frac{\pi}{2}}\begin{bmatrix} -1 \\ 1 \end{bmatrix} = \frac{\sqrt{2}}{2}e^{-i\frac{\pi}{2}}\,v \tag{5.81}$$

Hence, $v$ is also an eigenvector of matrix $T$ with eigenvalue of $\frac{\sqrt{2}}{2}e^{-i\frac{\pi}{2}}$.

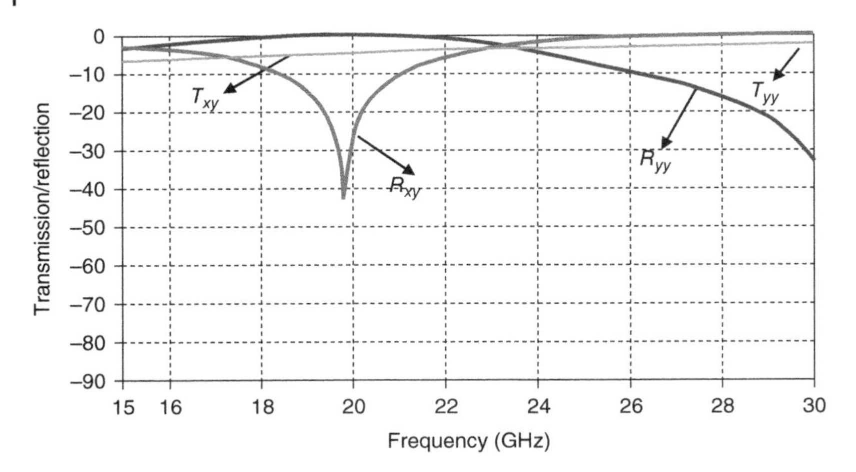


**Figure 5.30** *UV*-components for both reflection and transmission responses.

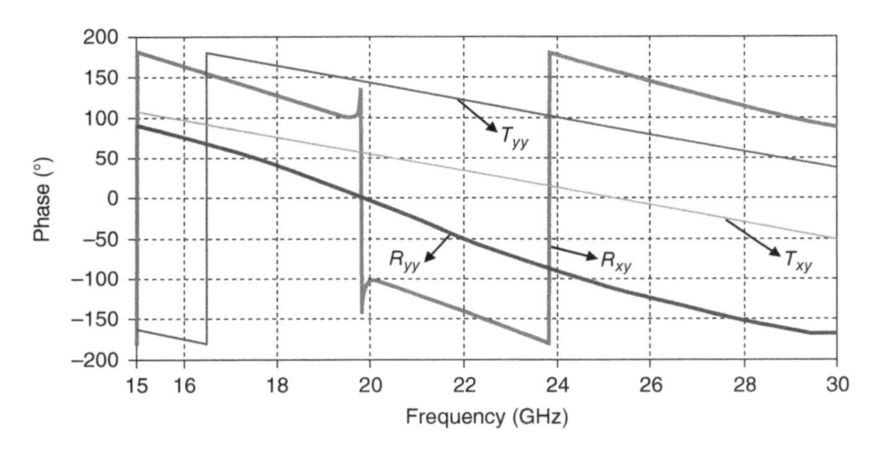


**Figure 5.31** Phase response.

The *uv*-components for both reflection and transmission responses are shown in Figure 5.30. The co-component of reflected and transmitted waves achieving −3 dB (i.e., $\approx\frac{\sqrt{2}}{2}$) at 22.5 GHz which is as per aforementioned mathematical formulations. The proposed metasurface may be optimized by varying its dimensions to achieve linear- to circular-polarization conversion (i.e., QWP) at 22.5 GHz. In this way, the metasurface produces both beam splitting and QWP operation simultaneously at 22.5 GHz. The phase response of the metasurface is shown in Figure 5.31. It can be seen that there exists a phase difference of 90° at 22.5 GHz between reflection coefficients (i.e., $R_{yy}$ and $R_{xy}$) and between transmission coefficients (i.e., $T_{yy}$ and $T_{xy}$), which means that in addition to beam splitting, the linear- to circular-polarization conversion is also achieved in both transmission and reflection modes. The handedness of the wave can be decided by observing the PER as shown in Figure 5.32. It can be seen from Figure 5.32 that the PER is above +20 dB for

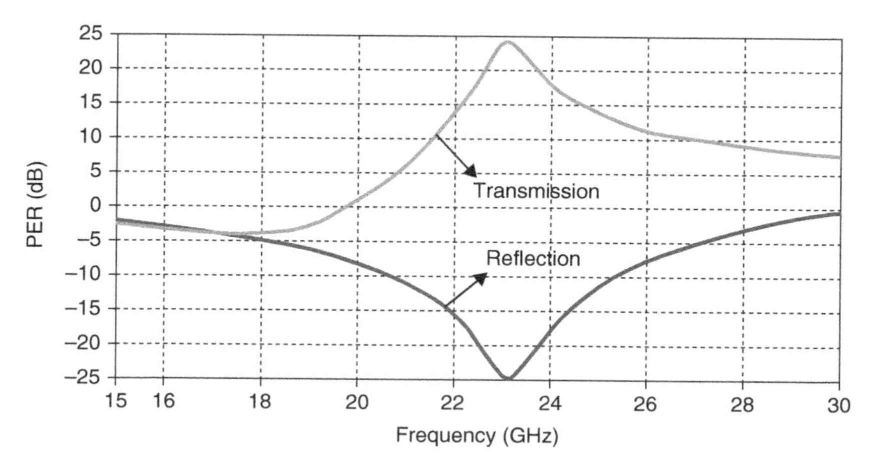


**Figure 5.32** PER response.

transmission at 22.5 GHz, i.e., producing RHCP conversion and −20 dB for reflection at same frequency i.e., producing left-hand circularly polarization conversion (i.e. LHCP).

## 5.4 Absorbers

A perfect absorber is one, which fully absorbs the incident wave within the desired frequency band. Due to various applications of absorbers in sensors, emitters, and stealth technology, the absorbers have received special attention in the field of engineering and applied physics. Few examples of conventional absorbers are ferrite and wedge absorbers. The optically transparent microwave absorbers, i.e. Jaumann or Salisbury screen, have also been of great interest for scientific community. However, due to their bulky size and narrow absorption band, the metamaterial absorbers are preferred. *Landy* et al. became pioneers of metamaterial absorbers after publishing their research findings in 2008 [27]. Since then, several metasurface-based absorbers have been presented in the literature [28–30]. The basic requirement of absorption is the existence of isotropy in the structure or anisotropy with 90° rotation symmetry (C4). To design a broadband absorption, the resistive sheets and multilayered structure may be used. In the subsequent sections, the mathematical background and design examples for the metasurface-based absorbers will be discussed.

### 5.4.1 Mathematical Background

Mathematically, the absorption can be characterized using the following equation [28]:

$$A(\omega) = 1 - R(\omega) - T(\omega) \tag{5.82}$$

where,

$$R(\omega) = \left|\frac{Z_{in} - Z_o}{Z_{in} + Z_o}\right|^2 \tag{5.83}$$

and

$$\frac{1}{Z_{in}(\omega)} = \frac{1}{Z_M(\omega)} + \frac{1}{Z_d(\omega)} \tag{5.84}$$

$$\frac{1}{Z_M(\omega)} = \frac{1}{R_n + j\omega L_n + \frac{1}{j\omega C_n}} \tag{5.85}$$

$$Z_d(\omega) = j\sqrt{\frac{\mu_r\mu_o}{\varepsilon_r\varepsilon_o}}\ \tan(kh) \tag{5.86}$$

$$k = \frac{k_o}{\sqrt{\varepsilon_r\mu_r}} \tag{5.87}$$

where $Z_o$ is the characteristic impedance of free space which is equal to 377 Ω and $n = 1$, 2, 3.... Moreover, $\mu_o$, $\varepsilon_o$, and $k_o$ are the permeability, permittivity, and wavenumber in the free space, respectively. Similarly, $\mu_r$, $\varepsilon_r$, and $k$ represent relative permeability, permittivity, and wavenumber in the material, respectively. The $T(\omega)$ is equal to zero which is due to the fact that absorbers are typically made of metasurfaces with complete metallic sheet, acting as ground plane, on its back side which causes no transmission.

Absorption may also be defined by the following mathematical formulation:

$$A(\omega) = 1 - |R_{xx}|^2 - |R_{yx}|^2 - T \tag{5.88}$$

where $|R_{xx}|^2$ and $|R_{yx}|^2$ are co-polarized and cross-polarized reflected powers, respectively. On the other hand, the transmitted power, i.e., $T = |T_{xx}|^2 + |T_{yx}|^2$ is zero because of the metallic ground plane. For absorption, the co- and cross-components of the reflected wave should be zero. However, few authors mistakenly consider only the co-component of the reflected wave and neglecting the cross-component of reflection by using the C2-symmetery concept. In that case, the structure behaves like a cross-polarizer, as in polarization conversion the co-component always remains below −10 dB, thus mistakenly claiming the polarizer as an absorber. Since absorbers find number of applications in EM compatibility/interference and stealth technology, therefore, a complete absorption is required and a cross-polarizer shall not be confused with an absorber. For example, a CP radar can easily detect the target which is so-called co-polarization absorber. In addition, in the anechoic chamber, one cannot install co-polarized absorbers when there is a requirement of complete absorption in the chamber. In the following, a clear distinction between cross-polarizer and an absorber will be presented.

### 5.4.2 Design Examples

To demonstrate the distinction between cross-polarizer and an absorber, two metasurfaces are proposed as shown in Figures 5.33 and 5.36. The metasurface, designed for absorber, has an array of circular rings placed on front side while having metallic ground plane on the back side shown in Figure 5.33. The FR-4 lossy substrate is used. It can be seen from Figure 5.34 that both co- (i.e., $R_{yy}$) and cross-components (i.e., $R_{xy}$) of reflection coefficients are below −10 dB, which means all the incident energy is absorbed. The $R_{yy}$ and $R_{xy}$ are also presented in tabular form in Table 5.4. In addition, absorption remains above 90%, as shown in Figure 5.35, which is a requirement for perfect absorption. Therefore, it can be concluded that the aforementioned metasurface is actually an absorber.

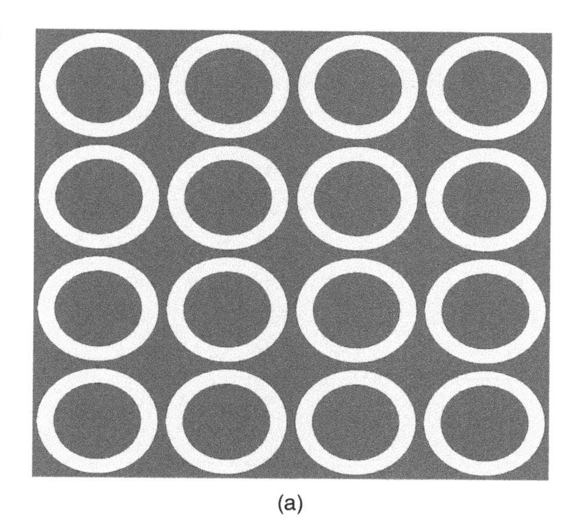

(a)

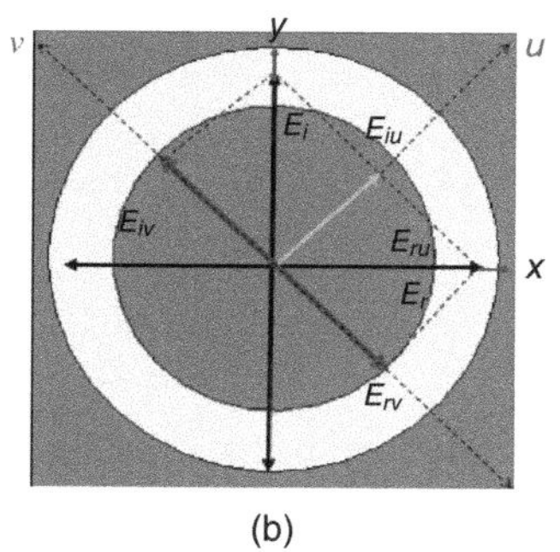


(b)

**Figure 5.33** (a) Proposed metasurface for absorber. (b) Unit cell configuration.

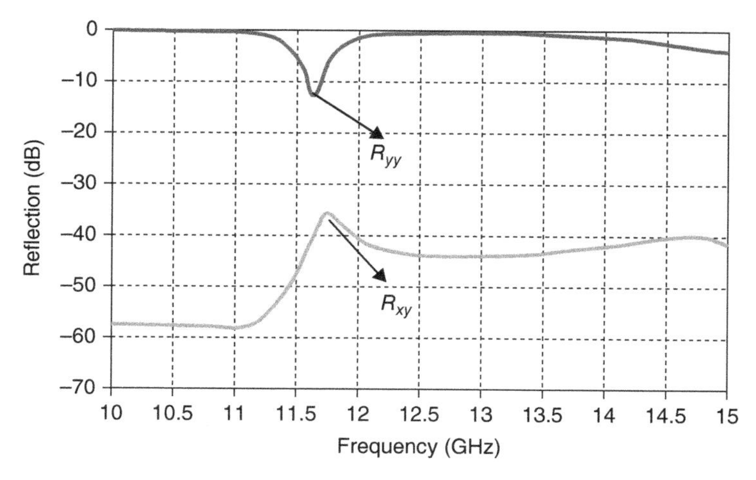


**Figure 5.34** Reflection coefficient response.

**Table 5.4** Reflection coefficients of metasurface.

| Frequency (GHz) | $R_{yy}$ (dB) | $R_{xy}$ (dB) |
|---|---|---|
| 10 | −0.14 | −57.46 |
| 11 | −0.54 | −58.08 |
| 11.65 | −13.00 | −37.00 |
| 12 | −1.49 | −40.74 |
| 13 | −0.59 | −43.90 |
| 14 | −1.33 | −41.98 |
| 15 | −4.02 | −41.36 |

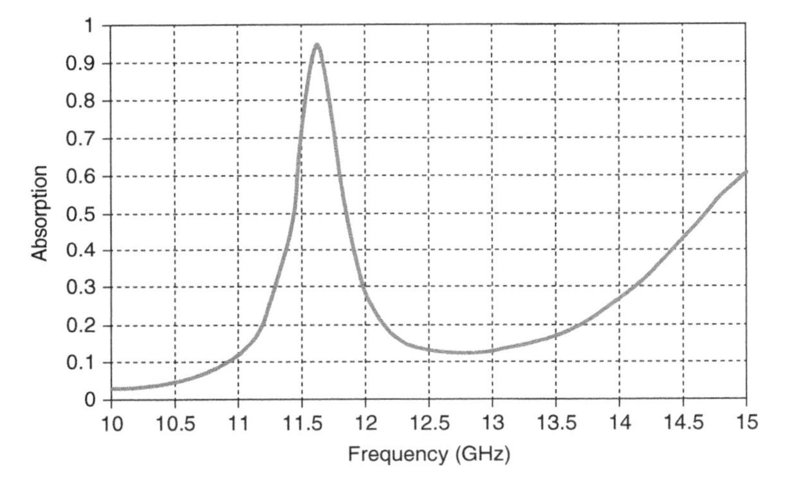


**Figure 5.35** Absorption response.

To demonstrate the behavior of a polarizer, a metasurface is proposed as shown in Figure 5.36. It has an array of circular rings with slits at the angle of 45° placed on the front side while having metallic ground plane on the back side. The slit in the unit cell is introduced to create anisotropy and mirror symmetry.

It can be observed from reflection coefficient response, shown in Figure 5.37, that the co-component of reflection coefficient (i.e., $R_{yy}$) is much lower than −10 dB, while the cross-component of reflection coefficient (i.e., $R_{xy}$) is almost zero dB at two frequency bands, i.e., 5.52–5.75 and 9.51–9.74 GHz. It means that the PCR is more than 90%; therefore, it can be concluded that the cross-polarization conversion is achieved at two frequency bands, i.e., 5.52–5.75 and 9.51–9.74 GHz. On the other hand, circular polarization conversion is achieved at four frequency bands, i.e., 5.34–5.43, 5.85–5.98, 9.25–9.39, and 9.84–9.92 GHz, as shown in Figure 5.38, since PER is either greater than +20 dB or lower than −20 dB at these frequency bands. At frequency bands i.e., 5.34–5.43 and 9.84–9.92 GHz, the PER is lower than −20 dB, i.e., producing LHCP conversion. While the

**Figure 5.36** (a) Proposed metasurface for polarizer. (b) Unit cell configuration.

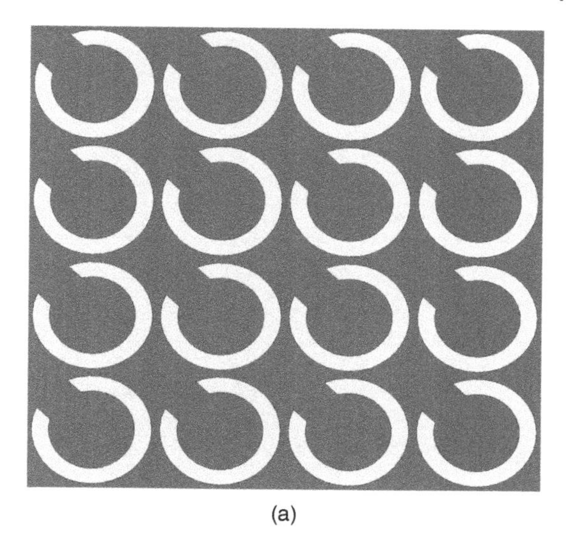


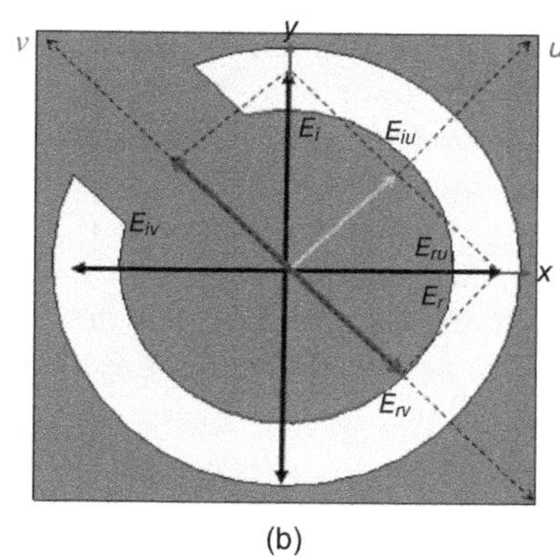


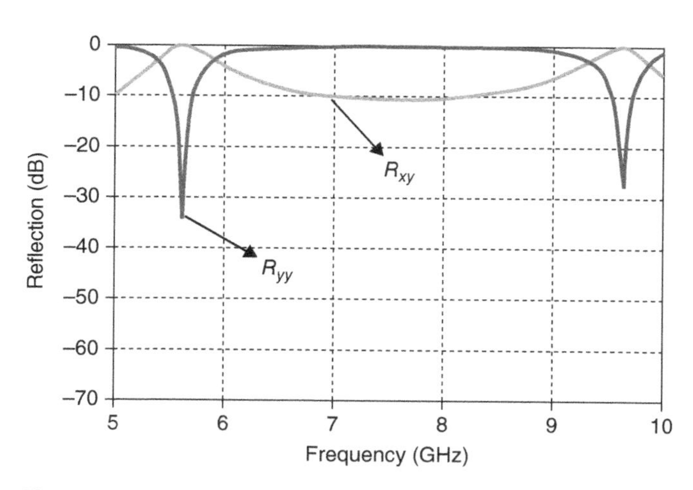


**Figure 5.37** Reflection coefficient response.

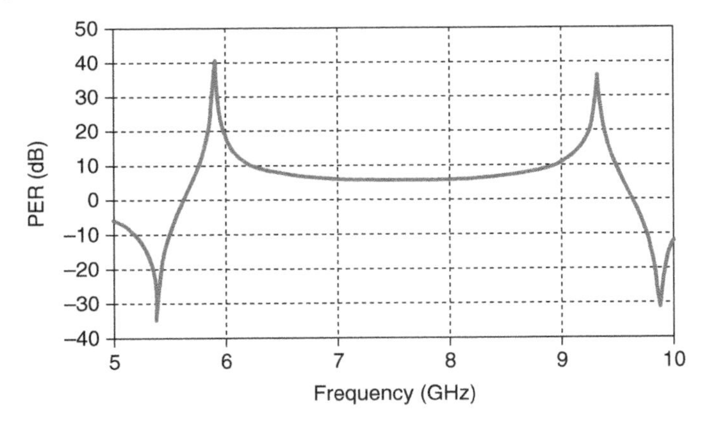


**Figure 5.38** PER response.

RHCP conversion is achieved at frequency bands i.e., 5.85–5.98 and 9.25–9.39 GHz since PER is above +20 dB at these frequency bands.

It is pertinent to mention that this metasurface has co-component of reflection coefficient (i.e., $R_{yy}$) much lower than −10 dB, while the cross-component of reflection coefficient (i.e., $R_{xy}$) is almost zero dB at two frequency bands, i.e., 5.52–5.75 and 9.51–9.74 GHz (as shown in Figure 5.37); therefore, it shall be treated as a polarizer. For absorber, it is necessary that both co- (i.e., $R_{yy}$) and cross-components (i.e., $R_{xy}$) of reflection coefficients are below −10 dB. Therefore, this metasurface shall not be confused as an absorber.

The authors are confident that this section has clearly demonstrated the distinction between cross-polarizer and an absorber with the help of two metasurfaces (shown in Figures 5.33 and 5.36).

## 5.5 Summary

In summary, this chapter has presented the underlying concepts related of polarizers, beam splitters, and absorbers. The theoretical concepts and mathematical analysis for the aforementioned operations are also presented. The main aim of this chapter is to design different unit cells to achieve different functionalities. Furthermore, to avoid confusion among absorbers and polarizers, the C2 and C4 symmetry as well as mirror symmetry concept along with their mathematics are shown. The design requirements for polarizers in reflection/transmission mode, the beam splitting along with linear- to circular-polarization conversion through single design, and the perfect absorbers are also discussed. The terms PCR, PER, and absorptivity are elaborated along with their basic criterion.

The authors are confident that the readers will get a detailed insight into the metasurface research domain after reviewing this chapter. It will pave the way for researchers to link and support their advanced research in metasurface domain with basic underlying concepts.

## References


1 Yu, N. and Capasso, F. (2014). Flat optics with designer metasurfaces. *Nature Materials* 13 (2): 139–150.

2 Gau, J.R., Burnside, W.D., and Gilreath, M. (1997). Chebyshev multilevel absorber design concept. *IEEE Transactions on Antennas and Propagation* 45 (8): 1286–1293.

3 Michielssen, E., Sajer, J.M., Ranjithan, S., and Mittra, R. (1993). Design of lightweight, broad-band microwave absorbers using genetic algorithms. *IEEE Transactions on Microwave Theory and Techniques* 41 (6): 1024–1031.

4 Ahmed, F., Ahmed, A., Tahir, F.A., and Chattha, H.T. (2019). An ultrathin flexible metasurface for half mirror and QWP operation. In: *2019 IEEE International Symposium on Antennas and Propagation and USNC-URSI Radio Science Meeting*, 439–440. IEEE.

5 Zheng, Y., Zhou, Y., Gao, J. et al. (2017). Ultra-wideband polarization conversion metasurface and its application cases for antenna radiation enhancement and scattering suppression. *Scientific Reports* 7 (1): 1–12.

6 Ahmed, A., Ahmed, F., and Tahir, F.A. (2019). Metasurface design for cross-polarization conversion and absorption applications. In: *2019 IEEE International Symposium on Antennas and Propagation and USNC-URSI Radio Science Meeting*, 1821–1822. IEEE.

7 Lee, G.Y., Sung, J., and Lee, B. (2019). Recent advances in metasurface hologram technologies. *ETRI Journal* 41 (1): 10–22.

8 Lawrence, M., Barton, D.R. III,, and Dionne, J.A. (2018). Nonreciprocal flat optics with silicon metasurfaces. *Nano Letters* 18 (2): 1104–1109.

9 Liu, N., Mesch, M., Weiss, T. et al. (2010). Infrared perfect absorber and its application as plasmonic sensor. *Nano Letters* 10 (7): 2342–2348.

10 Zhang, P. and Li, J. (2017). Compact UWB and low-RCS Vivaldi antenna using ultrathin microwave-absorbing materials. *IEEE Antennas and Wireless Propagation Letters* 16: 1965–1968.

11 Yu, N., Genevet, P., Aieta, F. et al. (2013). Flat optics: controlling wavefronts with optical antenna metasurfaces. *IEEE Journal of Selected Topics in Quantum Electronics* 19 (3): 4700423–4700423.

12 Yang, Y., Wang, W., Moitra, P. et al. (2014). Dielectric meta-reflectarray for broadband linear polarization conversion and optical vortex generation. *Nano Letters* 14 (3): 1394–1399.

13 Akhmedzhanov, I.M. and Bozhevolnyi, S.I. (2018). Scanning differential microscopy for characterization of reflecting phase-gradient metasurfaces. *Optics Communications* 427: 603–608.

14 Rodrigues, S.P., Lan, S., Kang, L. et al. (2014). Nonlinear imaging and spectroscopy of chiral metamaterials. *Advanced Materials* 26 (35): 6157–6162.

15 Ahmed, F. and Tahir, F.A. (2019). A multiband cross- and circular polarizer. In: *2019 International Symposium on Antennas and Propagation (ISAP)*, 1–2. IEEE.

16 Ahmed, F., Hassan, T., and Shoaib, N. (2020). Comments on "an ultrawideband ultrathin metamaterial absorber based on circular split rings". *IEEE Antennas and Wireless Propagation Letters* 19 (3): 512–514.

**17** Ahmed, F., Ahmed, A., Tamoor, T., and Hassan, T. (2019). Comment on "dual-band perfect metamaterial absorber based on an asymmetric H-shaped structure for terahertz waves". *Materials* 12 (23): 3914. https://doi.org/10.3390/ma11112193.

**18** Chirikjian, G.S., Kyatkin, A.B., and Buckingham, A.C. (2001). Engineering applications of noncommutative harmonic analysis: with emphasis on rotation and motion groups. *Applied Mechanics Reviews* 54 (6): B97–B98.

**19** Freistuhler, H., Szepessy, A., and Horie, Y. (2002). Advances in the theory of shock waves. Progress in nonlinear differential equations and their applications, Vol. 47. *Applied Mechanics Reviews* 55 (4): B63–B64.

**20** Bao, W., Sun, F., and Wei, G.W. (2003). Numerical methods for the generalized Zakharov system. *Journal of Computational Physics* 190 (1): 201–228.

**21** Kuhfittig, P.K. (2013). *Introduction to the Laplace Transform*, vol. 8. Springer Science & Business Media.

**22** Mustafa, M.E., Tahir, F.A., and Amin, M. (2019). Broadband waveplate operation by orthotropic metasurface reflector. *Journal of Applied Physics* 126 (18): 185108.

**23** Khan, M.I. and Tahir, F.A. (2017). An angularly stable dual-broadband anisotropic cross polarization conversion metasurface. *Journal of Applied Physics* 122 (5): 053103.

**24** Amin, M., Siddiqui, O., and Tahir, F.A. (2018). Quasi-crystal metasurface for simultaneous half-and quarter-wave plate operation. *Scientific Reports* 8 (1): 1–10.

**25** Tamayama, Y., Yasui, K., Nakanishi, T., and Kitano, M. (2014). A linear-to-circular polarization converter with half transmission and half reflection using a single-layered metamaterial. *Applied Physics Letters* 105 (2): 021110.

**26** Khan, M.I. and Tahir, F.A. (2017). Simultaneous quarter-wave plate and half-mirror operation through a highly flexible single layer anisotropic metasurface. *Scientific Reports* 7 (1): 1–9.

**27** Landy, N.I., Sajuyigbe, S., Mock, J.J. et al. (2008). Perfect metamaterial absorber. *Physical Review Letters* 100 (20): 207402.

**28** Kazemzadeh, A. and Karlsson, A. (2010). Multilayered wideband absorbers for oblique angle of incidence. *IEEE Transactions on Antennas and Propagation* 58 (11): 3637–3646.

**29** Huang, X., Yang, H., Yu, S. et al. (2013). Triple-band polarization-insensitive wide-angle ultra-thin planar spiral metamaterial absorber. *Journal of Applied Physics* 113 (21): 213516.

**30** Ahmed, F., Hassan, T., and Shoaib, N. (2020). A multiband bianisotropic FSS with polarization-insensitive and angularly stable properties. *IEEE Antennas and Wireless Propagation Letters* 19 (10): 1833–1837. https://doi.org/10.1109/LAWP.2020.3020949.

# 6

# The Role of Smart Metasurfaces in Smart Grid Energy Management

*Islam Safak Bayram[1], Muhammad Ismail[2], and Raka Jovanovic[3]*

[1]*Department of Electronic and Electrical Engineering, University of Strathclyde, Glasgow, United Kingdom*
[2]*Department of Computer Science, Tennessee Tech University, Cookeville, TN, USA*
[3]*Qatar Environment and Energy Research Institute, Hamad Bin Khalifa University, Doha, Qatar*

## 6.1 Introduction

For more than a century, the business of power utilities has been straightforward and focused on delivering electricity to end-users via passive power grids. However, over the last decade, concerns about climate change and political steps toward sustainable development goals have pushed grid operators to integrate more renewable energy resources, accommodate energy storage units, and serve increasing numbers of plug-in electric vehicles (PEVs). Such advances compel utilities to transform existing infrastructure into smart power grids that are envisaged to be cleaner, decentralized, and customer centric. To accomplish such goals, there is a strong need for a supporting communication infrastructure to extensively monitor, control, and enable automation of various grid assets. In this context, communication networks play a crucial role in providing connectivity needed to support the power grid services in a scalable, flexible, and seamless way.

To further elucidate the role of communication networks in supporting power grid services, we highlight a number of relevant case studies. For instance, electricity load forecasting is a critical task in balancing supply and demand based on data collected from smart meters and Internet of Things (IoT)-based sensors [1]. While the delivery of high volumes of information is a relatively straight forward task, the analysis of big data in a systematic fashion is a complex one. In specific, cloud-based infrastructure provides a scalable, flexible, and dynamic solution to handle high volumes of data traffic with minimal latency, thanks to the concept of edge and Fog cloud nodes [2]. Another critical application is smart energy management at residential units in which electricity consumption is minimized while satisfying customer comfort constraints [3]. In this application, there is a demand for a ubiquitous, large scale, and secure communication network where large number of sensors and actuators exchange information. Typical environment of energy management has little to no human interaction, hence machine-to-machine communication systems are commonly used [4]. Moreover, as the share of renewable energy sources increases, power system state estimation becomes a key functionality for energy management systems. With phasor

*Backscattering and RF Sensing for Future Wireless Communication*, First Edition.
Edited by Qammer H. Abbasi, Hasan T. Abbas, Akram Alomainy, and Muhammad Ali Imran.

measurement units (PMUs), power system states (e.g., voltage levels and phase angles) are measured and transmitted to a central entity for decision making. In such applications, emerging 5G networks could play a key role as a facilitator of advanced state estimation systems. Last but not least, wide area communication networks are instrumental in empowering a group of PEVs to participate in a vehicle-to-grid (V2G) ancillary services market program. In such applications, PEVs are parked across a wide region with diverse signal strength and latency requirements. Hence, overall efficiency of the aforementioned applications and services is linked to the performance of the underlying communication systems.

The above discussion paragraph outlines the importance of communication networks and its performance will have a direct impact on the operation of power systems. Since the introduction of the first generation (1G) wireless communication systems, the wireless environment between the communicating end points is assumed to be fixed and cannot be modified. Therefore, wireless network operations suffer from degradation in the received signal quality due to interaction with a number of physical objects/obstacles (e.g., doors, walls, buildings) that cause scattering of wireless signals in multiple paths in an uncontrollable fashion. Scattered signals arrive at the receiving side, making it challenging to differentiate from the original signal. When added up in a destructive manner, the received signal quality is highly degraded. Smart metasurfaces or smart radio environments are proposed as emerging new technologies designed to provide uninterrupted wireless connectivity by sensing ambient environments via sensors and adaptively adjusting electromagnetic functionalities via unmanned sensing feedback system [5]. Smart metasurfaces have the potential to improve communication system performance such as lowering delays and increasing communications reliability and support power grid operations. By integrating this advanced technology in smart rooms and buildings, reliable communication support can be provided to support the novel functionalities of the smart power grids.

In this chapter, first, we present emerging smart metasurfaces technologies and explain how they can be instrumental in enhancing power system performance. Second, we present a systematic overview of emerging smart grid applications that rely on communication network. Third, we present real-world signal measurements of 4G networks supporting smart grid services and show how smart metasurfaces can improve the overall system performance. In what follows, note that terms metasurface, smart surface, intelligent surface, and smart radio environment are used interchangeably.

## 6.2 Smart Metasurfaces

A smart metasurface is an artificially engineered surface that is composed of a large number of sub-wavelength unit cells deployed periodically on a two-dimensional surface. This way the wireless environment is transformed into a smart programmable environment whose operation is cultivated to support continuous connectivity with quality of service guarantee by recycling existing scatterd wireless signals [6]. The primary motivation for the use of smart metasurfaces is the increase in data traffic. Over the next decade, the global internet traffic is expected to grow by 55% each year, and there is a need to utilize physical environments as reconfigurable intelligent surfaces to fulfill the challenging requirements of future wireless networks [7]. Moreover, as the number of IoT devices increases, there is a need to redesign wireless communication between a Wi-Fi access point, and IoT devices as the IoT units consume high amounts of energy to transmit data to the Wi-Fi access points. Smart metasurfaces could be instrumental in reflecting radio waves in a way that IoT devices can preserve energy and hence operate on battery rather than wall power [8].

It is noteworthy that in Greek the prefix meta means "beyond," which refers to a material that shows special exotic properties that a natural material or surface does not posses. For instance, the thickness of the surface could be smaller or larger than the communication wavelength, andthe distance between neighboring unit cells could be smaller than the wavelength or size of each cell can be smaller than the wavelength [7]. These design metrics are considered during the manufacturing phase of the smart metasurface to produce the desired properties such as wave refraction, absorption, and focusing of incoming wireless signals (depicted in Figure 6.1). Over the last years, smart metasurfaces have been manufactured in the form of frequency selective surfaces [9], smart reflective mirrors [10], embedded smart antennas on walls [11], and coating of environmental objects [12]. In all cases, metasurfaces steer radio signal propagation in a customized way.

An increasing number of prototyping and testing studies are being carried out to unlock potential benefits offered by this technology. For instance, in Japan, a prototype trial for a transparent dynamic metasurface using 5G signals was conducted [?]. The purpose of this testing was to show how transparent dynamic metasurface such as building glasses or billboards can be used in places which are not well suited to install base stations such as high-security areas or indoor areas that require selective blocking of wireless reception. In another study [13], concurrent use of a metasurface imager and recognizer is proposed to remotely monitor healthcare of elderly people staying indoors using radio frequency (RF) signals. In Arun and Balakrishnan [14], a smart wall concept is developed with 3000 small antennas controlled by a software to maximize the signal strength at indoor environments. The proposed surface contains a two-dimensional surface with an array of simple RF switch elements. Each switch either reflects the signal or lets it through. It is further advocated that such a system could be instrumental in network connected smart homes where a number of IoT devices and sensors are in use. In Dunna et al. [8], a smart surface is presented to boost wireless network throughput by creating additional spatial stream to receiving node. This application is particularly designed to boost MIMO Wi-Fi networks to increase Wi-Fi throughput.

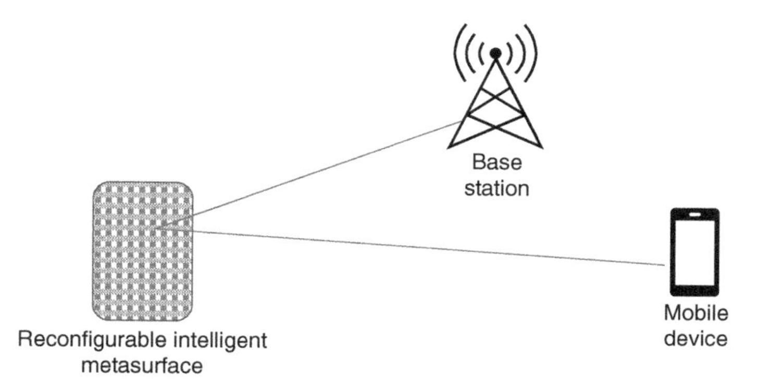


**Figure 6.1** Reconfigurable intelligent metasurface directs signal to mobile device.

## 6.3 Communication Support in Smart Power Grids

Electric power grids are going under a major transformation to support growing energy demands in a greener way. Smart electricity grids are envisioned to be decentralized, more efficient, resilient, and customer-responsive networks that could also support electrification of transportation. To realize this transformation, smart grids will employ state-of-the-art information and communication technologies such as 4G/5G networks for wide area communications and standards such as Zigbee, Wi-Fi, Ethernet, and Bluetooth for indoor communications. These technologies will carry data generated from monitoring and measurement devices such as smart meters, phasor measurement units (PMUs), IoT sensors, and supervisory control and data acquisition systems (SCADA) [3]. An overview of communication system requirements for various smart grid applications is presented in Table 6.1. It is noteworthy that while required data rates are relatively slow, latency requirements are tight and smart metasurfaces could play a key role in reducing latency. In this section, we present an overview of smart grid applications in detail and show their performance can be improved by smart metasurface technology.

### 6.3.1 Demand Response and Energy Management

Demand response (DR) is an emerging smart grid application that aims to shape electricity usage in a way that produces desired changes in load profiles [16]. The importance of DR applications is further linked to the higher use of renewable energy sources since by using DR applications, utility operators can create the demand-side flexibility needed to handle the intermittency associated with green energy generators. Overall, DR can be achieved by one of the following ways: (i) an economic approach in which financial incentives or differential pricing are offered to motivate consumers to shift and lower their electricity usage levels and (ii) a technological approach in which remote control devices such as smart thermostats, IoT sensors, and load switches are used to manage the load in an automated fashion in coordination with the utility company. As depicted in Figure 6.2, residential units host a number of smart appliances such as lights, washing machines, and air

**Table 6.1** Communication system requirements for various smart grid applications [15].

| Application | Data rate | Data size | Latency |
|---|---|---|---|
| AMI[a] | 10–100 kbps/node | 100B–1000B | 2–10 s |
| DR[b] | 14–100 kbps per device | 100B | 500 ms |
| HEM[c] | 9.6–56 kbps | 10–100B | 300–2000 ms |
| Trans. montr. | 9.6–56 kbps | 25B | 15–200 ms |
| Substation auto. | 9.6–56 kbps | 25B | 15–200 ms |
| PMU state est. | 9 kbps | 52B | 10 ms |

a) Advanced metering infrastructure.
b) Demand response.
c) Home energy management.
Source: Modified from Ghorbanian et al. [15].

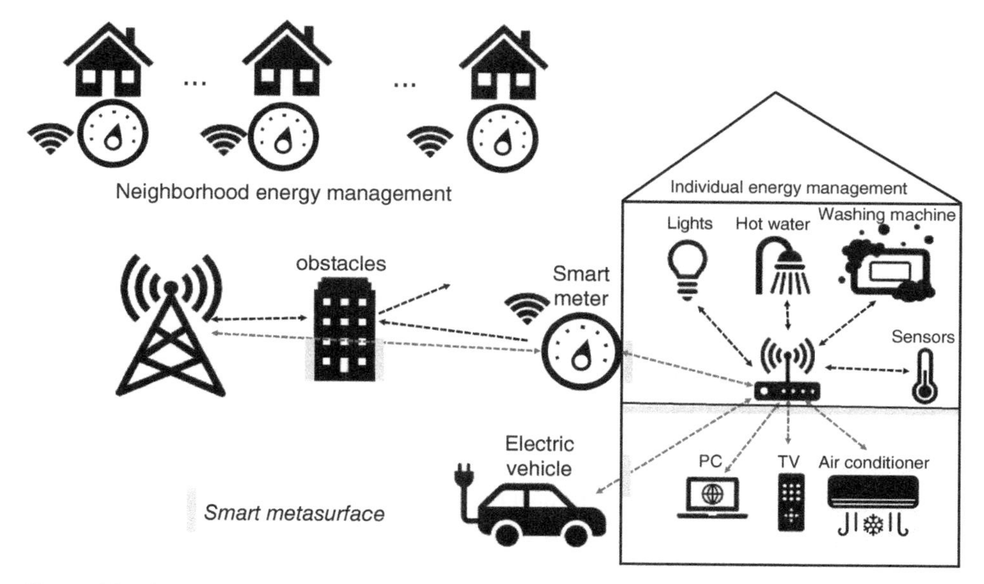


**Figure 6.2** An overview of demand response application with smart metasurface. Signal improvements with smart metasurfaces and wireless signals in the absence of metasurfaces are depicted.

conditioner units. The overall goal of energy management schemes is to schedule appliance usage in accordance with factors such as real-time electricity prices, user preferences and comfort levels, and appliance deadlines [3, 17]. Efficient management of residential loads is further critical to combat climate change and lower carbon emissions since one-third of the electricity consumption takes place at the residential sector [16]. It is important to note that residential units host a number of obstacles (e.g., walls, stairs, furniture) which negatively impact the signal quality of the appliances. For instance, in a demonstration study presented in Van Cutsem et al. [18], a wireless mesh network is developed to access all indoor climate sensors located around a two story villa. In addition, Wi-Fi extenders were used to boost signals for various appliances and energy monitors. This demonstrates that special measures need to be taken to enhance indoor signal quality to support DR strategies.

A key enabler of DR programs is smart meters or smart energy monitors which are typically located at hard to reach places such as basement or storage rooms of the residential units. Obstacles and poor signal coverage prevent utility companies from collecting data out of such smart meters, hence, smart functionalities offered by utilities diminishes. According to a survey study conducted in the United Kingdom [2], majority of the smart meter issues are related to the poor signal quality between the meter and the wireless network for meter reading and communication losses between the meter and the in home display. Smart metasurfaces can be instrumental in redirecting wireless signals in a way to support connectivity between home energy management equipments. Another use case is related to overcoming outdoor obstacles. As shown in Figure 6.2, smart metasurfaces can be employed to reduce signal attenuation especially in city centres to reach smart meters.

Although DR applications have the potential for efficient energy management in residential units, independently taken management actions can lead to unexpected results such as rebound effect in peaks, network instabilities, and contingencies. Therefore, coordination mechanisms are further needed to manage aggregated loads in a neighborhood area [19, 20]. In this case, there is a need to communicate with hundreds to thousands of appliances in near real time, and smart radio environment could be used to improve overall optimisation performance by creating dedicated line of sight paths between critical endpoints.

### 6.3.2 Plug-In Electric Vehicle Load Management

The plug-in electric vehicle (PEV) industry continues to grow rapidly in line with the recently introduced environmental sustainability and net-zero emission goals. The push toward PEVs is supported by legislation and regulations to encourage PEV uptake. For instance, a number of countries including the United Kingdom, France, and Norway plan to phase out fossil fuel cars by introducing a ban on the sale of such vehicles and increase the coverage of charging network within the next two decades [16]. However, reaching net zero emission goals means an exponential growth in PEV sales. For instance, in the United Kingdom, there are currently 0.2 million PEVs and this number needs to be around four million by 2030 to reach government targets in reducing carbon emissions. In a similar manner, the State of California has a mandate to push one and half million PEVs by 2025. Nevertheless, power grids are supporting backbone of the PEV networks and grid assets are becoming more congested due to growing number of PEV loads. Even though, the long-term remedy is the upgrade of the network capacity, a practical solution for the short term is to develop smart charging and control techniques to alleviate the stress on the grid. To realize such techniques communication networks will play a key role in enabling the interaction between drivers and grid operators in accordance with varying grid conditions [21].

One of the most important applications is smart PEV charging which refers to the adaptive adjustment of PEV charging power according to a number of factors such as power network congestion, electricity prices, and customer needs. It should be highlighted that smart charging mechanisms provide grid operators with the load flexibility needed during contingencies and peak hours. In Figure 6.3, different charging methodologies are compared according to provided load flexibility and communication system requirement. For instance, in the time of use, the pricing static charging rates are assigned at each time period, while in dynamic pricing with automated control charging rate is adjusted more frequently according to the aforementioned factors. Therefore, the amount of information needs to be exchanged and the associated delay constraints increase as the complexity of the control scheme increases. In that sense, PEV communication controller is a key component that supports information exchange between vehicles and PEV chargers. Since PEVs are typically parked in closed parking spaces with poor signals quality and limited communication system infrastructure, smart surfaces to be deployed at parking lots can increase the effectiveness of PEV smart load management.

### 6.3.3 Grid Monitoring and State Estimation

As the percentage of renewable energy generators increases, regional utility operators experience increased numbers of energy exchange and renewable curtailments. At the

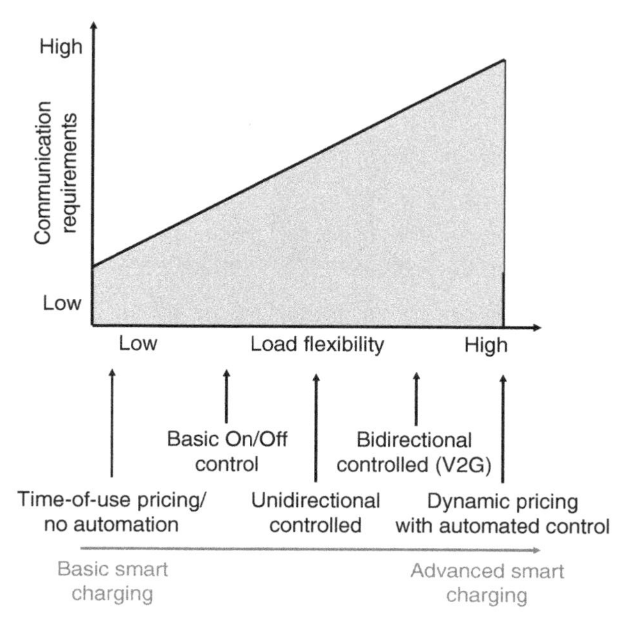


**Figure 6.3** PEV smart charging and communication needs.

distribution level, the share of renewable energies is increasing inline with renewable energy and sustainability goals [22]. Higher penetration of intermittent renewables combined with declining ratio of stabilizing rotational mass (e.g., coal, natural gas power plants) leads to stability issues. To ensure system stability and lower monetary losses, variations in system state should be monitored in a timely manner. In power system operations, state estimation plays a key role in monitoring the system status and taking necessary actions to maintain acceptable operation conditions and detect potential cyber threats. In state estimation, PMU and SCADA devices are used to measure voltage magnitude and phase angle at selected buses of the power network. While SCADA systems are considered as legacy systems and provide low-resolution (e.g., eight c06s per second) measurements, PMU systems are expanding at a fast rate and measure system state at high sampling rates such as 50–60 c06s per second. To utilize PMU networks, there is a need for a novel communication network that would enable wide area monitoring in a reliable and ultralow latency manner. As depicted in Figure 6.4, most of the delays occur during network transmission. Therefore, minimizing propagation delays provides a wider response time for utility operator to take actions [23, 24].

Wide area power system monitoring is composed of several hierarhical levels. At the first level, PMU devices are located at various parts of the power grid such as substations, main generator sites, and major interconnection points. In recent years, with the falling cost of hardware, PMUs are also deployed at distribution networks with high penetration of renewables [25]. In addition, various field sensors and actuators are connected to the power grid along with PMUs. As shown in Figure 6.5, phasor data concentrators (PDC) aggregate data from first level devices and send them to a central control center, also known as super PDC, where control decisions are taken. As discussed above, smart radio interfaces will improve

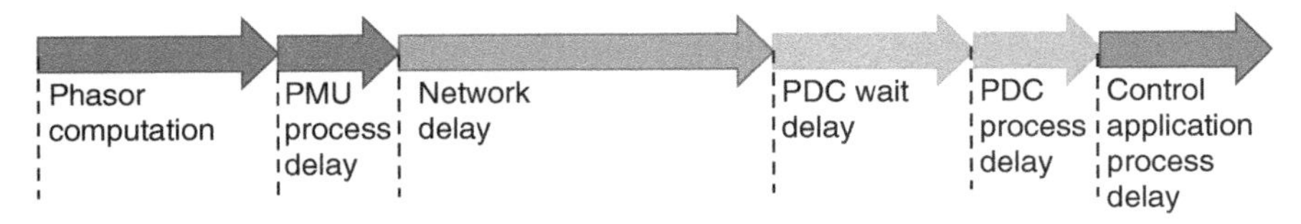


**Figure 6.4** Wide area power system monitoring delay stages. Source: Modified from Hojabri et al. [22].

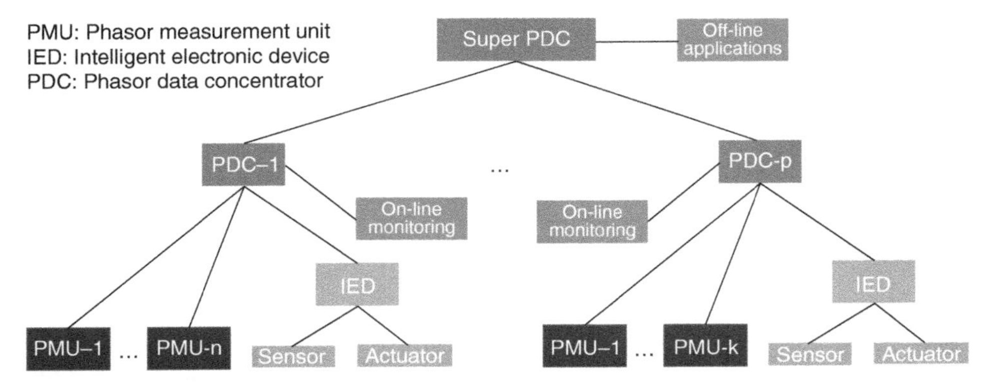


**Figure 6.5** Wide area power system monitoring system architecture. Source: Modified from Hojabri et al. [22].

overall network performance by intensifying associated wireless signals and/or creating line of sight between critical nodes (e.g., PMU to PDC).

Smart metasurfaces and 5G+ networks will provide an ideal avenue for development of accurate state estimation applications. Smart metasurfaces will improve system performance since most PMU devices are located indoors with significant isolation from line of sight.

### 6.3.4 Peer-to-Peer Energy Trading

As the distributed renewable energy generation and storage capabilities are becoming economically viable, energy trading is gradually becoming a profit making option for end-users. In energy trading applications, users participate in a decentralized bilateral energy trading market to purchase or sell locally generated electricity without stressing the power grid. An overview of peer-to-peer energy trading system is depicted in Figure 6.6. To successfully implement energy trading applications, there is a need to employ communication networks to monitor energy generation, consumption, and storage in real time and enable reliable information dissemination which are critical for energy transactions. Smart metasurfaces have the potential to facilitate and improve communication network performance to obtain necessary readings from generation and storage assets and enable transactions.

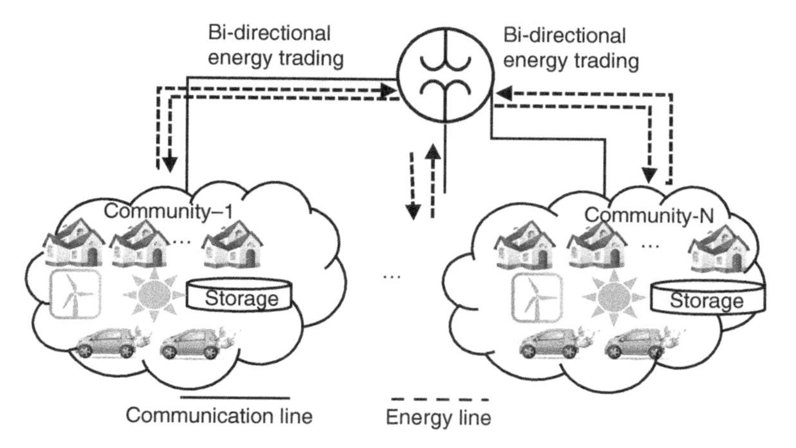


**Figure 6.6** Schematic overview of peer-to-peer energy trading in smart communities. Source: Bayram et al. [21]. IEEE, 2014.

### 6.3.5 Potential Applications of Intelligent Surfaces in Smart Energy Grids

Previous sections summarized smart grid applications and how smart metasurfaces can be used to improve the overall system performance. We summarize below these potential applications as:

- Smart surfaces can be instrumental in boosting connectivity among various sensors, devices, and smart meters in smart homes.
- Smart surfaces can create a dedicated line of sight between critical nodes such as PMUs to a PDC to ensure reliable communication.
- Delay sensitive applications such as vehicle-to-grid ancillary services participation requires strong signal reception. Smart metasurfaces can boost signal strength and reduce end-to-end communication delays.

## 6.4 Case Study: Communication System Performance Improvement in Vehicle-to-Grid Networks

In power systems, ancillary services market is a crucial element in balancing the demand with the supply in real time with tight tolerance bounds. Typically, dedicated generators that can provide fast response are used to respond according to market signals within seconds to minutes. Over the last years, V2G applications are gaining popularity as the use of stationary battery packs enhances system efficiency while providing financial benefits to PEV owners. In V2G, a group of PEVs managed by an aggregator charge or discharge their battery according to market signals [26]. To this end, the PEV aggregator needs to acquire information such as PEV departure and arrival times, battery state of charge information, and billing. The complexity of the communication network is also related to the number of participants and aggeregators. As the number of participants increase, coordination of all PEVs becomes complicated. As shown in Figure 6.7, smart metasurfaces will play a key role

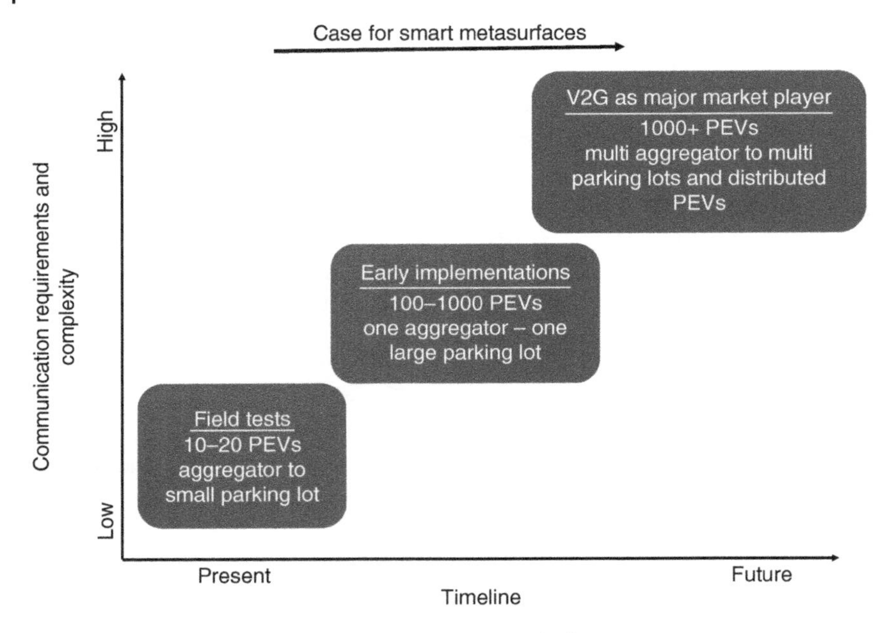


**Figure 6.7** V2G and communication networks complexity.

in future V2G networks. One of the main reasons is the that PEVs typically stay parked in multifloor lots located at dense areas such as city centers, hospitals, or university campuses where signal strength could play a key role. Communication delays particularly important as ancillary services like frequency regulation has tight response delays (typically between 500 ms and two seconds) [27]. Hence, metasurfaces can reduce overall system delays by strengthening wireless signals.

To present the impacts of signal strength on communication delays, a real-world implementation of a 4G communication network testbed is presented [28]. The market operator is emulated with a laptop computer, while an aggregator is emulated with a Raspberry Pi 3B device. To enable secure communication, a virtual private network is created between server and client. To mimic market signals, different packet sizes including 100, 500, and 1000 bytes are created using both transport control protocol (TCP) and user diagram protocol (UDP). Recall that TCP provides high reliability due to its guaranteed delivery scheme, while UDP enables faster delivery by sacrificing delivery guarantees. Moreover, to mimic real-world cases, three different signal levels are chosen: strong (−69 dBm), medium (−105 dBm), and poor (−118 dBm) strengths. Note that poor and medium signal strengths are very common in closed parking spaces.

In communication system performance modeling, cumulative distribution functions (CDFs) enable holistic comparisons between two systems. CDFs for end-to-end communication delays are presented in Figures 6.8, 6.9, and 6.10 for all cases. It is note worthy that in poor signal environments, end-to-end delay for TCP is significantly higher than the UDP due to more frequent retransmission of packets, while in strong signal case the gap between latency performance of two protocols is small. Smart radio environments present

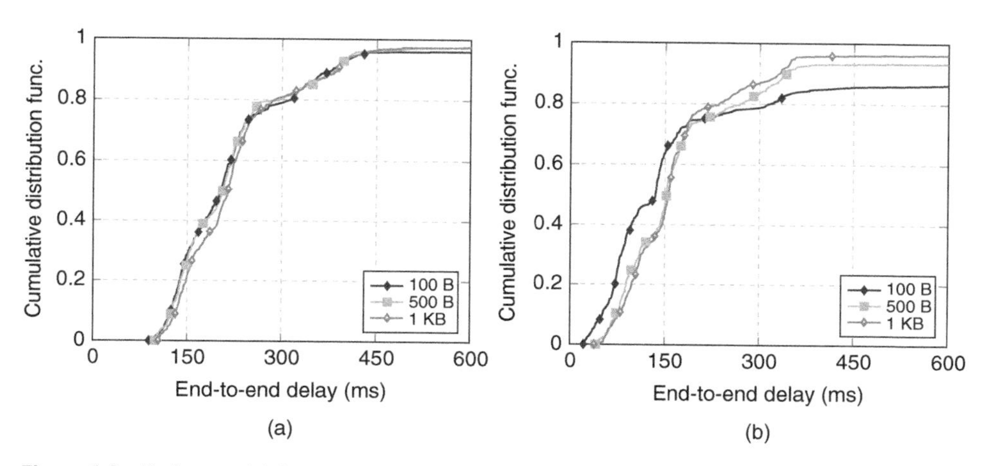


**Figure 6.8** End-to-end delay comparison for strong signal case. Source: Zeinali et al. [28].

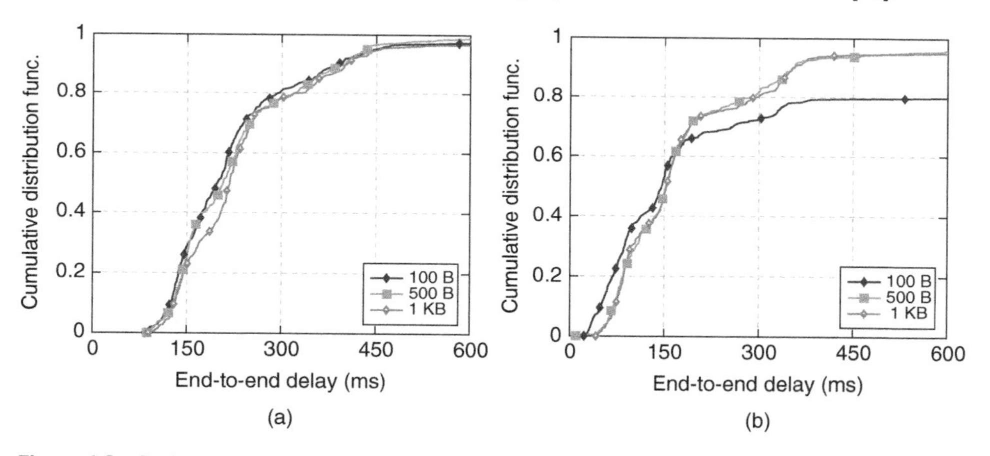


**Figure 6.9** End-to-end delay comparison for medium signal levels. Source: Zeinali et al. [28].

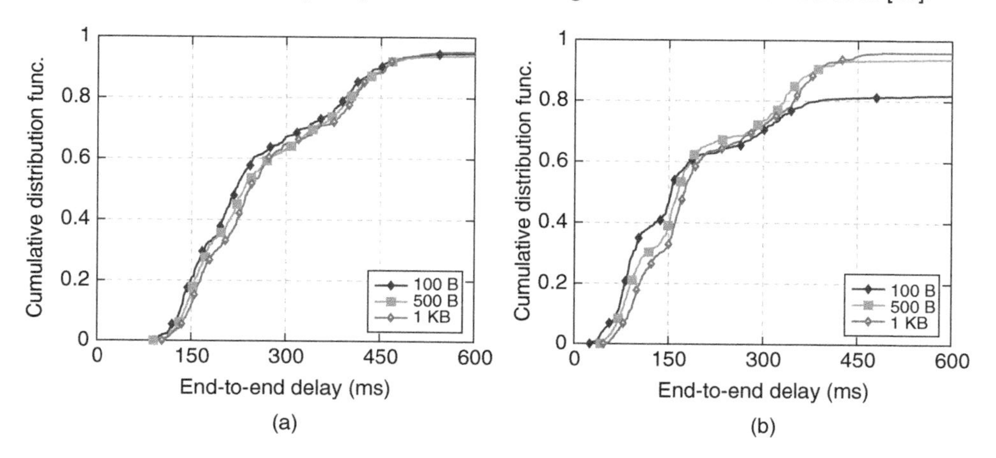


**Figure 6.10** End-to-end delay comparison for poor signal level. Source: Zeinali et al. [28].

**Table 6.2** Potential communication delay improvements with smart radio environments.

| Packet size (Bytes) | Trans. protocol | % Latency improvement |
|---|---|---|
| 100 | TCP | 16.10 |
| 500 | TCP | 15.70 |
| 1000 | TCP | 15.06 |
| 100 | UDP | 23.70 |
| 500 | UDP | 10.18 |
| 1000 | UDP | 14.21 |

TCP, transmission control protocol; UDP, user datagram protocol.

an opportunity to improve communication system performance. To present potential benefits, we calculate the 90% confidence intervals for communications and calculated the delay difference between strong and poor signals to reflect the improvements offered by metrasurfaces. As presented in Table 6.2, smart metasurfaces have the potential to enhance delay performance significantly and hence improve power system operations.

## 6.5 Conclusions

In this chapter, we have presented the potential role of smart metasurfaces in enhancing the performance of smart grid energy management systems. First, we provided an overview of research efforts and early demonstration studies related to intelligent surfaces. Next, we identified potential use cases in DR, PEV load management, state estimation, and peer-to-peer energy trading. In the last section, we have presented actual 4G network end-to-end communication delays and showed that communication performance can be improved by 15–24% by smart metasurfaces depending on packet size and transmission protocol.

## References


1 Li, L., Ota, K., and Dong, M. (2017). When weather matters: Iot-based electrical load forecasting for smart grid. *IEEE Communications Magazine* 55 (10): 46–51.

2 Atat, R., Ismail, M., Shaaban, M.F., Serpedin, E., and Overbye, T. (2019). Stochastic geometry-based model for dynamic allocation of metering equipment in spatio-temporal expanding power grids. *IEEE Transactions on Smart Grid* 11 (3): 2080–2091.

3 Bayram, I.S. and Ustan, T.S. (2017). A survey on behind the meter energy management systems in smart grid. *Renewable and Sustainable Energy Reviews* 72:1208–1232.

4 Zhou, Z., Gong, J., He, Y., and Zhang, Y. (2017). Software defined machine-to-machine communication for smart energy management. *IEEE Communications Magazine* 55 (10): 52–60.

5 Di Renzo, M., Debbah, M., Phan-Huy, D.-T., et al. (2019). Smart radio environments empowered by reconfigurable AI meta-surfaces: an idea whose time has come. *EURASIP Journal on Wireless Communications and Networking* 2019 (1): 1–20.

6 Liaskos, C., Nie, S., Tsioliaridou, A., Pitsillides, et al. (2018). A new wireless communication paradigm through software-controlled metasurfaces. *IEEE Communications Magazine* 56 (9): 162–169.

7 Navarro-Ortiz, J., Romero-Diaz, P., Sendra, S., et al. (2020). A survey on 5G usage scenarios and traffic models. *IEEE Communications Surveys & Tutorials* 22 (2): 905929.

8 Dunna, M., Zhang, C., Sievenpiper, D., and Bharadia, D. (2020). Scattermimo: enabling virtual mimo with smart surfaces. *Proceedings of the 26th Annual International Conference on Mobile Computing and Networking*, London, UK (21–25 September 2020), 1–14.

9 Zhang, B., Jornet, J.M., Akyildiz, I.F., and Wu, Z.P. (2019). Mutual coupling reduction for ultra-dense multi-band plasmonic nano-antenna arrays using graphene-based frequency selective surface. *IEEE Access* 7: 33214–33225.

10 Holloway, C.L., Kuester, E.F., Gordon, J.A., et al. An overview of the theory and applications of metasurfaces: the two-dimensional equivalents of metamaterials. *IEEE Antennas and Propagation Magazine* 54 (2): 10–35.

11 Hu, S., Rusek, F., and Edfors, O. (2018). Beyond massive mimo: the potential of data transmission with large intelligent surfaces. *IEEE Transactions on Signal Processing* 66 (10):2746–2758.

12 Liaskos, C., Tsioliaridou, A., Pitsillides, A., Ioannidis, S., and Akyildiz, I. (2018). Using any surface to realize a new paradigm for wireless communications. *Communications of the ACM* 61 (11): 30–33.

13 Li, L., Shuang, Y., Ma, Q., et al. (2019). Intelligent metasurface imager and recognizer. *Light: Science & Applications* 8 (1): 1–9.

14 Arun, V. and Balakrishnan, H. (2020). Rfocus: beamforming using thousands of passive antennas. *17th {USENIX} Symposium on Networked Systems Design and Implementation ({NSDI} 20)*, Santa Clara, CA (25–27 February 2020), 1047–1061.

15 Ghorbanian, M., Hacopian Dolatabadi, S., Masjedi, M., and Siano, P. (2019). Communication in smart grids: a comprehensive review on the existing and future communication and information infrastructures. *IEEE Systems Journal* 13 (4): 4001–4014.

16 Sahin, E.S., Bayram, I.S., and Koc, M. (2019). Demand side management opportunities, framework, and implications for sustainable development in resource-rich countries: case study qatar. *Journal of Cleaner Production* 241: 118332.

17 Jovanovic, R., Bousselham, A., and Bayram, I.S. (2016). Residential demand response scheduling with consideration of consumer preferences. *Applied Sciences* 6 (1): 16.

18 Van Cutsem, O., Bayram, I.S., Maher, K., and Fosse, J.-C. (2019). Demonstration of a smart villa energy monitoring platform in qatar. *2019 UK/China Emerging Technologies (UCET)*, 1–4, IEEE.

19 Celik, B., Roche, R., Suryanarayanan, S., Bouquain, D., and Miraoui, A. (2017). Electric energy management in residential areas through coordination of multiple smart homes. *Renewable and Sustainable Energy Reviews* 80: 260–275.

**20** Jovanovic, R., Tuba, M., and Voß, S. (2016). An ant colony optimization algorithm for partitioning graphs with supply and demand. *Applied Soft Computing* 41: 317–330. ISSN 1568-4946.

**21** Bayram, I.S., Shakir, M.Z., Abdullah, M., and Qaraqe, K. (2014). A survey on energy trading in smart grid. *2014 IEEE Global Conference on Signal and Information Processing (GlobalSIP)* (3–5 December 2014), 258–262. IEEE.

**22** Hojabri, M., Dersch, U., Papaemmanouil, A., and Bosshart, P. (2019). A comprehensive survey on phasor measurement unit applications in distribution systems. *Energies* 12 (23): 4552.

**23** Appasani, B. and Mohanta D.K. (2018). Co-optimal placement of PMUs and their communication infrastructure for minimization of propagation delay in the wams. *IEEE Transactions on Industrial Informatics* 14 (5): 2120–2132.

**24** Cosovic, M., Tsitsimelis, A., Vukobratovic, D., Matamoros, J., and Anton-Haro, C. (2017). 5G mobile cellular networks: enabling distributed state estimation for smart grids. *IEEE Communications Magazine* 55 (10): 62–69.

**25** Danielson, C.F.M., Vanfretti, L., Almas, M.S., Choompoobutrgool, Y., and Gjerde, J.O. (2013). Analysis of communication network challenges for synchrophasor-based wide-area applications. *2013 IREP Symposium Bulk Power System Dynamics and Control-IX Optimization, Security and Control of the Emerging Power Grid* (25–30 August 2013), 1–13. IEEE.

**26** Han, S., Han, S., and Sezaki, K. (2010). Development of an optimal vehicle-to-grid aggregator for frequency regulation. *IEEE Transactions on Smart Grid* 1 (1): 65–72.

**27** Hong, Q., Karimi, M., Sun, M., et al. (2020). Design and validation of a wide area monitoring and control system for fast frequency response. *IEEE Transactions on Smart Grid* 11 (4): 33943404.

**28** M. Zeinali, Bayram, I.S., and Thompson, J. (2020). Performance assessment of UK's cellular network for vehicle to grid energy trading: opportunities for 5G and beyond. *2020 IEEE International Conference on Communications Workshops (ICC Workshops)*, Dublin, Ireland, 1–6

# 7

# Passive UHF RFID Tag Antennas-Based Sensing for Internet of Things Paradigm

*Abubakar Sharif[1,4], Jun Ouyang[1], Kamran Arshad[2], Muhammad Ali Imran[3], and Qammer H. Abbasi[3]*

[1] *School of Electronic Science and Engineering, University of Electronic Science and Technology of China, Chengdu, China*
[2] *College of Engineering and IT, Ajman University, Ajman, United Arab Emirates*
[3] *James watt school of Engineering, University of Glasgow, Glasgow, United Kingdom*
[4] *Department of Electrical Engineering, Government College University Faisalabad, Faisalabad, Pakistan*

## 7.1 Introduction: Healthcare Provision and Radar Technology

In recent years, the Internet of Things (IoT) and radiofrequency identification (RFID) foresee a revolution toward the way people interact with physical objects and surrounding environment by transforming these physical objects into smart devices using RFID tags/printable wireless stickers [1]. RFID is being considered as a last mile solution for IoT that uses electromagnetic waves backscattering phenomenon for identification/sensing proposes. Nowadays, applications of RFID technology are covering every aspect of daily life, such as supply chain, access to buildings and transportation, animal tracking, patient monitoring, personal identification, facilitating the inventory, shipping of good, assembly lines, tagging food and retail items, localization, even assisting visually impaired people and so on (as shown in Figure 7.1) [2, 3].

In the last decade, RFID technology has accomplished a significant development mainly due to the reduction in the cost of RFID chips. The remarkable developments in microelectronics and RF domains are leading toward tremendous advances in wireless communication systems, in particular RFID systems [4]. The ultra-high frequency (UHF) RFID tags have proven their superiority among other competitors low frequency (LF) and HF band RFID as shown in Figure 7.2 in many applications such as retail management and supply chain management. The UHF tags have not only provided low-cost solutions regarding labeling/tagging of daily life items and product but also emerged as low-cost, inkjet-printed sensors for numerous applications due to their multiple tags reading capabilities, long read range, and low-cost inkjet printable structures [5]. The sensitivity of UHF tags toward different environment surfaces is a major weakness. On the other hand, the sensitivity of UHF tags can be exploited to transform these tags into versatile, low-cost sensing devices that fact opens a new paradigm in collaboration with IoT [6, 7].

*Backscattering and RF Sensing for Future Wireless Communication,* First Edition.
Edited by Qammer H. Abbasi, Hasan T. Abbas, Akram Alomainy, and Muhammad Ali Imran.

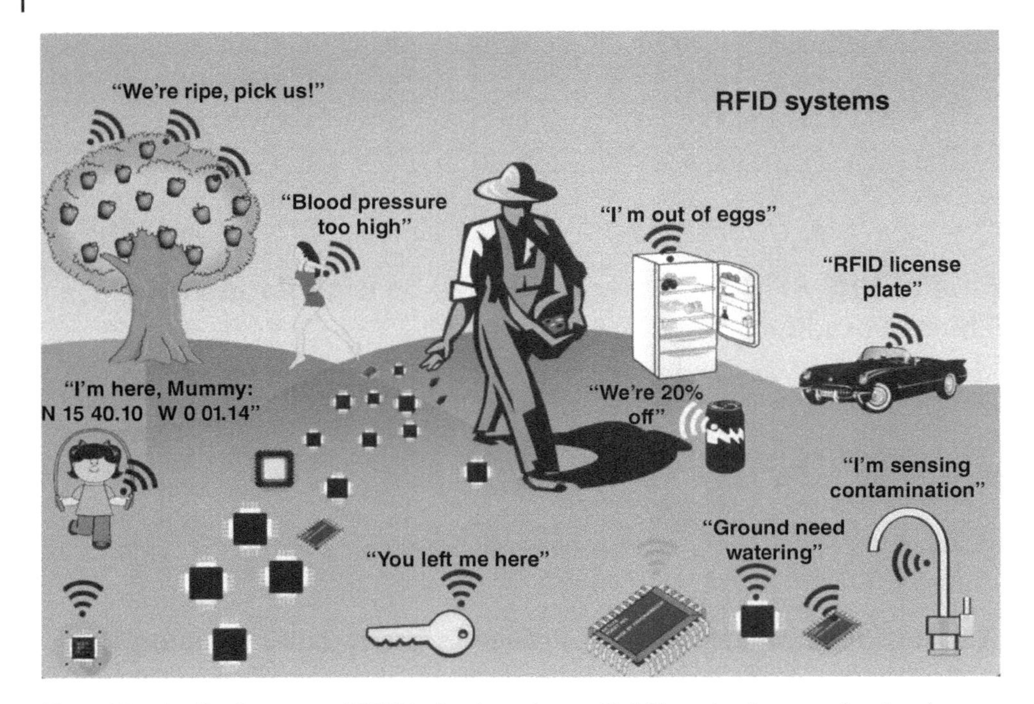


**Figure 7.1** Application areas of RFID technology along with IoT ranging from sensing, tagging, labeling, personal healthcare, etc. Sources: Dobkin [2] and Finkenzeller [3].

Depending on their power sources, the RFID tags can be categorized into three types. The tags are passive, semi-passive, and active. Table 7.1 describes a brief comparison of passive, battery-assisted semi-passive, and active tags. The passive tags use backscattering phenomenon for transmitting the information stored in microchip. Precisely, the passive tag intercepts power from the incoming signal transmitted by the reader to support the operation of their circuitry and transmits back this signal to the reader by modulating it with information stored in chips. Passive tags are inexpensive, easy to manufacture, and suitable for use at large scales (can be comparable to barcodes). Passive RFID tags pose enormous advantages because they require no independent power source (battery), no power amplifier, no crystal frequency reference, and no maintenance cost (for replacing battery or other). Semi-passive, also known as battery-assisted passive (BAP) tags, have a small battery for a power source for its internal circuitry like sensors. However, it uses the backscatter signal to communicate with the reader (for transmitting the signal from tag to reader). BAPs find their application mostly in sensors. Active RFID tags are equipped with a power source, a battery, to communicate with the interrogating device. So, they are expensive. However, the active tag can perform complex tasks with communication capability at very large distances up to 100 m.

In this context, passive UHF RFID tags are the most suitable candidates for sensing, especially for IoT domain, where every physical device must have to tell the story of themselves and surrounding environments.

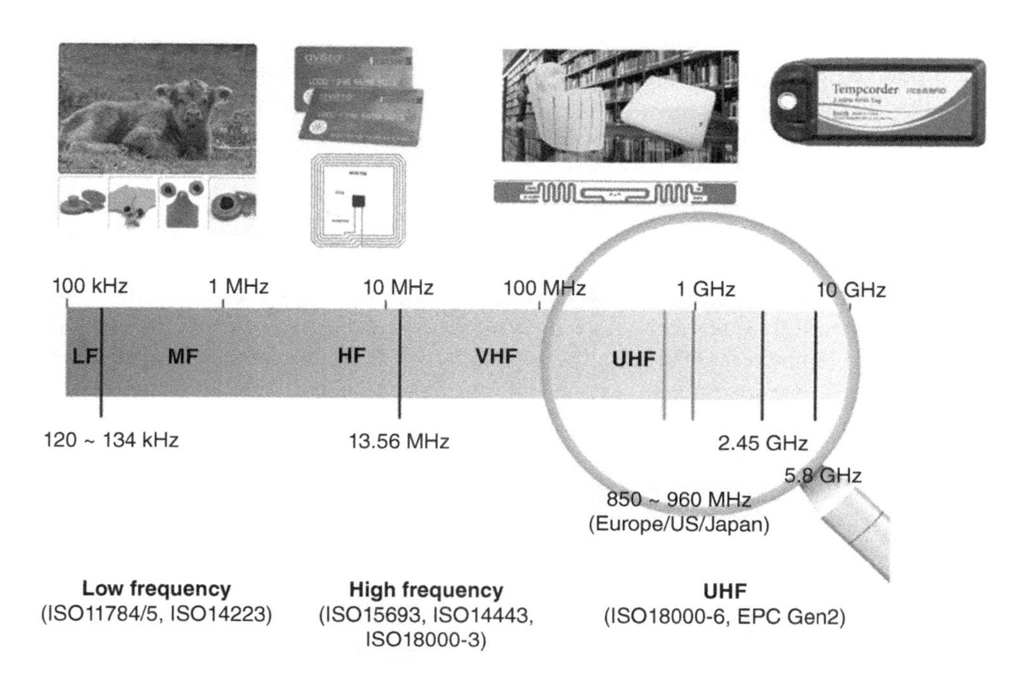


**Figure 7.2** RFID systems with respect to the operational frequency band.

**Table 7.1** Comparison of passive, Battery-Assisted Passive (BAP) and active RFID tag.

| | Passive | Semi passive or Battery-Assisted Passive "BAP" | Active |
|---|---|---|---|
| **Power source** | Use backscatter signal from reader at reading time | Internal battery (small) | Internal battery |
| **Power availability** | Only on reading | On reading for reader query (continuous for sensors internal circuitry) | Continuous transmit |
| **Read range** | Short (up to 10 m) | Medium (10 m$^{+}$) | Very long (100 m$^{+}$) |
| **Data storage** | Small (100–512 bits) | Medium (K bits) | Huge (Mb's to MB's) |
| **Life time** | Indefinite (20 years) | One to four years | 1–12 months |
| **Cost** | A few pennies | Dollars | Tens/hundred $'s |
| **Size** | Very small, thin and inkjet printable | Large | Huge |

Several sensing techniques have emerged recently regarding passive UHF RFID-based sensing that includes tag antenna-based sensing [8–10], RFID chip-assisted sensing [11, 12], multi-port tag-based sensing [13, 14], nearby reference tag-assisted sensing [15], and loaded sensing [16]. Similarly, the growing body of literature shows the development of passive UHF tags for healthcare, wearable and other implantable applications such as temperature monitoring [17], glucose and blood pressure monitoring [18–20], and on-skin monitoring and breath anomalies discrimination [21]. Moreover, there have been numerous innovative IoT applications reported in literature covering industrial as well as daily life aspects such as humidity and temperature sensing [22], pressure sensing [23], strain sensing [24], concrete structure deformation sensing [25–27], corrosion, and crack detection for steel [28, 29]. These are some few exemplary applications regarding passive UHF RFID sensing, however, developing research shows the application of RFID tags are not limited. Therefore, this chapter summarizes an overview of sensing parameters, techniques, state-of-the art passive UHF RFID sensors design-based IoT solutions, and their applications in daily life environments.

The remainder of this chapter will provide some basics on UHF RFID principles, performance metrics, sensing techniques, and sensing applications in the IoT domain. The chapter is organised as follows. Section 7.2 describes some fundamentals about UHF RFID working principles, performance parameters, and RFID chip fundamentals. Section 7.3 aims to present a sensing methodology and techniques derived from a different state of the art and commercial solutions. Section 7.4 discusses the innovative applications of passive UHF tags in different IoT applications including healthcare, agriculture, food contaminations sensing, and other numerous applications. Section 7.5 concludes with a discussion of outstanding research challenges and trends on possible future developments.

## 7.2 UHF RFID Fundamentals and Performance Metrics

This section summarises the fundamental working principles of UHF RFID and performance metrics aiming to provide the readers with a background on this technology. For additional details and comprehensive explanations, the readers are referred to the following references [2, 3, 30].

Figure 7.3 illustrates a typical UHF RFID system. An RFID system comprises of a set of remote transponders known as RFID tags and a more complex device called a reader or interrogator. RFID tags include an antenna and an application-specific integrated circuit (ASIC) also known as a chip, containing the data about the tagged object.

The RFID reader generates a query signal toward RFID tags; the tag replies back an information signal back to the reader in response to that query signal. The readers are usually connected with some embedded systems, host computers having application software for data collection and sharing.

The passive UHF tags typically consist of three elements: (i) transponder (packed in ASIC) or simply an RFID chip, (ii) antenna, and (iii) dielectric substrate. Passive tags are usually very simple devices. Figure 7.4 shows a typical structure of a passive UHF tag, which consists only of a printed or etched metal antenna on a plastic substrate or plastic inlay and a single integrated circuit. Therefore, passive tags are much cheaper than other types of radio devices. UHF passive tag costs about $0,10, 500 times cheaper than mobile

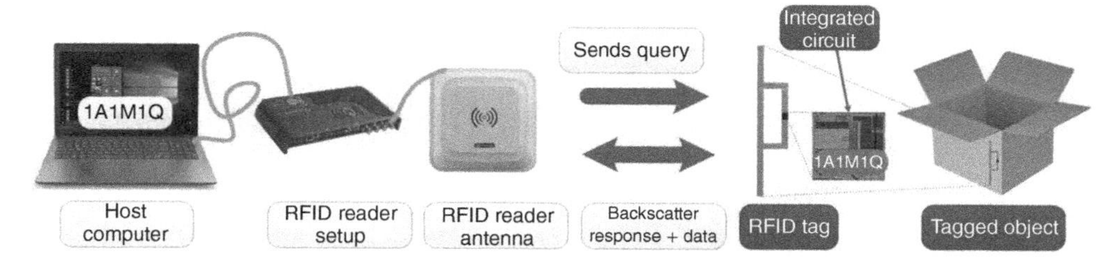


**Figure 7.3** Illustration of a typical RFID system.

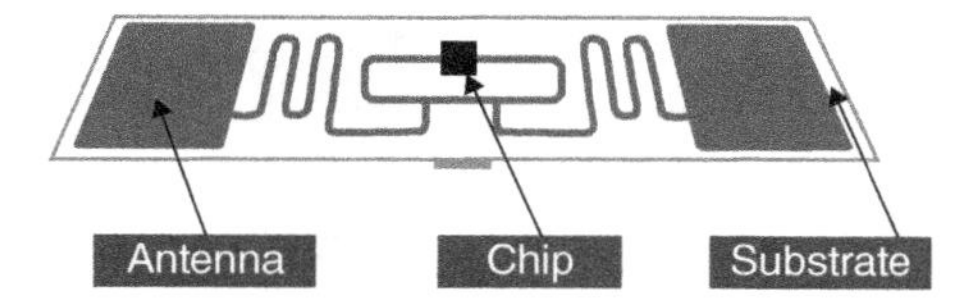


**Figure 7.4** A typical structure of passive UHF RFID tag.

phones, which are sold hundreds of millions per year. Passive tags usually have longer life span and do not require any maintenance. Moreover, the life of passive tags is depend on quality and degradation profile of label material rather than use of batteries. Therefore, the passive tags are expected to be readable for 10–20 years in different environments.

### 7.2.1 UHF RFID Microchips Insight

The main element of an RFID tag is a microchip, which incorporates an integrated circuit. The RFID chip is an essential element of a conventional RFID system because it contains a unique identifier that gives meaning to an RFID application. The performance of label in terms of reading distance will depend mainly on the sensitivity of the chip. Figure 7.5 shows the architecture of an RFID chip. It is divided into three main blocks: analog RF interface, digital control block, and nonvolatile memory. In recent years, tremendous advances in low-cost CMOS technology have allowed the production of micro-transponders with relatively less cost and energy consumption. A single monolithic chip contains the whole transponder used for passive RFID tags. Currently, the RFID chips consume only a few tens of μW of electric power and cover a quarter mm$^2$ of die area.

### 7.2.2 Performance Parameters for Passive UHF RFID Tag Antennas

The important tag performance parameters include the following:

i. Input impedance match and bandwidth
ii. Radar cross section
iii. Read range

#### 7.2.2.1 Input Impedance and Bandwidth

The basic principle of tag operation consists of switching from one impedance state to another. In fact, when a tag receives a signal from the reader antenna with sufficient power, the tag response back by switching the chip impedance between two states $Z_{c1}$ and $Z_{c2}$. The $Z_{c1}$ and $Z_{c2}$ correspond to the impedance of tag chip in receiving modes or downlink

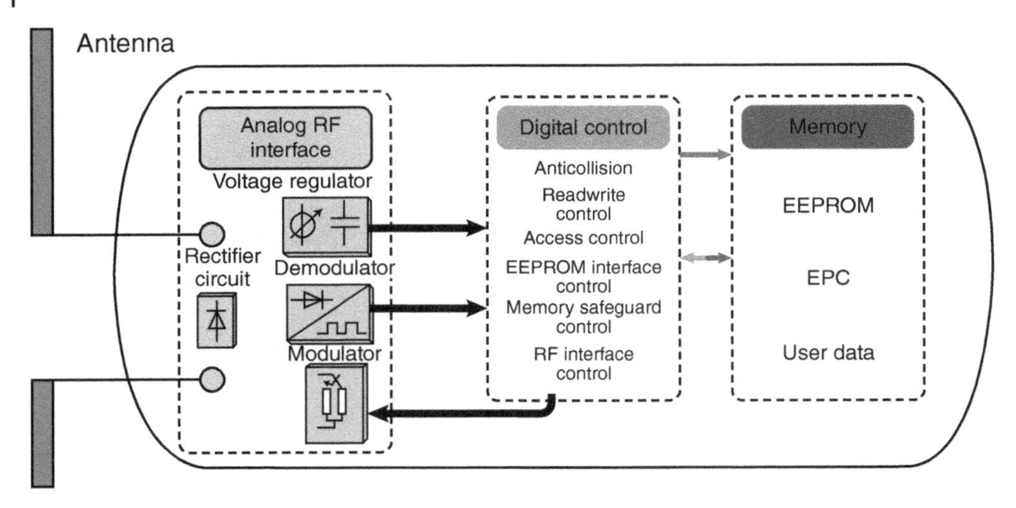


**Figure 7.5** Schematic of an RFID tag with functional blocks of the chip.

(from reader to tag communication) and transmitting mode or uplink (from tag to reader communication), as shown in Figure 7.6a and b, respectively. One impedance state required to match with antenna impedance $Z_a = Z_{c1}$ in order to achieve maximum power for internal circuitry operation of the chip.

The other matching state $Z_a = Z_{c2}$ is necessary for achieving maximum power transfer from chip toward tag antenna and ultimately required for uplink. Therefore, the proper impedance matching is an important parameter for passive UHF RFID tags. Since the reader sensitivity is good enough to read the tag in uplink transmission, so the most important impedance match condition for the passive UHF RFID tag is an impedance matching with the chip ($Z_a = Z_{c1}$) in receiving mode (upon receiving the query from a reader) that is necessary to activate the chip's internal circuitry. To lower the cost of the passive tag, the matching must be achieved by proper tag antenna design without the intervention of any additional matching circuits. In order to find impedance matching condition, consider the Thevenin equivalent circuit of the passive RFID tag in receiving mode as expressed in Figure 7.6a, $Z_a = R_a + jX_a$ represent complex impedance of tag antenna, and $Z_c = R_c + jX_c$ represents impedance of RFID chip.

The current from the circuit can be calculated as follows:

$$I = \frac{V_a}{Z_a + Z_c} \tag{7.1}$$

Or

$$I = \frac{V_a}{(R_a + jX_a) + (R_c + jX_c)} \tag{7.2}$$

The power delivered to the chip can be written as

$$P_c = \frac{|I|^2 R_c}{2} = \frac{V^2 R_c}{2|Z_a + Z_c|}$$

Or

$$P_c = \frac{V^2 R_c}{2|(R_a + jX_a) + (R_c + jX_c)|} \tag{7.3}$$

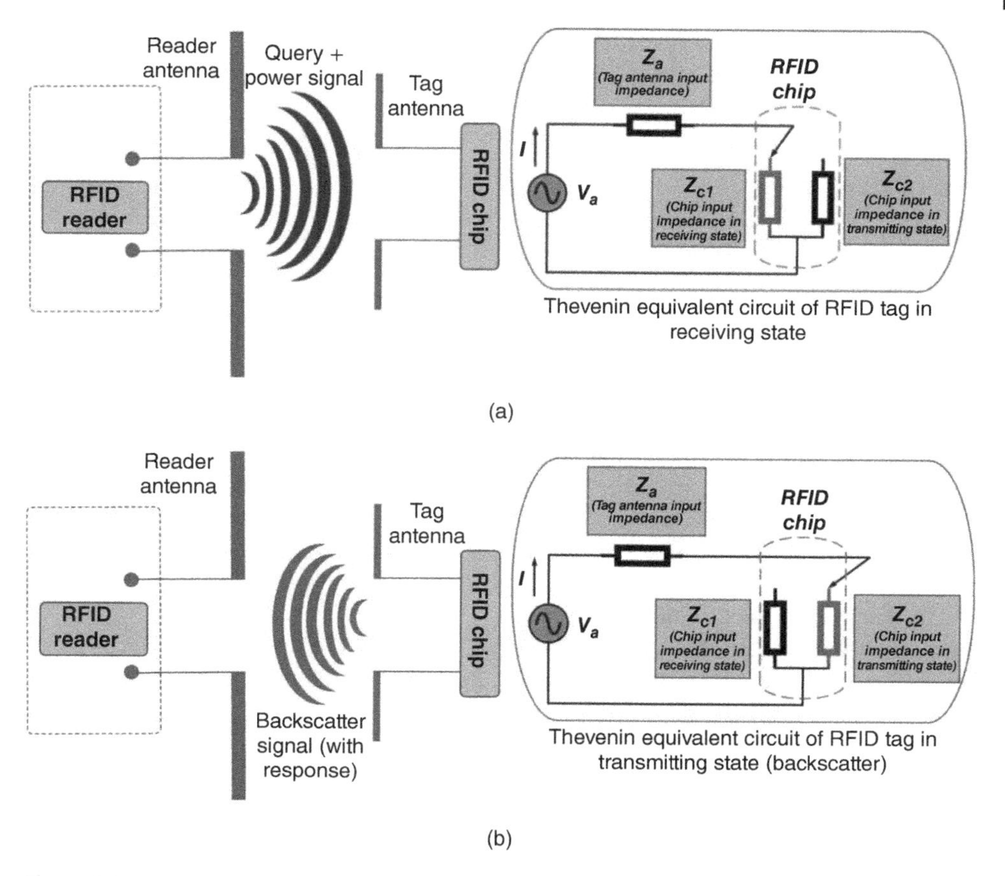


**Figure 7.6** Passive UHF tag antenna operation principle in (a) receiving mode (upon receiving query from reader), (b) transmitting mode (backscatter).

The maximum power will be delivered to chip only if the denominator of $P_c$ will be small, for real values of $R_a$ and $R_c$, the imaginary part must be zero. Therefore, the values of $X_c$ and $X_a$ must be the same with a negative sign.

So, the necessary condition to achieve a maximum power transfer from the tag antenna to chip can be expressed as

$$R_a = R_c$$
$$X_a = -X_c. \tag{7.4}$$

This condition is also termed as conjugate matching.

As mentioned in [1], another benefit of a good impedance match (conjugate matching) results in the form of voltage amplification expressed as follows:

$$\frac{|V_c|}{V_o} = \left| \frac{R_c + \frac{1}{j\omega C_c}}{2R_c} \right| \approx \frac{1}{2\omega C_c R_c} \gg 1 \tag{7.5}$$

where $V_c$ and $V_o$ are voltages at the IC terminal and open-circuit voltage of antenna, respectively, $R_c$ and $C_c$ are equivalent resistance and capacitance of chip.

Moreover, the power transferred to the chip can be expressed as

$$P_c = \tau P_a \tag{7.6}$$

where $\tau$ is the power transmission coefficient can be expressed as

$$\tau = 1 - |\Gamma_a| \tag{7.7}$$

Moreover, $\Gamma_a$ reflection coefficient of tag, $\Gamma_a = \frac{Z_c - Z_a}{Z_c + Z_a}$

$$\tau = \frac{4R_c R_a}{Z_c + Z_a} \leq 1 \tag{7.8}$$

The value of power transmission coefficient ranges between 0 and 1. Ideally, for maximum power transfer, $\tau = 1$.

The bandwidth of tag depends on impedance matching with a chip. It is good to have a wide bandwidth for a tag antenna in order to cover more regional frequency bands (ranging from 860 to 960 MHz) and further to nullifies the effects of materials such as detuning of the tag. Therefore, a tag antenna with wide bandwidth is less sensitive toward environment surfaces, and it will operate in wider range of UHF RFID frequency bands.

#### 7.2.2.2 Radar Cross Section (RCS)

The measure of backscatter power captured by passive UHF RFID reader is termed as radar cross section (RCS) of the tag. The backscattered power received by the reader from the tag antenna can be expressed as

$$P_{backscatter} = KP_{tag}G_{tag} \tag{7.9}$$

where $P_{tag}$ is the power received by tag antenna from the reader, and $G_{tag}$ is the gain of tag antenna. $K$ is term as backscattering coefficient as

$$K = \frac{4R_a}{|Z_c + Z_a|^2}$$

The reader antenna transmits and receives simultaneously. The tag takes power and query data from the reader's transmitted signal. The tag modulates its response signal by changing its antenna impedance state. By varying impedance, the tag varies its RCS and thus changes the backscatter power reflected by tag, as illustrated in Figure 7.7a. This is quite similar to radar's working principle, as shown in Figure 7.7b.

#### 7.2.2.3 Read Range

The read range is sometimes considered as the ultimate parameter/goal in determining the performance of passive UHF tags. The read range is the maximum distance between the tag and reader at which the tag can be successfully read by an RFID reader. The range equation can be determined using parameters as described in Figure 7.8.

Suppose $R_{read\_range}$ of RFID tag antenna. The parameters of reader setup are known, such as input power $P_t$ and reader antenna gain $G_{Reader}$. Also, $P_{tag}$ is the power received by the tag antenna, and $G_{tag}$ is the gain of tag antenna. Therefore, the Friis equation can be written as

$$P_{tag} = \frac{P_t G_{Reader}}{4\pi R_{read\_range}} * \frac{\lambda^2}{4\pi} G_{tag} \eta_p \tag{7.10}$$

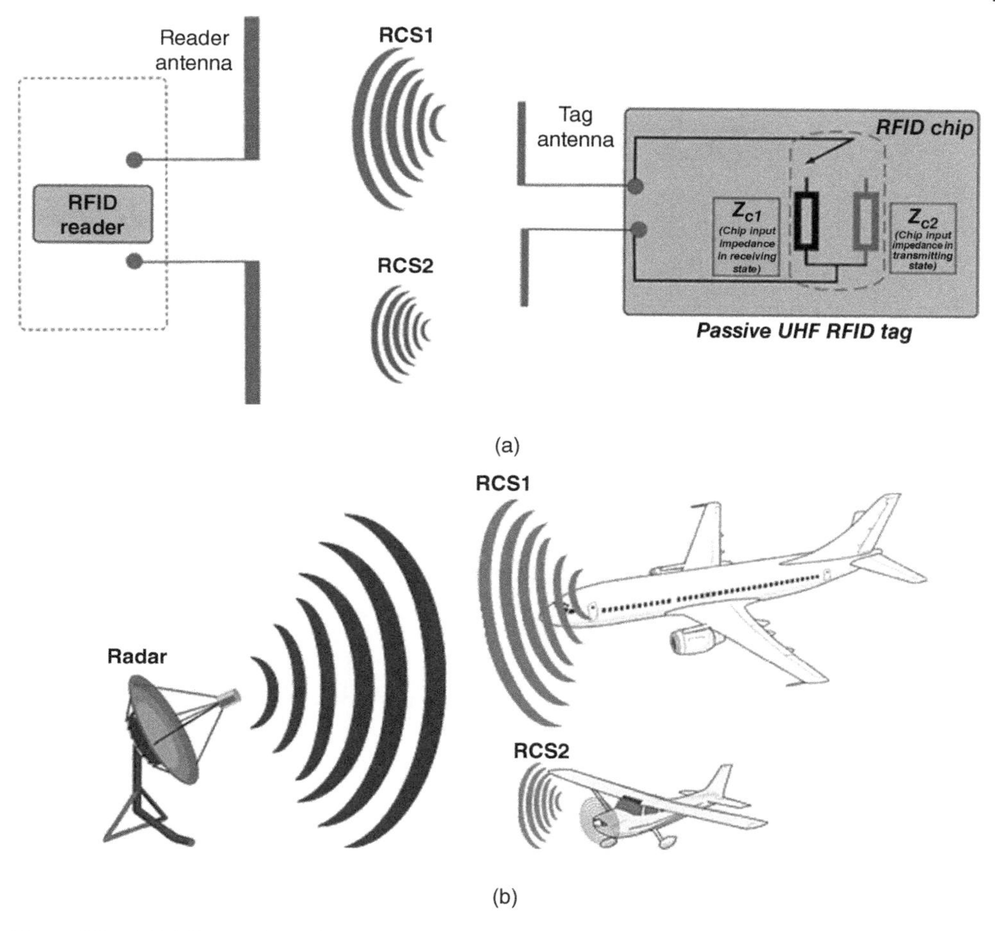


**Figure 7.7** (a) RFID working principle by switching RCS1 and RCS2 depends on impedance state. (b) equivalent radar principle similar to operation of passive RFID tag.

where $\eta_p$ is the polarization mismatch between tag and reader antennas. Let $\eta_p = 1$ so tag and reader antenna has no polarization mismatch.

To find maximum read range, the Equation (7.2–7.10) can be rearranged as:

$$R^2_{read\_range} = \left(\frac{\lambda}{4\pi}\right)^2 * \frac{P_t G_{Reader} G_{tag}}{P_{tag}}$$

Or

$$R_{read\_range} = \left(\frac{\lambda}{4\pi}\right) * \sqrt{\frac{P_t G_{Reader} G_{tag}}{P_{tag}}} \tag{7.11}$$

Whereas $P_{Tag}$ received by the tag can be expressed in term of $P_{th}$, which is known as the minimum threshold power to turn on RFID chip.

$$P_{th} = P_{tag} * \tau$$

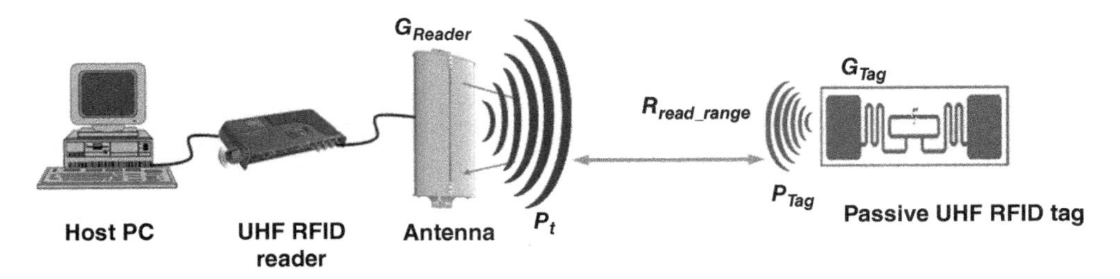


**Figure 7.8** Estimating the read range of passive tags.

where $\tau$ is the power transmission coefficient, can be determined using Equation (7.2–7.8) and finally tag read range of tag can be estimated as

$$R_{read\_range} = \left(\frac{\lambda}{4\pi}\right) * \sqrt{\frac{P_t G_{Reader} G_{tag}}{P_{tag}}} \tag{7.12}$$

Moreover, sometimes it is necessary to find backscatter power from a tag, especially in case of sensing applications, so we can also determine the backscatter power of RFID tag as

$$P_{backscatter} = P_{tag} G_{tag} * |\Gamma_m|^2 \tag{7.13}$$

$$P_{backscatter} = \left(\frac{P_t G_{Reader}}{4\pi R^2_{read\_range}} * \frac{\lambda^2}{4\pi} G_{tag}\right) * G_{tag} * |\Gamma_m|^2 \tag{7.14}$$

Or

$$P_{backscatter} = \left(\frac{P_t G_{Reader}}{4\pi R^2_{read\_range}} * rcs_T\right)$$

where $rcs_T$ is RFID tag's RCS, $rcs_T = \frac{\lambda^2}{4\pi} G^2_{tag} \cdot |\Gamma_a|^2$

The reader's sensitivity can also be calculated using the backscatter signal as:

$$P_{Reader\_th} = P_{backscatter} * \frac{1}{4\pi R^2_{read\_range}} * \frac{\lambda^2}{4\pi} G^2_{Reader} \tag{7.15}$$

The maximum distance $R_{Reader}$ shows the maximum read range of reader for detecting backscatter signal from tag.

$$R_{Reader} = \sqrt{\frac{G_{Reader}}{4\pi} * \sqrt{\frac{P_t * rcs_T}{4\pi P_{Reader\_th}}}} \tag{7.16}$$

As referred to Equation (7.2–7.12), the read range of tag depends on threshold power of the chip $P_{th}$, Gain of tag $G_{tag}$, impedance matching factor of tag $\tau$, and maximum allowed transmit power $EIRP = P_t G_{Reader}$, also known as effective isotropically radiated power, which depends on local countries or regional regulations, e.g., 4 W for the USA. Therefore, the task of UHF tag designer is to design tag antennas with optimized gain and impedance matching for desired performance on environmental surfaces, including some other important factors such as cost.

## 7.3 Sensing Methodology and Techniques

### 7.3.1 Single Tag Antenna-Based Sensing

Figure 7.9 shows a single tag antenna-based sensing. This sensing technique is based on a single tag antenna connected to a simple RFID chip, which is usually used for tagging or labeling proposes. The tag antenna was designed and optimized with certain environmental parameters. The change in environment parameter is recognized in terms of increasing or decreasing in dielectric loss or tangent loss of tagged surfaces of the material. The single tag-based sensing technique is low-cost and portrays a real picture of IoT sensing. Many researchers employed single tag antenna-based sensing; however, there are many challenges in order to get acceptable results from this sensing technique. An inkjet-printed tag antenna was proposed to detect salinity and sugar contents of water [8]. The sensor tag antenna was based on a slot-match technique, which provides a good imaginary impedance both in free space and on the surface of the water bottle.

The tag antenna characteristics were optimized to achieve maximum read range after mounting on water bottle surfaces without adding any salt or sugar contents. The salt and sugar contents were added in water and sensed by measuring and comparing the backscatter power from the tag in terms of received signal strength indicator (RSSI). In [9], a sensor tag antenna fabricated on an FR-4 substrate was designed for humidity sensing of the soil in flowerpots. The theoretical read range and RSSI were measured by adding 20, 40, 60, and 80 mL water in a soil-filled flowerpot. Food quality and contamination were sensed using batteryless wireless stickers or sensors [31]. Baby formula and alcohol adulteration were detected with 96 and 97% accuracy, respectively. The sensing methodology utilized in this research work was known as two-frequency excitation. In the two-frequency excitation technique, the first frequency was used to power up the wireless sticker in the ISM frequency band. The second frequency was used to sense the alteration of the sensor's

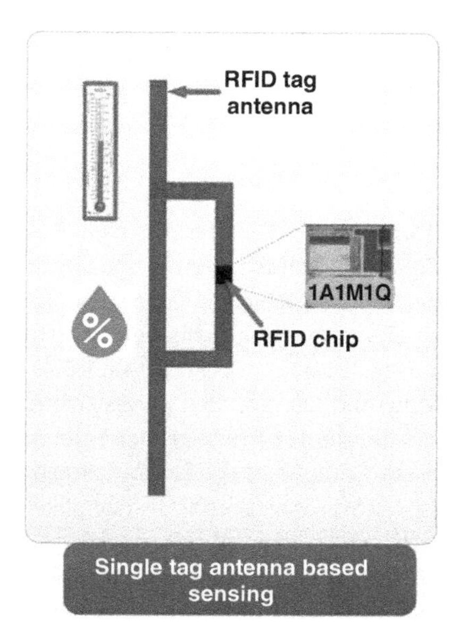


**Figure 7.9** Sensing technique based on a single tag antenna.

**Table 7.2** Sensitivity and impedance of some commercial RFID chips.

| RFID chip | $P_{th}$ (dBm),typical | $R_C$ (Ω) | $C_C$ (pF) | $Z_C$ (Ω) at 915 MHz |
|---|---|---|---|---|
| **Impinj Monza R6** | −20 | 1200 | 1.23 | 16 − 140j |
| **Impinj Monza R6-P** | −20 | 1200 | 1.44 | 12 − 120j |
| **Impinj Monza 4** | −17.4 | 1650 | 1.21 | 11 − 143j |
| **Alien Higgs H4** | −20.5 | 1800 | 0.95 | 18 − 181j |
| **Alien Higgs H3** | −20 | 1500 | 0.95 | 27 − 200j |
| **Alien Higgs H3 (SOT)** | −15 | 1500 | 0.9 | 24 − 190j |
| **NXP UCODE 8** | −23 | 4400 | 0.69 | 19 − j234 |
| **NXP UCODE 7** | −21 | 5500 | 0.68 | 12.8 − j248 |

response over a wide bandwidth. Furthermore, the machine learning algorithm (XGBoost) was employed to accuracy improvement in contamination detection and degree of purity. In [32], the simple UHF RFID tags were used as sensors for detecting the presence of the user and their daily physical activities by utilizing signal processing and machine learning techniques. Different objects were tagged, and user activities and presence were detected by intercepting the RSSI and received the signal's phase. User presence and daily activities were detected with F1 score of 96.7 and 82.8%, respectively.

A passive meandered dipole-based UHF RFID sensor antenna was proposed in [33] for sensing dielectric properties of aqueous solutions and organic liquids. The tag was mounted underneath the borosilicate glass bottle. The sensor tag was able to detect the dielectric properties of both known and unknown solutions. In [10, 34], a wearable strain sensor has been proposed for biomedical monitoring of respiration. The proposed system was based on backscatter power variations measurements resulted from the physical stretching of tag during respiration. Moreover, the machine learning algorithm was also developed for the classification of breathing and nonbreathing states. A UHF RFID tag antenna was designed on the FR-4 substrate for intravenous (IV) fluid levels monitoring [35]. Another low-cost inkjet-printed RFID sensor was proposed for IV fluid levels monitoring [36] (Table 7.2).

#### 7.3.1.1 UHF RFID Backscattering-Based Sensing Methodology

Figure 7.10 illustrates the sensing methodology regarding the sensing of UHF RFID tag using backscatter power or RSSI. The reduction in gain and impedance mismatch due to changing the in the sensing factor (humidity, temperature, and etc.) can be sensed by measuring backscatter power from RFID tag. There is fixed known distance "$d$" between reader setup and tag mounted on water filled plastic bottle. The parameters of reader set up such as input power $P_t$ and reader antenna gain $G_{Reader}$ are already known.

Let "g" denotes the factor or substance, which has to be sensed. Where $P_r$, $P_{rmc}$, $P_{bs}$, and $P_{RSSI}$ are power received by tag, power input to RFID chip, backscatter power, and backscattered power measured by a reader in terms of RSSI, respectively.

Therefore, the equations expressed in [2] can be rewritten accordingly in free space to obtain the sensing parameters.

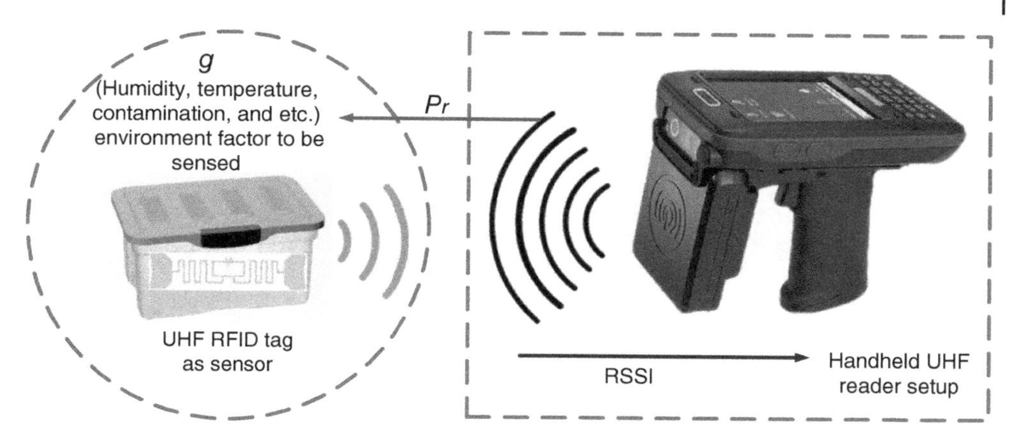


**Figure 7.10** Sketch of sensing methodology employed to use the proposed tag as sensor.

The power received by tag antenna can be written as

$$P_r = \frac{P_t G_{Reader}}{4\pi d^2} * \frac{\lambda^2}{4\pi} G_{Tag}[g]\eta_p \tag{7.17}$$

where $G_{Tag}[g]$ and $Z_{Tag}[g]$ are gain and input impedance parameters of the tag estimated with respect to substance g. Also, $\eta_p$ is polarization mismatch between tag and reader antennas.

The power transferred to microchip can be expressed by

$$P_{rmc} = P_r * \tau[g] \tag{7.18}$$

where $\tau[g] = 1 - |\Gamma_m(g)|^2$ is the power transmission coefficient, $|\Gamma_m(g)|$ is reflection coefficient of tag, which measures impedance mismatch between RFID chip impedance $Z_{chip} = R_{chip} + jX_{chip}$ and antenna impedance $Z_{Tag}[g] = R_{Tag}[g] + jX_{Tag}[g]$.

The power transmission coefficient and reflection coefficient with respect to substance contents g can also be expressed in terms of antenna and chip impedance as follows:

$$|\Gamma_m(g)| = \frac{Z_{chip} - Z_{Tag}[g]}{Z_{chip} + Z_{Tag}[g]} \tag{7.19}$$

$$\tau[g] = \frac{4R_{chip}R_{Tag}}{|Z_{chip} + Z_{Tag}|^2} \tag{7.20}$$

Moreover, the power transferred to microchip can be written as

$$P_{rmc} = \frac{P_t G_{Reader}}{4\pi d^2} * \frac{\lambda^2}{4\pi} G_{Tag}[g]\eta_p * \tau[g] \tag{7.21}$$

The backscatter power from the tag is

$$P_{bs} = P_r \cdot G_{Tag}[g] \cdot |\Gamma_m(g)|^2$$

$$P_{bs} = \frac{P_t G_{Reader}}{4\pi d^2} * \frac{\lambda^2}{4\pi} G_{Tag}[g]\eta_p \cdot G_{Tag}[g] \cdot |\Gamma_m(g)|^2$$

$$P_{bs} = \frac{P_t G_{Reader}}{4\pi d^2} * rcs_T[g] \tag{7.22}$$

where $rcs_T[g]$ is the RFID tag's radar cross section

$$rcs_T[g] = \frac{\lambda^2}{4\pi} G^2_{Tag}[g] \cdot |\Gamma_m(g)|^2 \quad (7.23)$$

Similarly, the backscatter power from the tag can be measured by a reader in terms of RSSI:

$$P_{RSSI} = \frac{1}{4\pi d^2} * \frac{\lambda^2}{4\pi} G_{Reader} * P_{bs} \quad (7.24)$$

$$P_{RSSI} = \frac{1}{4\pi d^2} * \frac{\lambda^2}{4\pi} G_{Reader} * \frac{P_t G_{Reader}}{4\pi d^2} * rcs_T[g]$$

$$P_{RSSI} = \frac{1}{4\pi} \left( \frac{\lambda}{4\pi d} \right)^2 P_t G^2_{Reader} rcs_T[g] \eta_p^2 \quad (7.25)$$

### 7.3.2 Single Tag Antenna with On-chip Circuitry-Based Sensing

This sensing technique utilized on-chip sensor circuitry of RFID chips for sensing of temperature, humidity, and other parameters (as shown in Figure 7.11). Some of these tag solutions operate in fully passive and other work in semi-passive mode. However, we will explore passive tags with the on-chip sensor circuit and also discuss the commercial solution. An epidermal temperature sensor tag was realized in [37] using the EM4325 IC and a sub-millimeter bio-compatible membrane. This sensor tag achieved a read range of 0.7–2.3 m in battery-less and battery-assisted mode. The overnight temperature monitoring and surveillance gate crossing testing showed the robustness of this temperature sensor and

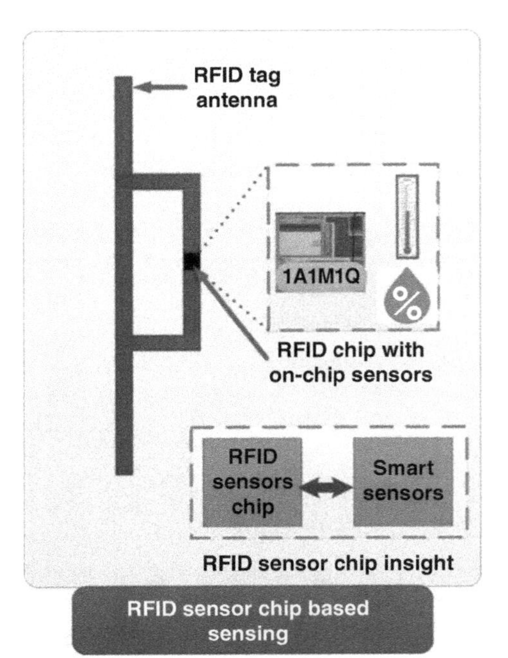


**Figure 7.11** Sensing technique based on chip integrated sensors.

**Table 7.3** RFID IC chip with on-chip sensors.

| RFID chip | Manufacturer | RF band | Digital interfaces | RF sensitivity | On-chip sensors |
|---|---|---|---|---|---|
| **EM4325** | EM microelectronic | UHF/EPC Class-3 Gen2 | SPI | −8.3 dBm | Temperature |
| **SL900A** | AMS | UHF/EPC Class-3 | SPI | −7 dBm | Temperature |
| **Rocky100** | Farsens | UHF/EPC Class-1 Gen2 | SPI | −13 dBm | External sensors |
| **WM72016** | Ramtron | UHF/EPC Class-1 Gen2 | DSPI | <−6 dBm | External sensors |
| **Magnus-S3 M3D IC** | Axzon (formerly RFMicron) | UHF/EPC Class-1 Gen2 | SPI | NA | Temperature moisture proximity |
| **Monza X-2K/X-8K** | Impinj | UHF/EPC Class-1 Gen2 | I2C | −19.1 dBm | External sensors |

open the doors toward applications such as temperature sensing during mass-screening at airports and country border crossing terminals. The temperature measurement was also done by mounting this sensors tag at different positions on the human body [38]. In [39], a finger-augmented UHF RFID sensor-based system was proposed for restoring peripheral thermal feeling. The tag was fabricated on the U-shaped bio-silicone substrate, and a reader antenna was integrated with a sports wristband. The EM4325 microchip IC chip was used for temperature sensing of touched objects (Table 7.3).

A flexible pH sensor was designed for sweat monitoring to predict the early signs of some psychophysical diseases [11]. This sensor was realized using three simple components multilayer pH sensor, SL900A, and a sensor antenna optimized for human skin. The sensor antenna was fabricated on a Kapton substrate and can also able to sense the temperature due to the on-chip sensor of SL900A. This sensor can be readable up to 1 and 2 m after mounting on the human arm for battery-less and battery-assisted mode. A graphene oxide(GO)-based wearable sensor was developed using Kapton substrate for wireless monitoring of breath and breath anomalies [21, 40–43]. The proposed wearable sensor consists of a flexible antenna integrated into a face mask and a humidity sensitive GO sensor. The whole breath monitoring system was experimentally tested and demonstrated the ability to detect inhalation/exhalation cycles along with the abnormal respiration patterns by measuring the change in resistance of graphene oxide. In [12], a tag antenna screen printed on a cardboard box was proposed with multiple sensing capabilities. The sensor antenna consists of a commercial force sensor printed opening detector to continuously monitor the force applied on the package along with the opening detection throughout the supply chain. The sensor antenna was based on SL900A's RFID chip. A UHF RFID sensor was proposed on a flexible Kapton substrate with multiple optical sensing capabilities [44]. The sensor composed of SL900 A RFID microchip and five photodiodes to cover a wide spectral range (from near-infrared to visible and ultraviolet spectral regions). This tag antenna can sense the spectral range of incident electromagnetic energy ranging from ultraviolet to infrared. Furthermore, an idea has been developed using this sensor tag by applying illuminants in order to sense fruit ripening and cards from a color chart.

### 7.3.3 Multiport Tag Antenna-Based Sensing

Figure 7.12 illustrates the sensing technique based on sensor having multiple ports. In this configuration, one port was usually connected to the sensor node, while the other port was considered as a reference node, as shown in Figure 7.13. A metamaterial inspired tag antenna was proposed in [14] with a multi-port configuration for sensing applications. The two-port sensor antenna was integrated with the two RFID chips. One RFID chip was connected to a passive sensor, while the other chip was considered as a reference node. In [45], a long-range solar-powered multi-port sensor antenna with thin-film solar cells was proposed. This sensor used I2C based Impinj Monza X-Dura RFID chip, which was further connected to the external microcontroller unit, temperature, and humidity sensor. A two-port tag antenna was realized in [46] using electromagnetic band gap structures (EBG). The EBG-based tag antenna achieved a size reduction of 75% as compared to conventional antennas. A single tag antenna with two excitation ports was designed in [47] for tagging and sensing applications. The two ports are matched with Texas RFID chips (RI-UHF-IC116-00) and one chip in interfaces with a solar-powered resistive sensor in order to get enhanced read range. In [48], a multiport UHF RFID sensor tag antenna with the circular polarization (CP) was designed. The 20% size reduction was achieved with four stars etched on a patch antenna. The sensor tag was matched with two RFID chips using an inductive loop matching technique. One sensor port was connected to the resistive sensor, while the other port acts as a reference node.

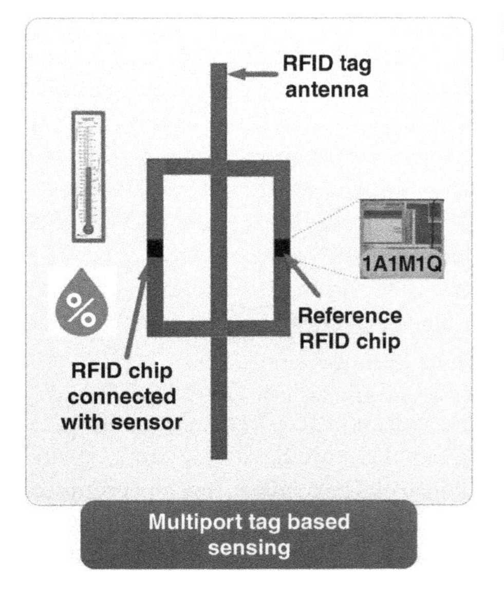


**Figure 7.12** Sensing technique-based multiport sensors.

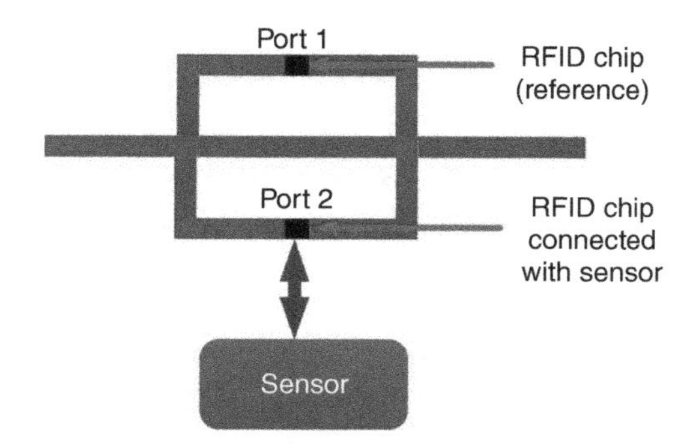


**Figure 7.13** Sensing methodology of multiport sensor antennas.

### 7.3.4 Reference Multiple Tag-Based Sensing

Figure 7.14 shows the sensing technique that is based on nearby reference sensor tag antennas. This sensing technique utilized some of the tags placed nearby as a reference node, while others as a sensor node. A low-cost, inkjet printed tag antenna with capacitive sensing capability was proposed in [49] for calibration-free sensing with dual tag configuration. In [50], a printed humidity sensor with memory function was proposed. The sensor operates as the change in resistive memory was detected from 2 kΩ to 50 Ω with the exposure of humidity or water. This sensor also worked as a twin tag configuration. Two ordinary tags were exploited as a moisture sensor [51, 52]. One tag was left open, and another tag was integrated inside absorbent material. The level of moisture was determined by comparing the backscattered signal strength. In [15], two tags were placed on a dry and wet soil, respectively, for soil moisture sensing. Similarly, a system was realized for wirelessly

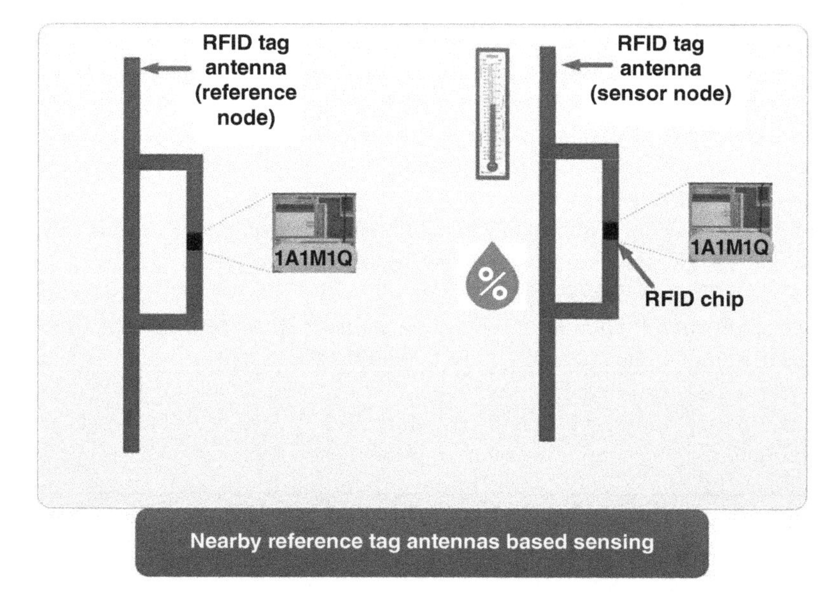


**Figure 7.14** Sensing technique based on nearby reference sensor antennas.

crack monitoring using coupled RFID tags. The cracked was detected by placing two passive tags on top of the crack and further by detecting the phase of the backscattered field.

## 7.4 Conclusion and Outstanding Challenges

This chapter described some of the latest findings regarding passive UHF RFID tags, their sensing methodologies, and innovative applications in the context of IoT paradigm ranging from healthcare to environmental sensing solutions. The sensitivity of UHF RFID tags was exploited to transform them into low-cost, versatile sensors for a wide range of applications. The communication channel parameter such as backscatter RSSI, and the RF phase was used as sensing metrics for the effects of sensing environments and user activities. Moreover, machine learning and signal processing techniques have also been developed to add more fascination and accuracy in UHF RFID-based passive sensing.

Although, there are many passive sensing solutions were developed and tested in the lab environment. However, there were very few commercial solutions regarding passive UHF RFID sensors. Some outstanding research challenges regarding the passive UHF RFID sensors still needed to be addressed and are briefly discussed later:

- Since the operations of the passive sensors depend on power harvesting from RFID source or reader. The passive sensors utilized RF energy to power the controller, or sensor to respond for reading/write commands. Antennas are basic units regarding efficient energy harvesting. So, the design of efficient, low-cost, inkjet printable tags is still challenging. Although most of RFID ICs are power efficient, however, the operation logic of the sensors is more complex and time consuming. Therefore, it is still very challenging to provide power and logic operations with RF energy harvesting only.
- The sensing techniques regarding passive sensors show a high degree of heterogeneities regarding IC functionalities, sensor antennas, sensing methodologies, and techniques. Therefore, these heterogeneities hinder to develop any standardization and cross-platform integration. Most sensor solutions, including antenna design and RFID microchip, were developed for particular scenarios. Due to variations of IC chip impedance, every scenario needs a custom-designed sensor antenna in order to get a conjugate impedance match.
- The reliability of passive sensors is another important challenge due to their sensitivity to environmental factors. So, there is a need for high-performance sensors in order to cope with harsh environmental effects. Therefore, the reliability of passive RFID-sensing devices still poses significant challenges in obtaining stable RFID sensor information. Figure 7.15 shows the main factors that should be taken into account by tag designers to design a good performance tag for environmental applications. First of all, some application specifications are very important such as cost and required read range. Second, IC with suitable parameters such as input impedance, memory requirement, size, and cost of IC are also key factors. Moreover, it is worth to mention that RF regulations are not the same for UHF RFID tags worldwide, at least there three different frequency bands are used for different reader power level. The antenna is vital and main driving part of an RFID tag, especially when passive RFID tags are used to transmit date back to the RFID reader. Tag antennas must be designed with good radiation efficiency to provide a

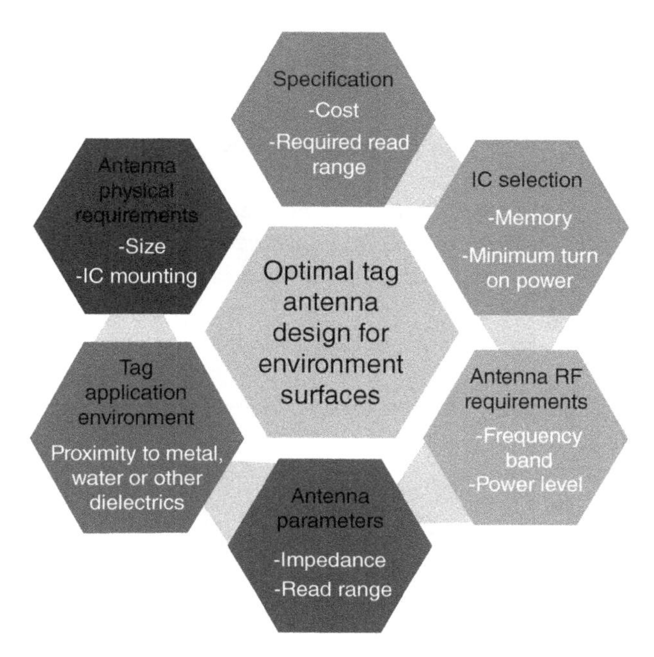


**Figure 7.15** Considerations for UHF RFID tag antenna designs for environment applications.

conjugate match with IC impedance. Furthermore, the tag antenna should also be optimized according to its application surface on which it would be mounted such as metal, water, and other dielectric material. Additionally, the size of the antenna must be considered, while optimizing to fulfil the specific applications requirements. Also, the IC mounting location must be carefully chosen, so that it will not be vulnerable to damage during operability [53].

## 7.5 Future Trends

RFID-based passive sensing has become prominent both in industry and academia, especially with the advent of the IoT, 5G, and beyond. The key research area regarding RFID sensor is the integration of new processes, techniques, and materials to improve the performance of RFID sensors. In addition, RFID sensor networks are also receiving significant attention in academia for wide-area and multiple object monitoring due to low-cost, battery-less, and light-weight sensors. The emergence of IoT has further generated wide range of innovative applications by utilizing the advantages associated with passive UHF RFID sensors ranging from baby diaper humidity sensing to pressure sensing in vehicles tyres and so on. Therefore, the sensing and communication techniques will revolutionize this world and transform it into a smart planet by providing innovation in traditional non-sensing RFID-based applications.

## References


1 Sharif, A., Ouyang, J., Yang, F. et al. (2019). Low-cost inkjet-printed UHF RFID tag-based system for internet of things applications using characteristic modes. *IEEE Internet of Things Journal* 6 (2): 3962–3975.

2 Dobkin, D.M. (2013). *The RF in RFID: UHF RFID in Practice*. Elsevier: Newnes.

3 Finkenzeller, K. (2010). *RFID Handbook: Fundamentals and Applications in Contactless Smart Cards, Radio Frequency Identification and Near-Field Communication*. Chichester: Wiley-Blackwell.

4 Chawla, V. and Ha, D. (2007). An overview of passive RFID. *IEEE Communications Magazine* 45 (9): 11–17.

5 Lakafosis, B.V., Rida, A., Vyas, R. et al. (2010). Progress towards the first wireless sensor networks consisting of inkjet-printed. *Sensor Tags* 98 (9): 1601–1609.

6 Perret, E., Tedjini, S., and Nair, R.S. (2012). Design of antennas for UHF RFID tags. *Proceedings of the IEEE* 100 (7): 2330–2340.

7 Roy, B.S., Jandhyala, V., Smith, J.R. et al. (2010). RFID: from supply chains to sensor nets. *Proceedings of the IEEE* 98 (9): 1583–1592.

8 Sharif, A., Ouyang, J., Raza, A. et al. (2019). Inkjet-printed UHF RFID tag based system for salinity and sugar detection. *Microwave and Optical Technology Letters* 61 (9): 2161–2168.

9 Alonso, D., Zhang, Q., Gao, Y., and Valderas, D. (2017). UHF passive RFID-based sensor-less system to detect humidity for irrigation monitoring. *Microwave and Optical Technology Letters* 59 (7): 1709–1715.

10 Patron, D., Mongan, W., Kurzweg, T.P. et al. (2016). On the use of knitted antennas and inductively coupled RFID tags for wearable applications. *IEEE Transactions on Biomedical Circuits and Systems* 10 (6): 1047–1057.

11 Nappi, S., Mazzaracchio, V., Fiore, L. et al. (2019). Flexible pH sensor for wireless monitoring of the human skin from the medimun distances. *FLEPS 2019 - IEEE International Conference on Flexible and Printable Sensors and Systems Proceedings*, Glasgow, United Kingdom, pp. 2–4.

12 Fernández-Salmerón, J., Rivadeneyra, A., Martínez-Martí, F. et al. (2015). Passive UHF RFID tag with multiple sensing capabilities. *Sensors (Switzerland)* 15 (10): 26769–26782.

13 Abdulhadi, A.E. and Denidni, T.A. (2017). Self-powered multi-port UHF RFID. *IEEE Journal of Radio Frequency Identification* 7281: 1–9.

14 Zaid, J., Abdulhadi, A.E., and Denidni, T.A. (2019). Miniaturized multi-port microstrip patch antenna using metamaterial for passive UHF RFID-tag sensor applications. *Sensors* 19 (9): 1982.

15 Pichorim, S.F., Gomes, N.J., and Batchelor, J.C. (2018). Two solutions of soil moisture sensing with rfid for landslide monitoring. *Sensors (Switzerland)* 18 (2): 452.

16 Manzari, S., Occhiuzzi, C., Nawale, S. et al. (2012). Humidity sensing by polymer-loaded UHF RFID antennas. *IEEE Sensors Journal* 12 (9): 2851–2858.

17 Milici, S., Amendola, S., Bianco, A., and Marrocco, G. (2014). Epidermal RFID passive sensor for body temperature measurements. *2014 IEEE RFID Technology and Applications Conference RFID-TA 2014*, Tampere, Finland, pp. 140–144.

**18** Xiao, Z., Tan, X., Chen, X., and Chen, S. (2015). An implantable RFID sensor tag toward continuous glucose monitoring. *IEEE Journal of Biomedical and Health Informatics* 19 (3): 910–919.

**19** Caldara, M., Nodari, B., Re, V., and Bonandrini, B. (2014). Miniaturized and low-power blood pressure telemetry system with RFID interface. *Procedia Engineering* 87: 344–347.

**20** Caldara, M., Nodari, B., Re, V., and Bonandrini, B. (2016). Miniaturized blood pressure telemetry system with RFID interface. *Electronics* 5 (4): 51.

**21** Caccami, M.C., Mulla, M.Y.S., Occhiuzzi, C. et al. (2018). Design and experimentation of a batteryless on-skin RFID graphene-oxide sensor for the monitoring and discrimination of breath anomalies. *IEEE Sensors Journal* 18 (21): 8893–8901.

**22** Oprea, A., Bârsan, N., Weimar, U. et al. (2008). Capacitive humidity sensors on flexible RFID labels. *Sensors and Actuators B: Chemical* 132 (2): 404–410.

**23** Rennane, A., Abdelnour, A., Kaddour, D. et al. (2018). Design of passive UHF RFID sensor on flexible foil for sports balls pressure monitoring. *IET Microwaves, Antennas and Propagation* 12 (14): 2154–2160.

**24** Occhiuzzi, C., Paggi, C., and Marrocco, G. (2011). Passive RFID strain-sensor based on meander-line antennas. *IEEE Transactions on Antennas and Propagation* 59 (12): 4836–4840.

**25** Caizzone, S. and DiGiampaolo, E. (2015). Wireless passive RFID crack width sensor for structural health monitoring. *IEEE Sensors Journal* 15 (12): 6767–6774.

**26** Caizzone, S. and DiGiampaolo, E. (2014). Passive RFID deformation sensor for concrete structures. *2014 IEEE RFID Technology and Applications Conference, RFID-TA 2014*, Tampere, Finland.

**27** Ikemoto, Y., Suzuki S., Okamoto H. et al. (2009). Force sensor system for structural health monitoring using passive RFID tags. *Sensor Review*. 29 (2): 127–136, doi: 10.1108/02602280910936237.

**28** Zhang, J., Tian, G.Y., and Zhao, A.B. (2017). Passive RFID sensor systems for crack detection & characterization. *NDT & E International* 86: 89–99.

**29** Yi, X., Cho, C., Cooper, J. et al. (2013). Passive wireless antenna sensor for strain and crack sensing - electromagnetic modeling, simulation, and testing. *Smart Materials and Structures* 22 (8): 085009.

**30** Marrocco, G. (2008). The art of UHF RFID antenna design: impedance-matching and size-reduction techniques. *IEEE Antennas and Propagation Magazine* 50 (1): 14.

**31** Ha, U., Ma, Y., Zhong, Z. et al. (2018). Learning food quality and safety from wireless stickers. *HotNets 2018 – Proceedings of 2018 ACM Workshop on Hot Topics in Networks*, Redmond, Washington, USA, pp. 106–112.

**32** Li, H., Wan, C.Y., Shah, R.C. et al. (2019). IDAct: towards unobtrusive recognition of user presence and daily activities. *2019 IEEE International Conference on RFID, RFID 2019*, Pisa, Italy, pp. 1–8.

**33** Makarovaite, V., Hillier, A.J.R., Holder, S.J. et al. (2019). Passive wireless UHF RFID antenna label for sensing dielectric properties of aqueous and organic liquids. *IEEE Sensors Journal* 19 (11): 4299–4307.

**34** Dion, G., Fontecchio, A.K., and Kurzweg, T.P. (2017). On implementing an unconventional infant vital signs monitor with passive RFID tags. *2017 IEEE International Conference on RFID (RFID)*, Phoenix, AZ, USA.

**35** Ting, S.H., Wu, C.K., and Luo, C.H. (2017). Design of dual mode RFID antenna for inventory management and IV fluid level warning system. *International Journal of Antennas and Propagation* 2017: Article ID 2470291. https://doi.org/10.1155/2017/2470291.

**36** Sharif, A., Ouyang, J., Yan, Y. et al. (2019). Low-cost inkjet-printed RFID tag antenna design for remote healthcare applications. *IEEE Journal of Electromagnetics, RF and Microwaves in Medicine and Biology* 3 (4): 261–268.

**37** Amendola, S., Bovesecchi, G., Palombi, A. et al. (2016). Design, calibration and experimentation of an epidermal RFID sensor for remote temperature monitoring. *IEEE Sensors Journal* 16 (19): 7250–7257.

**38** Miozzi, C., Amendola, S., Bergamini, A., and Marrocco, G. (2017). Reliability of a re-usable wireless epidermal temperature sensor in real conditions. *2017 IEEE 14th International Conference on Wearable and Implantable Body Sensor Networks, BSN 2017*, Eindhoven, Netherlands, pp. 95–98.

**39** Di Cecco, V., Amendola, S., Valentini, P.P., and Marrocco, G. (2017). Finger-augmented RFID system to restore peripheral thermal feeling. *2017 IEEE International Conference on RFID, RFID 2017*, Phoenix, AZ, USA.

**40** Caccami, M.C., Mulla, M.Y.S., Di Natale, C., and Marrocco, G. (2017). Wireless monitoring of breath by means of a graphene oxide-based radiofrequency identification wearable sensor. *2017 11th European Conference on Antennas and Propagation, EUCAP 2017*, Paris, France, pp. 3394–3396.

**41** Caccami, M.C., Mulla, M.Y.S., Di Natale, C., and Marrocco, G. (2018). Graphene oxide-based radiofrequency identification wearable sensor for breath monitoring. *IET Microwaves, Antennas and Propagation* 12 (4): 467–471.

**42** Caccami, M.C., Miozzi, C., Mulla, M.Y.S. et al. (2017). An epidermal graphene oxide-based RFID sensor for the wireless analysis of human breath. *2017 IEEE International Conference on RFID Technology and Application, RFID-TA 2017*, Warsaw, Poland.

**43** Occhiuzzi, C., Caccami, C., Amendola, S., and Marrocco, G. (2018). Breath-monitoring by means of epidermal temperature RFID sensors. *2018 3rd International Conference on Smart and Sustainable Technologies, SpliTech 2018*, Split, Croatia.

**44** Escobedo, P., Carvajal, M.A., Capitán-Vallvey, L.F. et al. (2016). Passive UHF RFID tag for multispectral assessment. *Sensors (Switzerland)* 16 (7): 1085.

**45** Abdulhadi, A.E. and Abhari, R. (2016). Multiport UHF RFID-tag antenna for enhanced energy harvesting of self-powered wireless sensors. *IEEE Transactions on Industrial Informatics* 12 (2): 801–808.

**46** Zaid, J., Farahani, M., Kesavan, A., and Denidni, T.A. (2017). Miniaturized microstrip patch antenna using electromagnetic band gap (EBG) structures for multiport passive UHF RFID-tag applications. *2017 IEEE Antennas and Propagation Society International Symposium, Proceedings*, San Diego, CA, USA.

**47** Abdulhadi, A.E. and Denidni, T.A. (2017). Self-powered multi-port UHF RFID tag-based-sensor. *IEEE Journal of Radio Frequency Identification* 1 (2): 115–123.

**48** Zaid, J., Abdulhadi, A., Kesavan, A. et al. (2017). Multiport circular polarized RFID-tag antenna for UHF sensor applications. *Sensors* 17 (7): 1576.

**49** Kim, S., Kawahara, Y., Georgiadis, A. et al. (2015). Low-cost inkjet-printed fully passive RFID tags for calibration-free capacitive/haptic sensor applications. *IEEE Sensors Journal* 15 (6): 3135–3145.

**50** Gao, J., Siden, J., Nilsson, H.E., and Gulliksson, M. (2013). Printed humidity sensor with memory functionality for passive RFID tags. *IEEE Sensors Journal* 13 (5): 1824–1834.

**51** Sidén, J., Zeng, X., Unander, T. et al. (2007). Remote moisture sensing utilizing ordinary RFID tags. *Proceedings of IEEE Sensors*, Atlanta, GA, USA.

**52** Gao, J., Siden, J., and Nilsson, H.E. (2011). Printed electromagnetic coupler with an embedded moisture sensor for ordinary passive RFID tags. *IEEE Electron Device Letters* 32 (12): 1767–1769.

**53** Mitchell, N. (2013). The art and science of UHF passive tag design: and selecting the tag that is best for your requirements. Alien Technology.

# 8

# RF Sensing for Healthcare Applications

*Syed Aziz Shah[1], Hasan Abbas[2], Muhammad Ali Imran[2], and Qammer H. Abbasi[2]*

[1] *Centre for Intelligent Healthcare Coventry University, Coventry, United Kingdom*
[2] *James watt school of Engineering, University of Glasgow, Glasgow, United Kingdom*

## 8.1 Introduction

The radio frequency (RF) sensing is an evolving technique in the field of healthcare sector [1]. However, the limitations in detection accurate and reliable human motion identification require more cooperative sensing techniques rather than a reliable system.

For its application in real-world intelligent healthcare facilities, human activity recognition and monitoring approaches exploiting a single RF sensing system might not provide the performance as and when required. To overcome this challenge, RF sensing technologies have been discussed here. This chapter primarily covers various RF sensing technologies as support for healthcare applications [2]. Initially, the detection, monitoring, and working function of this technology will be discussed, covering insight of signal processing of the various sensing devices and artificial intelligence-based algorithms. The results acquired for healthcare scenario will be delivered as well.

With an increase in life expectancy of an older adult, it has induced numerous multiple fatal conditions and critical events such as abnormality of gait, falling on the floor, and freezing of gait, heart disease, and so on. This is extremely vital to challenge the health of elderly individuals involved along with their friends and families, as well as presenting numerous other problems considering low-cost easily deployable healthcare system. Studies have shown that the timely intervention by healthcare professional after experiencing critical events such as falls could greatly decrease their fatal consequences. That continuous detection and monitoring among the target population can enable early intervention of potentially dangerous health provision [3–5].

Various RF sensing approaches have been introduced to address these open research areas and allow proactive detection of various fatal events in the remote healthcare domain. These technologies consist of acoustic sensors, passive infrared (PIR), pressure sensors on floors, video and depth cameras, wearable sensors including accelerometer, gyroscope, magnetic sensors, and radio frequency (RF) sensors. These various RF and wireless sensing alternatives have advantages and associated limitations as well [6]. These performance metrics include accuracy, F-measure, and sensitivity when identifying various disease

*Backscattering and RF Sensing for Future Wireless Communication,* First Edition.
Edited by Qammer H. Abbasi, Hasan T. Abbas, Akram Alomainy, and Muhammad Ali Imran.

symptoms, cost of the sensing system, and the detection of intricate body movements. This chapter gives details about the RF sensing technologies used for healthcare applications.

## 8.2 Basics of RF Sensing in Remote Healthcare

The applications of radio frequency wireless sensing considering remote healthcare sector include falls, FOG, breath detection, and walking behavior leverage movement induced by a subject to determine the abnormalities due to disease related to body motion. This is extremely vital as it totally depends on detecting unusual events.

The challenge of monitoring elderly people focusing on patient monitoring has opened new research areas of novel healthcare provisions that enhances for the classification of hospital central treatment and procedure to home-centric where various patients with different disease reside. This home-centric monitoring alternative solution can bring huge changes in the QoL of persons who suffer from disease disorders that will enable them to lead comfortable lives at homes. This can greatly mitigate challenges caused by various factors by the subjects as daily routine activities are being performed. Therefore, there is a huge requirement to tap various alternatives using radio frequency sensing.

Few of the major challenges to avoid the reliable healthcare monitoring systems that face high risk of health disorders exploiting radio frequency approaches provided enabling the patients to lead comfortable life residing at their residence. At the beginning, the radio frequency wireless sensing method need to determine anomaly considering movement symptoms. The high risk of fatal person movements and their corresponding disease disorders have very extensive effect on elderly populations. Numerous research works have identified that approximately 70% of elderly people having dementia disorder demonstrate that they highly experience pacing, lapping, and random walking. There is a high risk that patients with dementia disease will leave their homes if they are unattended and might experience fatal injury due to fall. Likewise, subjects on bed for a lot of period of time might develop bedsores as tissues get injured. Patients having respiratory disorders all of a sudden have breathing stoppage. Patients having Parkinson's disease experience freezing of gait episodes as mentioned earlier. Narcolepsy directly impacts the controlling of the sleeping pattern as described in Ref. [7]. Patients having this disease suddenly experience sleeping episodes when they perform intense activities of daily livings. These incidences rapidly increased when patients' age increases and the level of mobility slows down. Numerous alternatives and healthcare provisions have introduced to continuously check patients. These healthcare solutions are based on camera systems and are light-dependent at indoor environment. The constant recording of images and videos also raise privacy concerns as well. Systems exploiting acoustic sensors, pressure mats sensor, and PIR sensors are also noncontact systems that can be deployed at homes. Contact-based inertial sensors including ors such as accelerometer, gyroscope and magnetic sensors are installed to monitor patients with different disease. On the contrary, these sensing systems have been deployed on subjects' part of the body which discomforts disease patients on several occasions. Hence, both devices are expensive and had numerous overheads in terms of installing.

**Table 8.1** Advantages and limitations of existing technologies. Source: Syed Aziz Shah et el. in [3].

| Category | Technology | Advantages | Limitations |
|---|---|---|---|
| Sound-based | Ultrasonic acoustic audio [8, 9] | a. Very sensitive to motion<br>b. Objects and distances are typically determined precisely<br>c. Inexpensive (audio) | d. Work only directionally (ultrasonic)<br>e. Sensitive to temperature and angle of the target (ultrasonic)<br>f. Easily be influenced by other audio signals/noise<br>g. Prone to false detections |
| Motion-based | Accelerometer gyroscopes [10, 11] | a. No privacy issues<br>b. High detection accuracy<br>c. Lower cost | d. Raise physical discomfort issue<br>e. No direct linear or angular position information<br>f. Prone to false detection<br>g. Insensitive to very slow motions |
| Vision-based | Video camera [12, 13] | a. Effective security measure<br>b. Maintain records | c. Interfere with privacy<br>d. Ineffective in the dark<br>e. High computational cost |
| Sensor-based | Body-worn sensors [14, 15] | a. High detection accuracy<br>b. No privacy issue | c. Expensive devices (sensors)<br>d. Disturb or limit the activities of the users<br>e. Required sensors installation and calibration |

The application of RF sensing systems has been recently used for healthcare sector due to its noninvasive and noncontact nature. RF-based system devices can be used to continuously monitor patients suffering from various diseases as mentioned above. The radio frequency sensing devices use low-cost small wireless devices such as Wi-Fi signals. The radio frequency approaches include radar sensing and Wi-Fi sensing as well that radiate electromagnetic waves, where each body movement causes unique change on wireless medium. Table 8.1 (*Syed Aziz Shah et el. in [1]*) indicates the list of RF sensing systems considering remote healthcare along with their advantages and disadvantages.

## 8.3 Challenges in RF Sensing Technologies

Majority of the radio frequency sensing systems that encounter numerous challenges should be addressed in order to provide the reliable healthcare solutions. Some of them are as follows:

### 8.3.1 Robust RF System for Healthcare

The robustness is one of the major concerns faced by researchers and RF system designers that must be addressed to adequate sensing system delivery. Large number of RF sensing systems must be deployed and used as a ground truth scenario for considering a lab environment; therefore, it is very challenging in the context of the real-world applications. Moreover, the dataset acquired advance signal processing algorithms along with its applications should be applied using less computation analysis. This is specifically vital when targeting to identify important healthcare events that have a high probability of false alarms that can essentially jeopardize the deployment of RF-based technologies [16].

### 8.3.2 Reliability of RF Sensing for Remote Patient Monitoring

The reliability is also an another factor that also need to be addressed where a system can be deployed to monitoring patients using the noncontact method and examine whether it can be used for long-term vision. The RF sensing systems should also be designed in such a way that it should be easily used for monitoring elderly people. Adequate training, trials, and testing for older population are also very important for acceptance issues. The radio frequency systems for healthcare should also aid elderly in case of emergency situations, for instance, looking into vital signs.

### 8.3.3 Affordability in RF Systems for Healthcare

Numerous RF systems have been brought forward for decreasing the overall cost of wireless sensing systems for deployment purpose. On the contrary, affordability for single and multiple patients and their families is also a very challenging problem as some people might not be able to afford it. These systems must be of low cost while keeping reliability and robustness at high level.

### 8.3.4 Ethical Approval and Consent of Patients for Data Acquisition

Preserving privacy of a patient, obtaining information for security, and patient's personal data are few of the major things that should be kept in mind when laying out the radio frequency sensing systems. Although these systems do no directly take optical images such as vision-based systems, these should collect and record data so that no one can get access to it.

Another possible privacy problem faced by the patients is constant exposure to radio frequency waves day in and day out. Such kind of challenge is not usually faced by the passive radio frequency systems such as Wi-Fi systems, as it relies on very low transmitted power. Considering the active radio frequency wireless sensing systems, electromagnetic wave radiation device known as radio transmitter, majority of the existing works use low-power devices at license-free band. In this regard, careful research of possible risk of constant exposure needs to be studied in designing the RF system for healthcare applications.

## 8.4 Wireless Sensing Technologies for Healthcare Applications

This part provides a brief background into three widely used radio frequency sensing systems, namely Wi-Fi, radar, and RFID sensing systems. We will also discuss their different parameters.

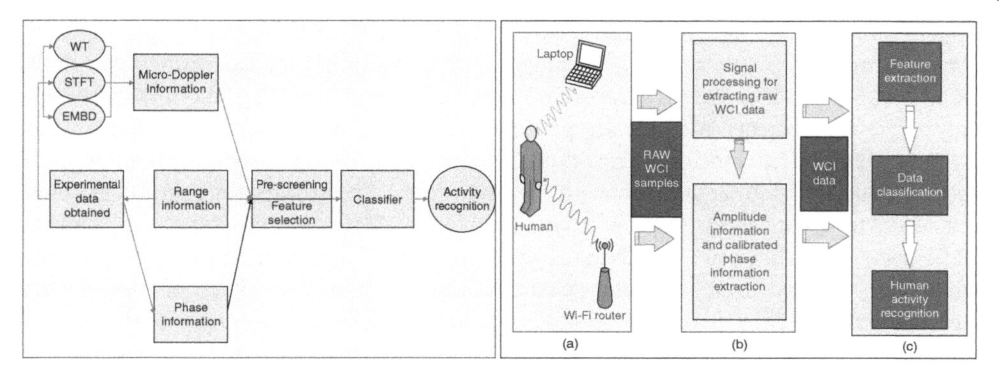


**Figure 8.1** Typical radar and Wi-Fi technologies for healthcare. (a) Data collection, (b) Data extraction, (c) Data processing. Source: Syed Aziz Shah et al. [3].

### 8.4.1 Radar Sensing Technologies for Healthcare Applications

The radio frequency sensing systems, including radar sensing approaches, consist of transceiver model and depend on the backscattered EM waves. On the contrary, the receiver along processing units operates advance signal processing to identify specific disease symptoms as indicated in Figure 8.1 (*Syed Aziz Shah et el.[3]*). The radio-frequency-transmitted waves when received back upon striking something have different aspects such as velocity, distance, and so on. The velocity of subject can be identified by exploiting micro-Doppler signatures. The radar systems can be divided into following two groups as discussed as follows:

### 8.4.2 Active Radar Sensing in Remote Healthcare

The active radar sensing systems are one type of radio frequency technique that carry two transceiver modules and are key component to active radar systems. These require particular dedicated hardware components that work at EM frequency range. Hence, the key limitations are that they cannot manipulate the transmitted RF signals. The radar sensing devices have the potential to obtain range profile when person is moving within radar range.

### 8.4.3 Passive Radar in Remote Healthcare

The passive radar systems depend on another source rather than relying on their own transmitter model. Their main plus point is that they are not very complex and no dedicated hardware is required which is expensive at times. Also, there is no need to give an adequate frequency range to them. The main limitation as far as these systems concerned is that the operating frequency cannot be changed, also the power level as well. The presence of subject within passive radar range is determined by calculating the delays occurred between third-party transmitter and receiving nodes.

The tradition passive radar sensing systems considering patient monitoring are shown in Figure 8.1a, for two types of radar sensing systems. The work function of radar sensing techniques starts with an application in healthcare and consists of the collection of data from different patients and manufacturing process. The passive radar sensing systems' data obtained for various patients are then plugged into advance signal processing algorithms using radar micro-Doppler signatures and range profile as well. The micro-Doppler signatures having subjects' motions are produced leveraging time-frequency dataset which is done using fourier transform algorithm. Classical micro-Doppler signatures are usually used to monitor various activities of daily livings to detect disease symptoms; however, some academics have used range-Doppler information as well. Following process has mathematical aspects which are usually elaborated in the context of time domain statistical features which are retrieved from raw radar sensor data. To mitigate the impact of false alarms when different machine learning and state-of-the-art machine learning algorithms are concerned, data pre-screening and radar dataset cleaning processes are performed to eliminate the noise.

## 8.5 RF Sensing Signal Processing for Patient Monitoring

The radio frequency sensing for healthcare works on the micro-Doppler signature obtained from various categories of subject that carry key imprints for signal processing. The micro-Doppler signatures that are obtained leveraging radar sensors use reflected RF signal on someone who is performing the activities of daily livings. This means that it can essentially get the speed of a person considering 2-D area. As the micro-Doppler data are obtained using the time-frequency dataset which is also known as STFT spectrograms, these minute body movements from different sources can possibly be elaborated in spectral plane as well. To obtain these particular data, the time-frequency or STFT method can be applied on the received signal strength indicator. Very small radio frequency waveform is formed in the context of overlapping small windows along their frequency range.

The STFT technique used the fast-fourier transforms using a small slide window technique that overlaps on panicle top of each to a particular percentage to form the spectrogram. On the contrary, time-frequency method comprises spectral components. The constant changes in the time-frequency method were due to the application of filtering algorithms that explain the frequency spectrum in the region of spectral components. This is important and is explained as there is a constant trade-off between the spectral and temporal components.

Figure 8.2 shows the traditions of the speed and micro-Doppler signature for the people performing activities when moving toward and far away from the sensor notes.

### 8.5.1 Feature Extraction from Single RF Sensor

For application where radar target is human subject having different sources of small-scale and large-scale body movements, the cumulative motions of hands and legs produce the time-frequency spectrograms, and it is needed to be in such a format that various activities can be classified using machine learning classifier.

Feature extraction is the process of retrieving particular properties of a given dataset that carries some patterns and is of highest value. Moreover, they are also a technique to reduce the dimensions of the data when the order of a dataset is smaller than a given input data.

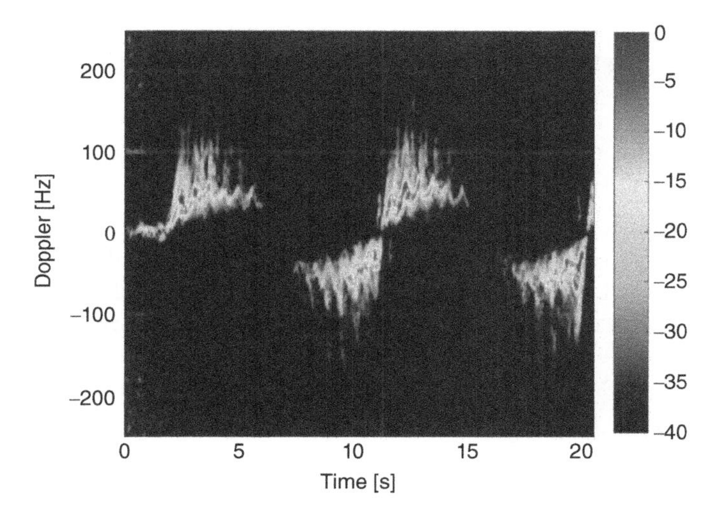


**Figure 8.2** Spectrogram of a patient walking back and forth while being observed in indoor settings.

### 8.5.2 Radar Features for Machine Learning Algorithms

Although it is humanly possible and can be visualized to transform this to a machine learning classifier, the properties of machine learning classifier need their features to be extracted from a specific dataset. In a nutshell, the variances required need to be correctly identified for each classifier.

There are two major sorts of feature extraction techniques as discussed, namely automatic and manual machine learning features. The first one was formed to get the reduction in dimensions such as principal component analysis (PCA) or SVD technique to decrease the input vector to higher ones.

### 8.5.3 Automatic Feature Selection in Machine Learning

These kinds of features are obtained using the available dataset where a maximum variance is determined for a given input machine learning algorithm. These features include PCA and independent component analysis (ICA).

PCA is used in exploratory data analysis and for making predictive models. It is commonly used for dimensionality reduction by projecting each data point only onto the first few principal components to obtain lower-dimensional data while preserving as much of the data's variation as possible. The first principal component can equivalently be defined as a direction that maximizes the variance of the projected data. On the other hand, ICA is a computational method for separating a multivariate signal into additive subcomponents. This is done by assuming that the subcomponents are non-Gaussian signals and that they are statistically independent from each other. ICA is a special case of blind source separation. A common example application is the "cocktail party problem" of listening in on one person's speech in a noisy room.

The time-frequency analysis is a tradition manual feature obtained from radar sensor. Treating the spectrogram in Figure 8.2 is an example of manual features that can directly be fed into a machine learning or deep learning classifier.

### 8.5.4 Multiple Sensor Combination

The single radar or Wi-Fi sensing can only record one type of dataset that can have several limitations. For instance, if a single senor node goes down, there is high likelihood that a particular activity will go undetected. Therefore, it is very important to use multiple radar sensor nodes so that we have several degrees of freedom. The sensor fusion is discussed as follows:

### 8.5.5 Working Function of Multiple Sensors

The sensor fusion means using two or more than two sensors for a specific purpose and is discussed as follows:

### 8.5.6 Multiple Sensor Architectures

### 8.5.7 Classification of Active Radar Sensor Node

It always depends on a data source where the data are acquired from, and also sensor fusion architecture is also very important which form a fusion network. The parameters set for sensor fusion include types of sensor used, micro-Doppler signatures, range information, and so on that provide more degree of freedom than single one. This sort of network exploits spatial diversity as presented in Ref. [17]. Sensor fusion is the combination of sensory data and the data derived from disparate sources such that the resulting information has less uncertainty than would be possible when these sources were used individually. The term uncertainty reduction in this case can mean more accurate, more complete, or more dependable, or refers to the result of an emerging view, such as stereoscopic vision (calculation of depth information by combining two-dimensional images from two cameras at slightly different viewpoints). The sensor fusion is shown in Figure 8.3. The data sources for a fusion process are not specified to originate from identical sensors. One can distinguish direct fusion, indirect fusion, and fusion of the outputs of the former two. Direct fusion is the fusion of sensor data from a set of heterogeneous or homogeneous sensors, soft sensors, and history values of sensor data, while indirect fusion uses information sources like a priori knowledge about the environment and human input. The radar sensing system data are essentially obtained using different tools that include MATLAB or sometimes Python as well where training and testing phases are performed. The data are usually divided into 70% for training and 30% for test purpose. The test is performed multiple times where several subjects are left out to put the classifier to the test.

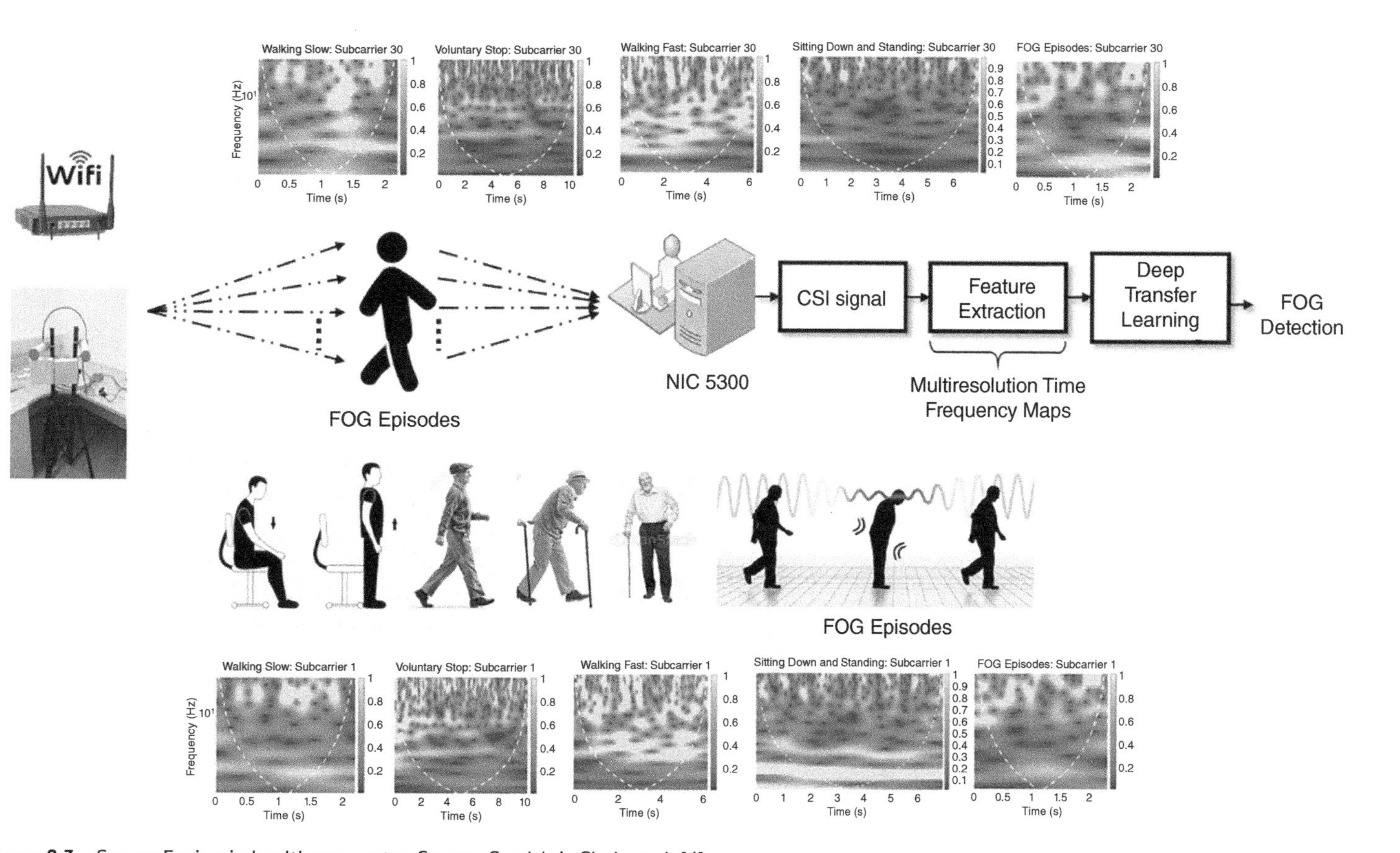


**Figure 8.3** Sensor Fusion in healthcare sector. Source: Syed Aziz Shah et al. [4].

## 8.6 Active Radar Sensing in Digital Healthcare

The radar sensors have different types of application in the context of healthcare applications. Nonstop and seamless monitoring of patient's chest and heartbeat can be done in the different fields such as security, search, rescue, and more precisely in digital healthcare. Numerous articles have been introduced covering the estimation of breathing rate and vital signs that are correlated with physiological symptoms. On the other hand, detecting, motions of torso, limbs, and so on, with high precision and accuracy are still a hard nut to crack. The variances amplitude and in-phase data are collected from micro-Doppler signature directly associated with a patients' lungs motions, and signal power change occurred when RF signal reflects back when it is hit by the patient's body. The overall RF signal reflected back due to different body parts can be easily estimated by the radar cross section (RCS). The RCS is directly related to the total area highlights by lung along with reflectivity and directivity of the radar at the operating frequency. Moreover, different sorts of human body motions on each side along vertical axes, the lung movement direction of both axes, and transceiver pair affect the RCS. Few of the applied use cases of RF sensing using radar are discussed as follows:

## 8.7 Posture Recognition on Bed

The simplest and common technique to determine posture on a bed is shown in Ref. [18]. The main work done shows various sorts of sleep postures onward the bed that exploited RCS, and it reveals that RCS for the back of the patient is estimated 15 times higher than a patient on the back side. The I-Q parts for the reflection of radar sensor are presented in Figure 8.4. In order to detect the RCS by removing the direct current values, the data are split into different segments so that each comprises a full breathing cycle. The component of the radar range radius for overall arc value was then calculated using center approximation techniques.

### 8.7.1 RF Sensing for Patients with Sleep Disorders

People suffering from sleep disease have usual abnormal sleep cycles which can eventually lead to physical and mental stress. Research has indicated various sorts of sleep diseases.

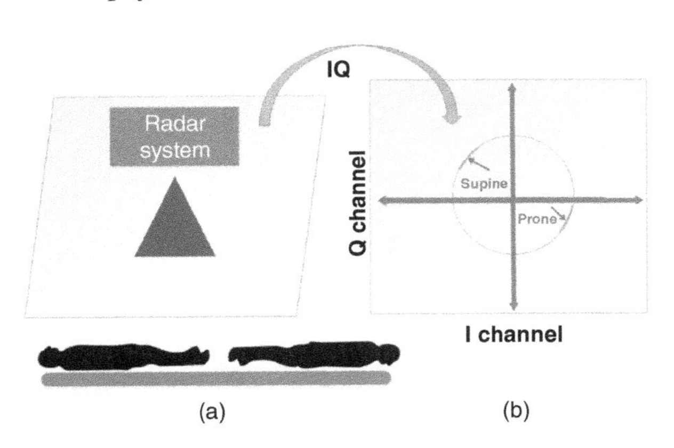


**Figure 8.4** (a) Patient' breathing position lying on back (b) and breathing identification in prone posture. Source: Modified from Kiriazi et al. [14].

Studies show that there are several kinds of sleep disorders. Polysomnography (PSG) is widely used to study and monitor patients with sleep conditions. The PSG is used to monitor muscle activity (EMG), eye movement (EOG), brain activity (EEG), and heart rates (ECG). Patients using this technology have to spend a night in a specialized lab, and all the physiological markers are obtained using a wearable sensor which is highly expensive. A noncontact physiological metric obtained using radar sensing is generally called physiological monitoring radar system (PRMS) that was launched recently. This sensing system is very cheap in price and easily deployable with detectable patient's sleep cycles. It primarily uses frequency-modulated continuous wave radar.

### 8.7.2 Radio Frequency Identification for Patients

Each patient poses a specific respiratory rate that is totally different and unique from another patient in the context of expiratory direction, lungs movements due to air influx, and tidal volume. Instead of these modalities, the collection of respiratory disorder in ventilation at the intensive care unit when the patient is at rest condition, normal respiratory rate, and abnormal one's episodes are determined in ventilator units. This is a huge variation in the condition that comprises a specific pattern and might be caused due to the central nervous system of a patient. It is quiet evident that each patient provides a particular respiratory rate between a large amount of combination of airflow. A wide number of the dataset is recorded for thousands of patients over time. The general architecture of the proposed system having training and implementation stage is shown in Figure 8.5.

### 8.7.3 Radar Sensing for Occupancy Monitoring in Healthcare Sector

Identifying single or more than one patient in indoor and outdoor settings using radar sensing systems is one of the challenging tasks researchers face. Using a different type of self-made and read-made antenna systems for radar sensing can enable caregivers to

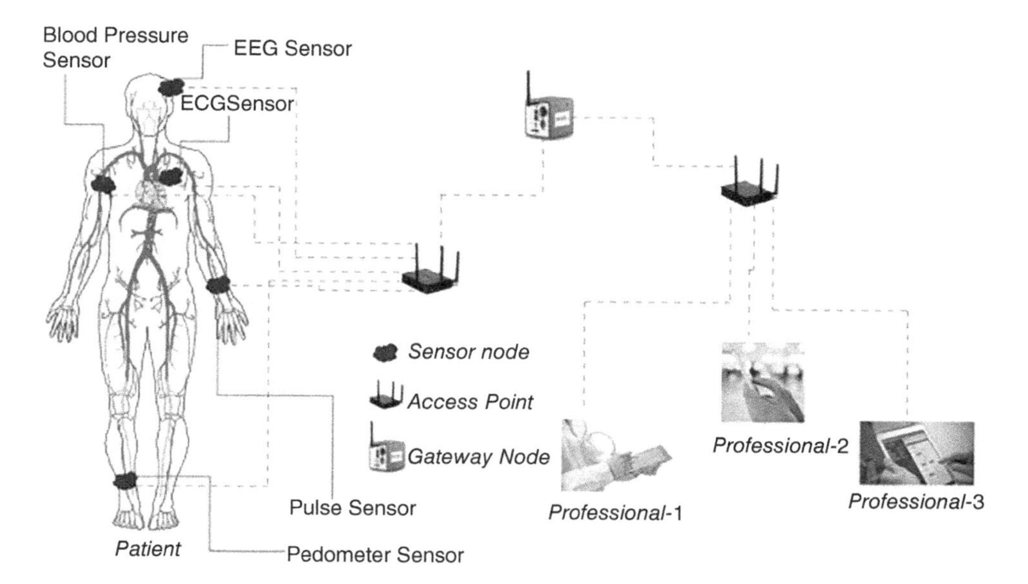


**Figure 8.5** Different stages of sensor combination as shoen in Ref. [6].

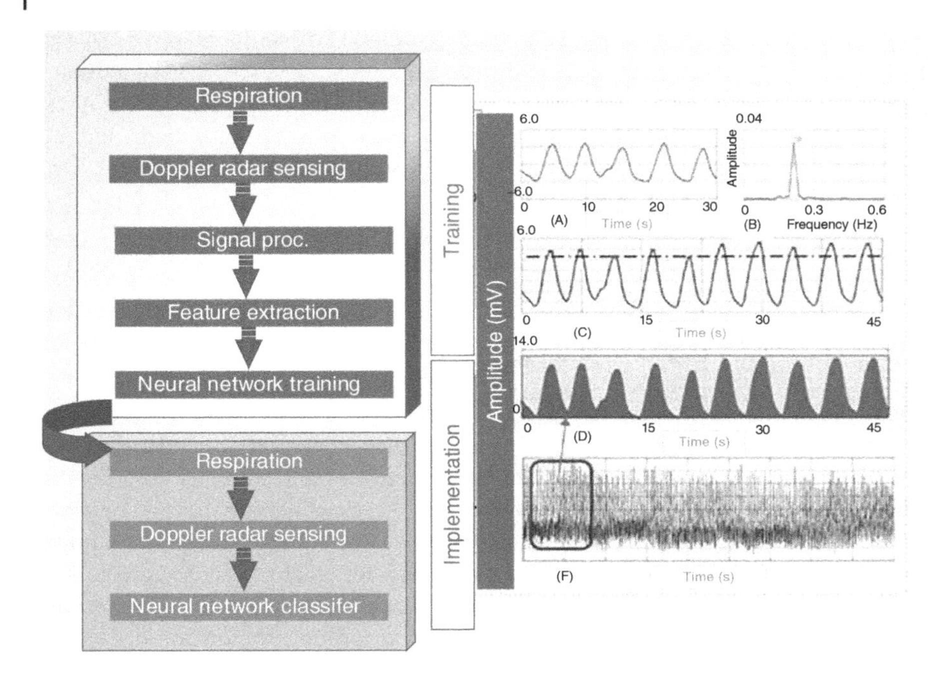


**Figure 8.6** General architecture for identifying people based on vital signs. Source: Syed Aziz Shah et al. [3].

determine a total number of patients and identify them exploiting various cardiovascular signature using micro-Doppler information. The occupancy detection sensing system using radar technology is only brought into reality when we examine the patient breathing and heart rate. By exploiting the angle of arrival of the RF signal, heart movement sequence can be identified as multiple antennas and a single antenna. Calculating the AoA of RF signals can aid in distinguishing and identifying patients' space and time in hospitals or care centers. Radar sensing system operating at 2.4 GHz using micro-Doppler signature exploiting coaxial parts of a leaky-wave cable has the potential to determine patients within radar range. The main research point that it produced was an accurate solution to tune the RF sensing device to be turned on and off automatically. This work saved much energy for long-term patient monitoring for occupancy purposes. Figure 8.6 shows general architecture for vital signs detection.

### 8.7.4 Activities of Daily Livings and Critical Events

Human activity identification and walking or gait approximation for a different kind of disease symptom can bring improvement in patients' life in terms of leading independent lives easily. The radar sensor using an RF signal can be exploited to determine various gait of patients suffering from different diseases including Parkinson's disease and dementia. This is done using Doppler and range information extracted from the radar sensor. This potential allows to estimate activities of daily living and to identify gait abnormalities such

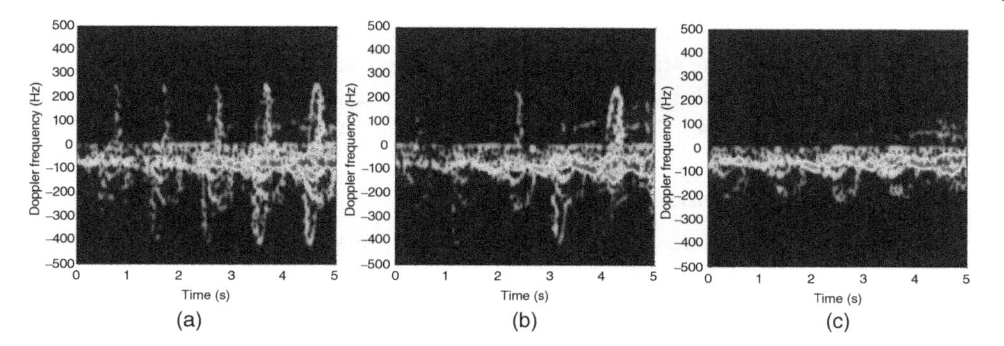


**Figure 8.7** Spectrograms of human gait: (a) walking with no handbag, (b) walking and holding handbag with one hand, and (c) walking and holding handbag with two hands. Source: Syed Aziz Shah et al. [3].

as walking back and forth, sitting on a chair, picking up the object, falling down, and so on. This is directly related to human actions and motions that can potentially be integrated to determine small activities such as doing work in the kitchen, brushing teeth, and other related activities. RF sensor fusion-based system can be used to demonstrate different human activities using radar sensing technologies along invasive wearable equipment to improve the overall accuracy. An FMCW radar that operates at 5.8 GHz frequency range is widely used to detect human activities.

Figure 8.7a shows the time-frequency images represented as spectrogram for the subject carrying nothing in hand. The two hands of the subject move along each other but in another direction as the person walks in indoor settings. As the subject was holding the handbag in his hands, one arm went back and forth, and the other arm was still due to holding the stuff. Hence, the peak values of positive and negative signature appeared alternately in the micro-Doppler as shown in Figure 8.7b. As the subject was walking within radar range by holding the bag with two hands rather than one, the huge portion of RF signal came from legs and torso motions as shown in Figure 8.7c. This RF radar dataset was collected and then plugged in into the support vector machine classifier to get the accuracy and precision performance metrics for all three subject walking patterns.

### 8.7.5 Noncontact Wi-Fi Sensing for Patient Monitoring

The presence of a patient doing movements in a certain area where moving within an area where Wi-Fi signals are present, they keep on disturbing wireless channel and produce a unique shift in frequency which is received by mobile phone or a computer. Each patient's body movement produced a certain change known as CSI. The CSI is exploited to show the radio frequency signal continuous propagation consisting of distortion, scattering, and reflection that formed by human motions. These technology and technique have created opportunities for biomedical and healthcare technology experts to develop a low-cost healthcare solution for monitoring patients with different diseases. The classical method includes RSSI recorded from orthogonal frequency division multiplexing RF signals to determine the activities of daily living and monitor particular disease symptoms related to body motions. However, the RSSI method does not provide an adequate result when monitoring chest movement and heartbeat.

On the contrary, CSI can estimate body motions below the centimeter level by analyzing various subcarriers. The process and procedure of data collection for CSI are similar to that of radar-based RF sensing systems. Numerous Wi-Fi routers manufactured by different companies allow us to record CSI when installed on a computer or laptop.

### 8.7.6 Wi-Fi-Based Activities of Daily Livings

A noncontact, least expensive person human activity detection using RF sensing was put forward recently known as the E-eyes. This RF sensing system used CSI collected from 30 OFDM subfrequency channels inside the room. The wireless sensing system delivered feasible and good results where the small number of wireless devices were used. Compared to the hand detection system used in RF sensing, the human activity recognition system is not strictly defined over a period of time, generally known as large-scale body motions. The post-processing of this system uses a smooth filtering technique at high RF frequencies while removing noise figure in a noisy environment. This system has two key parts: the first one is the formation of human activity recognition raw system and the second one is the data split for variances of amplitude information. The experimental campaign was done inside a large room and provided high classification. Something similar to this work, that is, gait identification work, was also introduced using CSI sensing where there are different wandering behaviors in dementia. The three walking patterns including random wandering, lapping wandering, and pacing wandering are presented in Figure 8.8a. The experiment was performed in a 12-by-12 m room at Xidian University, China, as shown in Figure 8.8b.

### 8.7.7 Vital Signs Monitoring using Wi-Fi Signals

The respiratory rate due to the lung movement occurs between 0.5 and 2.0 cm. This minute lung movement will cause a 0.25–1.0 cm phase variation at signal operating at S-Band. This enables us to exploit CSI obtained from Wi-Fi router to deliver noncontact monitoring system. The experimental data acquisition setup and outcomes indicate that the amplitude and phase information consisting of breathing waveforms can easily be extracted using this technology and are shown in Figure 8.9. The main goal of this filer is to mitigate unwanted

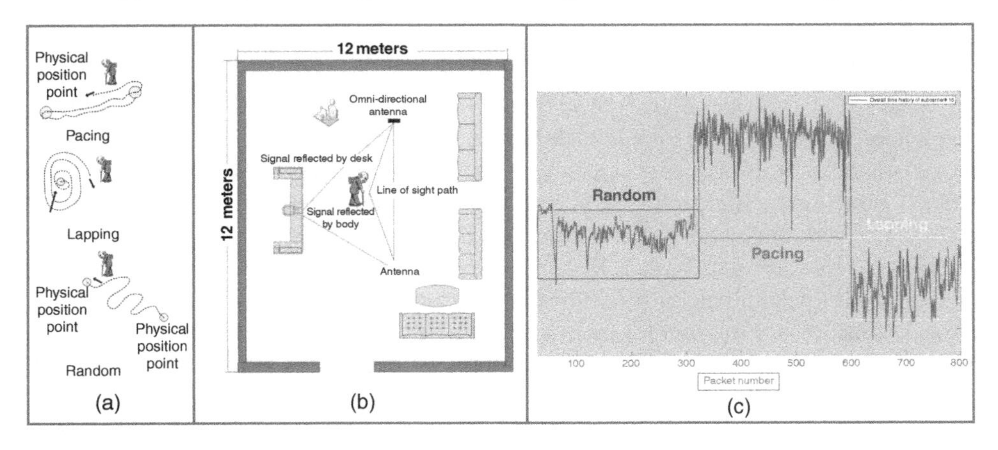


**Figure 8.8** A sketch for three wandering behaviors – (a) typical scenario for random, pacing, and lapping gaits, (b) experimental setup, and (c) time history of three gaits considering amplitude information. Source: Syed Aziz Shah et al. [3].

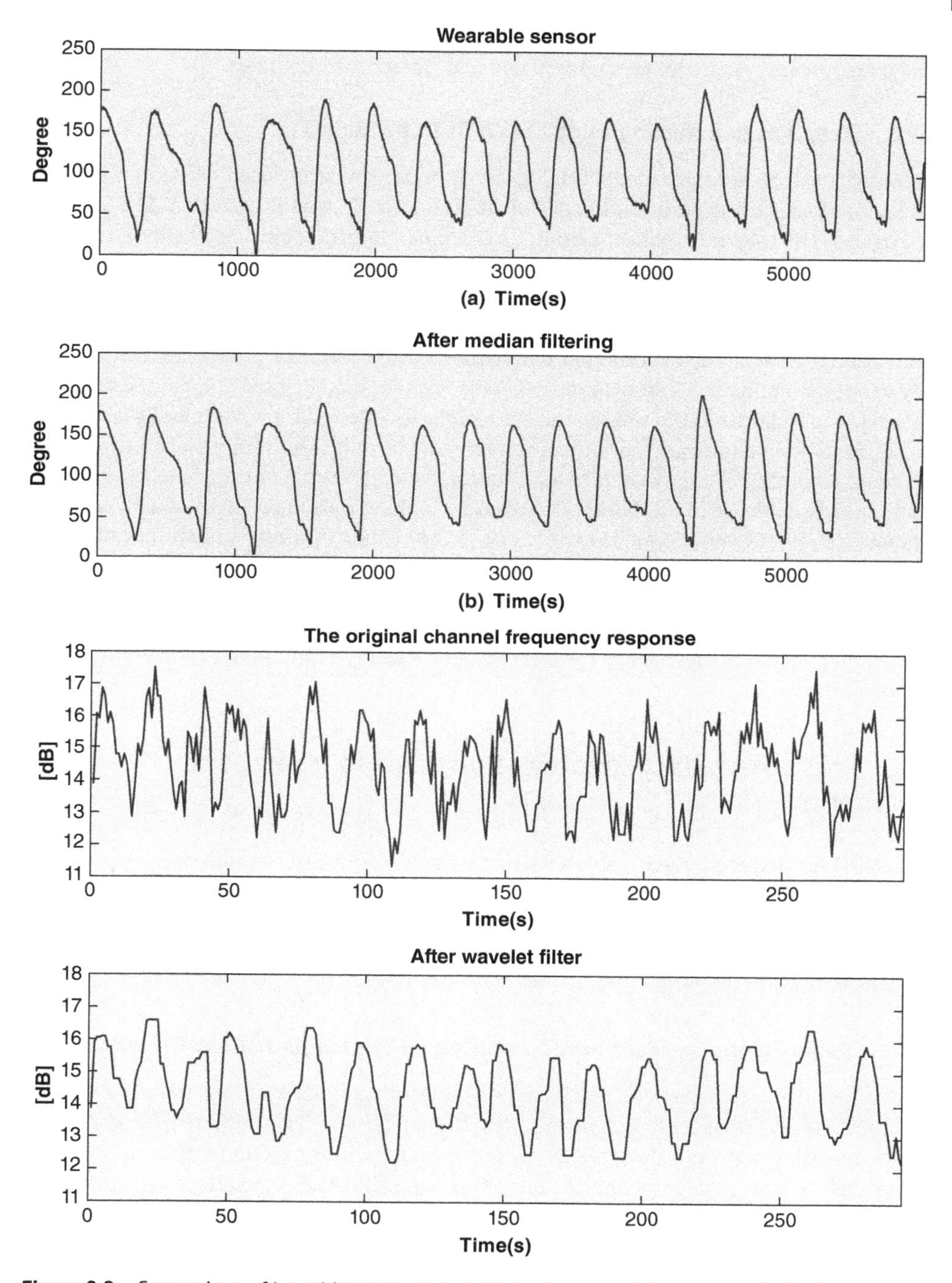


**Figure 8.9** Comparison of breathing pattern recorded using wearable sensor and wireless channel information. Source: Syed Aziz Shah et al. [3].

noise. The researchers have named this sensing system as S-band sensing system. The proposed healthcare system delivers normal and abnormal breathing rates.

### 8.7.8 Sleep Attack Detection using CSI Wi-Fi Technologies

Recently, network interface model Intel 5300 has been extensively used for various healthcare applications using perturbation of CSI amplitude and phase information. The CSI primarily uses OFDM data to collect amplitude and phase information. Specifically, it exploits the OFDM technique to get variances of amplitude and phase data in the form of subcarrier index. The main advantage of leveraging CSI collected from Wi-Fi signals over the RSSI technique is that single or more than that can be used for different healthcare applications. The host personal laptop or computer where the network interface card is installed is usually wirelessly connected to the ubiquitous Wi-Fi router and recorded CSI data in terms of packets. The instantized CSI phase information phase collected from commodity devices is mostly random in nature and not suitable for any activity or disease symptom detection. The seamless recording of channels' rate information data represented in terms of time allows us to analyze patients in a different direction by looking into their angle of arrival of RF signals from Wi-Fi router. That is mainly due to the change in wireless medium caused by the person moving over a period of time. When a patient stands stationary within wireless range, the minute person's body movement from various body parts also produced a small change that is detected by amplitude and phase information. For instance, classical activities of daily living are formed of various body parts leading to unique CSI imprint over a period of time.

## 8.8 Radio Frequency Identification Sensing for Patient Monitoring

The RFID sensing technique is also a leading solution for healthcare purposes.

The RFID technology determines patients via tags put on a person's body. Hence, there is not required of both transmitter tag and receiver tag to be in the line of sight to each other. This sensing solution only depends on the wireless medium between tag and reader and consists of three main parts.

### 8.8.1 Radio Frequency Identification Sensing for Patient for Patient Tracking

The supply chain management, detection and monitoring of patients in digital, and remote healthcare sector along with their inventories are few of the applications of RFID. The RFID wireless sensing method is presently going through an evolutionary and transitionary state, where before it was only applied for identified tasks, but now considered as alternative technology in the healthcare sector.

### 8.8.2 Radio Frequency Identification Sensing for Patient for Disease Detection

The third and most common application of RF sensing for healthcare application is that of the radio frequency identification used along other sorts of sensors to monitor patients for

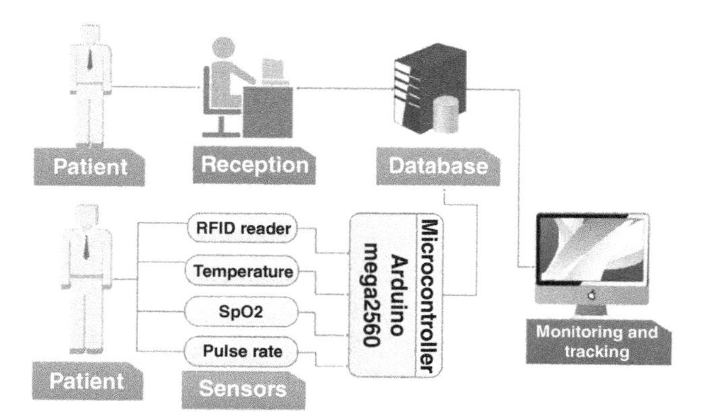


**Figure 8.10** Architecture of RFID-based tracking system. Source: Syed Aziz Shah et al. [3].

different purposes. Example includes implantable RFID sensing devices used in portable healthcare services [19]. The RFID wireless sensing system in conjunction with mobile communication networks can aid in identifying critical events such as abnormal breathing, rate, asthma, falls, and so on. The radio frequency identification enabled monitoring and detection system was recently introduced in Ref. [19]. Furthermore, Friesen et al [19] developed a smart system driven by smart watches using RFID technology to the observed patient after the operation being performed at care homes or hospitals. An article report [20] shows that the RFID sensing system for the healthcare sector and remote patient monitoring are at infant stages. However, there has been evidence that academics and researchers have paved the way for print ink on patient skin to monitor patients remotely. In addition, an RFID wireless sensing for heartbeat and respiratory monitoring system was introduced by Omar et al.[21]. The main objective of this smart system was to develop and introduce a low-cost and efficient vital sign monitoring system that can detect oxygen level and temperature, and track patients. This wireless system used microcontroller chip which is famously known as Arduino Mega 2560 to record data from sensors and transmit it through the Wi-Fi network as indicated in the following Figure 8.10.

Vora et al. [21] developed a smart vital signs detection system driven by RFID that has the potential to monitor abnormal respiratory rate. This research demonstrated that the detection of two critical and abnormal scenarios such as sleep apnea and bradycardia could only be brought in fruition leveraging RFID technology. It was showed a similarity of 99% when compared to the organic heart. The sleep apnea critical event that has a minimum of 15 seconds was identified. In addition, another work was introduced in Ref. [22] that used micro-sensing chips driven by RFID tags to identify heartbeat, the oxygen level in blood, and temperature. The main objective was to decrease the risk of life-threatening events in newly born babies. These RFID sensing devices were wirelessly connected to a centralized computer to notify the doctor or nurse for immediate intervention.

### 8.8.3 Radio Frequency Identification Sensing for Patient to Identify Falls

Critical event such as falls is one of the most life-threatening events encountered and experienced by elderly people at homes, care centers, and hospitals and has received considerable attention. Numerous wireless sensing devices have been designed and launched in the remote healthcare sector to observe fall events. The researchers who introduced work in Ref [23] provided a smart fall detector using RFID technology consisting of various readers and RFID sensor tags. This wireless sensing system was used in a way that nurse can get message and alerts in case of elderly person falls. The proposed sensing system was designed on the basis of variances of power level for various individuals and presented no results in terms of numbers. The second fall detector which was introduced as a prototype also used RFID technology where it was implemented using a movement detecting sensor. Another fall detection prototype was based on RFID technology where the design and implementation of a motion detection system were based on RFID tags termed as TagCare that was introduced in [24]. The key finding of this research work was micro-Doppler information that was used to identify activities of daily living and detect fall. A large amount of experiments were conducted to show the validity of TagCare. It was demonstrated that this technology shows promising results of up to 98%.

### 8.9.4 Radio Frequency Identification Sensing for Patient for Intricate Body Movement Observations

Breathing rate and heart rate monitoring are the categories of patients' vital signs group. Various RFID-based systems have been introduced to monitor infant vital signs such as the work in Refs [21] and [25] to identify potential life-threatening disease. This vital sign monitoring system can be deployed in home or hospital setup and can be integrated with another system to monitoring symptoms such as blood pressure and so on. To summarize, the RF sensing systems have been used in works such as in Refs [26–39].

## References

1 Haider, D., Shah, S.A., Shah, S.I., and Iftikhar, U. (2017). MIMO network and the alamouti, STBC (Space Time Block Coding). 5 (1): 23–27. https://doi.org/10.12691/ajeee-5-1-4.

2 Yang, X., Shah, S.A., Ren, A. et al. (2019). Freezing of gait detection considering leaky wave cable. *IEEE Transactions on Antennas and Propagation* 67 (1): 554–561. https://doi.org/10.1109/TAP.2018.2878081.

3 Shah, S.A. and Fioranelli, F. (1 Nov. 2019). RF Sensing Technologies for Assisted Daily Living in Healthcare: A Comprehensive Review. *IEEE Aerospace and Electronic Systems Magazine* 34 (11): 26–44. https://doi.org/10.1109/MAES.2019.2933971.

4 Toda, K. and Shinomiya, N. (2018). Fall detection system for the elderly using RFID tags with sensing capability. *2018 IEEE 7th Global Conference on Consumer Electronics (GCCE)*, pp. 475–478, doi: https://doi.org/10.1109/GCCE.2018.8574720.

5 Shah, S.A. et al. (1 Dec 2020). Sensor Fusion for Identification of Freezing of Gait Episodes Using Wi-Fi and Radar Imaging. *IEEE Sensors Journal* 20 (23): 14410–14422. https://doi.org/10.1109/JSEN.2020.3004767.

6 Hornyak, R., Lewis, M., and Sankaranarayan, B. (2015). Radio frequency identification-enabled capabilities in a healthcare context: an exploratory study. *Health Informatics Journal* 22 (3): 562–578. https://doi.org/10.1177/1460458215572923.

7 Shah, S., Ren, A., Fan, D. et al. (2018). Internet of things for sensing: a case study in the healthcare system. *Applied Sciences* 8 (4): 508. https://doi.org/10.3390/app8040508.

8 Rahman, A., Yavari, E., Lubecke, V.M., and Lubecke, O. (2016). Noncontact Doppler radar unique identification system using neural network classifier on life signs. *Proceedings of IEEE Topical Conference on Biomedical Wireless Technologies, Networks, and Sensing Systems (BioWireleSS)*, Texas, USA, pp. 46–48.

9 Høst-Madsen, A., Petrochilos, N., Boric-Lubecke, O. et al. (2008). Signal processing methods for Doppler radar heart rate monitoring. In: *Signal Processing Techniques for Knowledge Extraction and Information Fusion* (eds. D. Mandic, M. Golz, A. Kuh, et al.), 121–140. Boston, MA: Springer https://doi.org/10.1007/978-0-387-74367-7_7.

10 Massagram, W., Hafner, N.M., Park, B.K. et al. (2007). Feasibility of heart rate variability measurement from quadrature Doppler radar using arctangent demodulation with DC offset compensation. *Proceedings of 29th Annual International Conference of the IEEE Engineering in Medicine and Biology Society*, Lyon, France, pp. 1643–1646.

11 Droitcour, A.D. Seto, T.B., Park, B.-K. et al. (2009). Non-contact respiratory rate measurement validation for hospitalized patients. *Proceedingsof the Annual International Conference of the IEEE Engineering in Medicine and Biology Society*, Minneapolis, MN, USA, pp. 4812–4815.

12 Zhao, H., Hong, H., Sun, L. et al. (2017). Noncontact physiological dynamics detection using lowpower digital-IF Doppler radar. *IEEE Transactions on Instrumentation and Measurement* 66 (7): 1780–1788.

13 Zhao, H., Hong, H., Miao, D. et al. (2019). A noncontact breathing disorder recognition system using 2.4-GHz digital-IF Doppler radar. *IEEE Journal of Biomedical and Health Informatics* 23 (1): 208–217. https://doi.org/10.1109/JBHI.2018.2817258.

14 Kiriazi, J.E., Boric-Lubecke, O., and Lubecke, V.M. (2009). Radar cross section of human cardiopulmonary activityfor recumbent subject. *Proceedings of 2009 Annual International Conference of the IEEE Engineering in Medicine and Biology Society*, Minneapolis, MN, USA, pp. 4808–4811.

15 Baboli, M., Singh, A., Soll, B. et al. (2015). Good night: sleep monitoring using a physiological radar monitoring system integrated with a polysomnography system. *IEEE Microwave Magazine* 16 (6): 34–41.

16 Shah, F. and Fioranelli, S.A. (2019). Human activity recognition: preliminary results for dataset portability using FMCW radar. *2019 International Radar Conference (RADAR)*, TOULON, France, 2019, pp. 1–4. http://eprints.gla.ac.uk/184463/.

17 Chetty, K., Chen, Q., Ritchie, M., and Woodbridge, K. (2017). A low-cost through-the-wall FMCW radar for stand-off operation and activity detection. *Radar Sensor Technology XXI*, vol. 10188, p. 1018808, doi: https://doi.org/10.1117/12.2261680.

18 Shah, S.A., Zhao, N., Ren, A. et al. (2016). Posture recognition to prevent bedsores for multiple patients using leaking coaxial cable. *IEEE Access*, vol. 4, doi: https://doi.org/10.1109/ACCESS.2016.2628048.

19 Coustasse, A., Cunningham, B., Deslich, S. et al. (2015). Management of RFID systems in hospital transfusion services. *Business and Health Administration Association Annual Conference*, Chicago, IL, pp. 1–16. http://mds.marshall.edu/mgmt_faculty.

20 Ajami, S. and Rajabzadeh, A. (2013). Radio Frequency Identification (RFID) technology and patient safety. *Journal of Research in Medical Sciences: The Official Journal of*

*Isfahan University of Medical Sciences* 18 (9): 809–813. Available: https://www.ncbi.nlm.nih.gov/pubmed/24381626.

**21** Omar, H.Q., Khoshnaw, A., and Monnet, W. (2016). Smart patient management, monitoring and tracking system using radio-frequency identification (RFID) technology. *2016 IEEE EMBS Conference on Biomedical Engineering and Sciences (IECBES)*, pp. 40–45, doi: https://doi.org/10.1109/IECBES.2016.7843411.

**22** Yao, W., Chu, C.H., and Li, Z. (2012). The adoption and implementation of RFID technologies in healthcare: a literature review. *Journal of Medical Systems* 36 (6): 3507–3525. https://doi.org/10.1007/s10916-011-9789-8.

**23** Punin, C., Barzallo, B., Clotet, R. et al. (2019). A non-invasive medical device for parkinson's patients with episodes of freezing of gait. *Sensors (Switzerland)* 19 (3) https://doi.org/10.3390/s19030737.

**24** Chen, Y.C. and Lin, Y.W. (2010). Indoor RFID gait monitoring system for fall detection. *2010 2nd International Symposium on Aware Computing (ISAC 2010)*, pp. 207–212, doi: https://doi.org/10.1109/ISAC.2010.5670478.

**25** Vora, S.A., Mongan, W.M., Anday, E.K. et al. (2017). On implementing an unconventional infant vital signs monitor with passive RFID tags. *2017 IEEE International Conference on RFID (RFID)*, pp. 47–53, doi: https://doi.org/10.1109/RFID.2017.7945586.

**26** Lermthong, S., Suwanna, P., and Airphaiboon, S. (2017). Bedside patient monitoring by NFC. *BMEiCON 2016 - 9th Biomedical Engineering International Conference*, pp. 1–4, doi: https://doi.org/10.1109/BMEiCON.2016.7859641.

**27** Yang, X., Shah, S.A., Ren, A. et al. (2018). Monitoring of patients suffering from REM sleep behavior disorder. *IEEE Journal of Electromagnetics, RF and Microwaves in Medicine and Biology* 2 (2): 138–143. https://doi.org/10.1109/JERM.2018.2827705.

**28** Yang, X., Shah, S.A., Ren, A. et al. (2018). Detection of essential tremor at the $S$ -band. *IEEE Journal of Translational Engineering in Health and Medicine* 6: 1, 2000107–7. https://doi.org/10.1109/JTEHM.2017.2789298.

**29** Fan, D., Ren, A., Zhao, N. et al. (2018). Breathing rhythm analysis in body centric networks. *IEEE Access*, vol. 6, pp. 32507–32513, doi: https://doi.org/10.1109/ACCESS.2018.2846605.

**30** Khan, M.B., Yang, X., Ren, A. et al. Design of software defined radios based platform for activity recognition. *IEEE Access*, vol. 7, pp. 31083–31088, 2019, doi: https://doi.org/10.1109/ACCESS.2019.2902267.

**31** Shah, S.A., Zhang, Z., Ren, A. et al. (2017). Buried object sensing considering curved pipeline. *IEEE Antennas and Wireless Propagation Letters* 16: 2771–2775. https://doi.org/10.1109/LAWP.2017.2745501.

**32** Yang, X., Fan, D., Ren, A. et al. (2020). Diagnosis of the Hypopnea syndrome in the early stage. *Neural Computing and Applications* 32: 855–866. https://doi.org/10.1007/s00521-019-04037-8.

**33** Haider, D., Ren, A., Fan, D. et al. (2018). Utilizing a 5G spectrum for health care to detect the tremors and breathing activity for multiple sclerosis. *Transactions on Emerging Telecommunications Technologies* 29: e3454. https://doi.org/10.1002/ett.3454.

**34** Tanoli, S.A.K., Rehman, M., Khan, M.B. et al. (1983). An experimental channel capacity analysis of cooperative networks using Universal Software Radio Peripheral (USRP). *Sustainability* 2018: 10.

**35** Liu, L., Shah, S.A., Zhao, G., and Yang, X. (2018). Respiration symptoms monitoring in body area networks. *Applied Sciences* 8: 568.

**36** Dong, B., Ren, A., Shah, S.A. et al. (2017). Monitoring of atopic dermatitis using leaky coaxial cable. *Healthcare Technology Letters* 4 (6): 244–248. https://doi.org/10.1049/htl.2017.0021.

**37** Yang, X., Shah, S.A., Ren, A. et al. (2019). $S$ -band sensing-based motion assessment framework for cerebellar dysfunction patients. *IEEE Sensors Journal* 19 (19): 8460–8467. https://doi.org/10.1109/JSEN.2018.2861906.

**38** Zhang, Q., Haider, D., Wang, W. et al. (2018). Chronic obstructive pulmonary disease warning in the approximate ward environment. *Applied Sciences* 8: 1915.

**39** Masood, F., Ahmad, J., Shah, S.A. et al. (2020). A novel hybrid secure image encryption based on Julia set of fractals and 3D lorenz chaotic map. *Entropy* 22: 274.

# 9

# Electromagnetic Wave Manipulation with Metamaterials and Metasurfaces for Future Communication Technologies

*Muhammad Qasim Mehmood*[1], *Junsuk Rho*[2,3], *and Muhammad Zubair*[1]

[1] *NanoTech Lab, Electrical Engineering Department, Information Technology University (ITU) of the Punjab, Lahore, Pakistan*
[2] *Department of Mechanical Engineering, Pohang University of Science and Technology (POSTECH), Pohang, South Korea*
[3] *Department of Chemical Engineering, Pohang University of Science and Technology (POSTECH), Pohang, South Korea*

## 9.1 Introduction

Electromagnetic (EM) waves can be expressed mathematically by the expression of time-varying electric field, $\mathbf{E}(x, y, z, t) = \hat{\mathbf{P}}F(x, y, t)|E(x, t)|e^{k_0kz}e^{-ik_0nz}e^{iwt}$, as shown in Figure 9.1, where $\hat{\mathbf{P}}$ is the polarization vector, $F(x, y, t)$ is the spatial structure, $|E(x, t)|e^{k_0kz}$ is related to amplitude, $e^{-ik_0nz}$ is related to phase, $k$ is the imaginary part of refractive index, $n$ is the real part of refractive index, $k_0$ is the free-space wave number, $w$ is the angular frequency, and $t$ is time [1]. EM waves can be manipulated to achieve desired functionality by varying one or more physical dimensions of EM waves including wavelength or frequency, time, amplitude, phase polarization, and spatial structure. When an EM wave passes through a medium or material, some of its properties are altered. However, to take a full control of EM wave properties one can use artificially engineered materials known as metamaterials. The most important material properties affecting the propagation, radiation, or scattering of an EM wave are its permittivity $\epsilon$ and permeability $\mu$. Figure 9.2 illustrates the parameter space for $\epsilon$ and $\mu$ and some typical examples of materials for each category [2]. There are four possible sign combinations of ($\epsilon$, $\mu$) pair. Most natural materials lie in the three quadrants with signs (+,+), (+,−), or (−,+). However, the combination (−,−) with simultaneously negative permittivity and permeability corresponds to a class of materials known as "Left-Handed" materials that do not exist in nature and clearly fall in the category of EM metamaterials. EM metamaterials can be designed using artificial structures made of natural materials, whose array configuration provides unusual properties. Figure 9.3 shows a schematic illustration that natural materials are made of lattice arrangement of *atoms* ($p <<< \lambda$, where $p$ is the structural average size and

*Backscattering and RF Sensing for Future Wireless Communication*, First Edition.
Edited by Qammer H. Abbasi, Hasan T. Abbas, Akram Alomainy, and Muhammad Ali Imran.

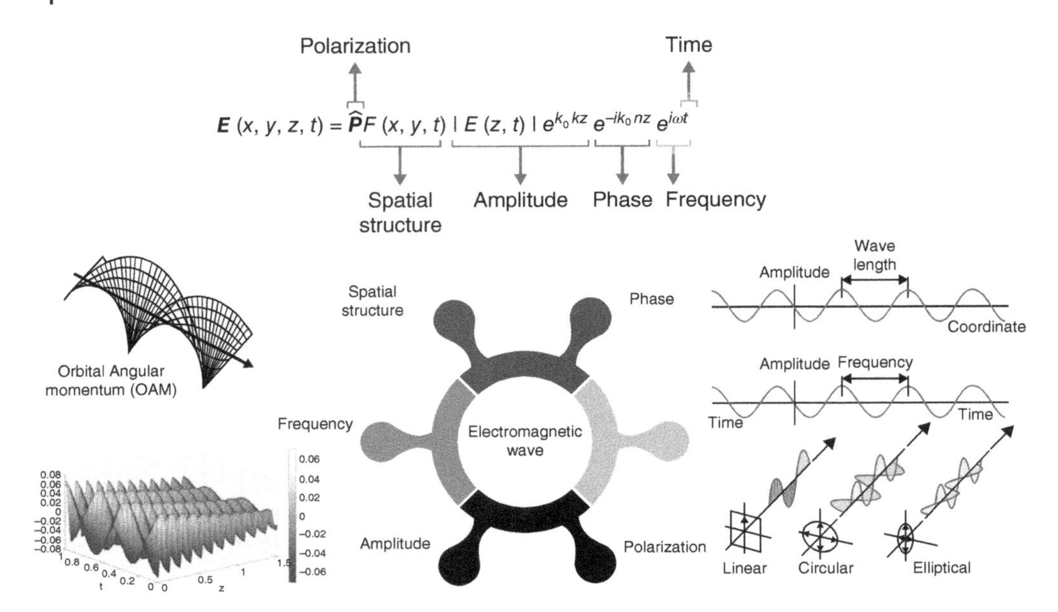


Figure 9.1 Schematic illustration of various dimensions of electromagnetic waves.

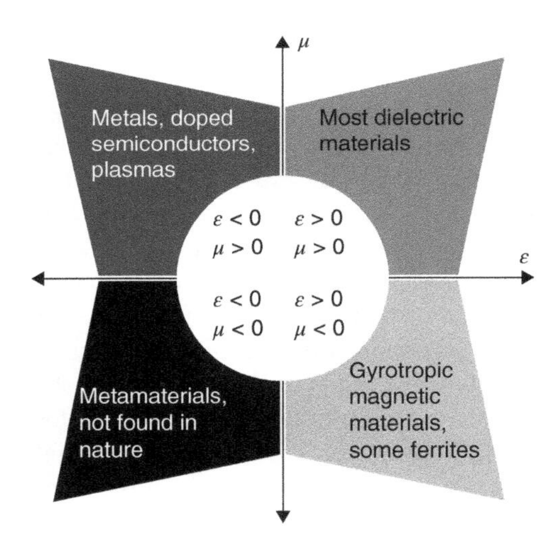


**Figure 9.2** Illustration of the parameter space for permittivity $\epsilon$ and permeability $\mu$ and the typical examples for each quadrant.

$\lambda$ is the guided wavelength), while *meta-atoms* are subwavelength geometric structures ($p << \lambda$) of natural materials and can be arranged in three-dimensional (3D) arrays to form *metamaterials* or in two-dimensional (2D) arrays to form *metasurfaces*.

Metamaterials and metasurfaces can be classified into different categories based on the frequency of operation and type of materials used to design primary meta-atoms. The meta-atoms have been designed using metals, dielectrics, two-dimensional (2D) materials like graphene, etc. The physical sizes of meta-atoms vary from micro- to nano-scale depending on the desired frequency of operation ranging from microwave to optical regime. The fabrication techniques of micro-scale metamaterials or metasurfaces include

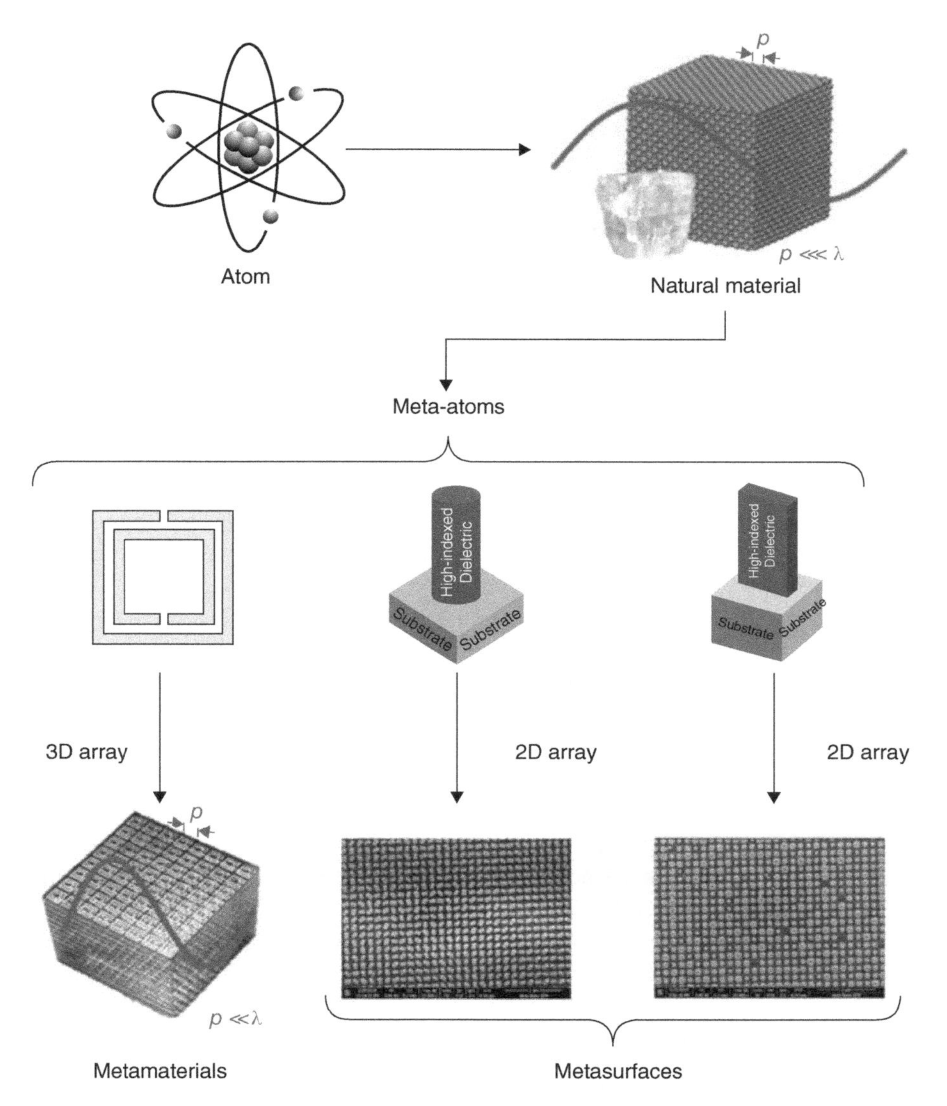


**Figure 9.3** Schematic illustration of atom, natural material, meta?atom, 2D array of meta-atoms called metasurface, and 3D array of meta-atoms called metamaterial.

printed-circuit board (PCB) technology [3], screen printing [4], inkjet printing [5, 6], and 3D printing technologies [7]. For nano-scale fabrications, one can use electron-beam lithography [8], focused ion beam [9], interference lithography [10], and nanoimprint lithography [11]. The EM properties of these periodic or aperiodic arrays of meta-atoms can be studied using computational EM methods like method of moments (MOM) [12, 13], finite difference time domain (FDTD) [14], and finite element method (FEM) [15]. There have been recent theoretical investigations for the description of metamaterials using homogenization theory [16]. Recently proposed fractional electrodynamics

[1, 17, 18, 19, 20]-based approaches can also be extended for the study of fast solutions of electrically large geometries. There are numerous novel phenomena that can occur in metasurfaces including novel ways of EM wave manipulation, anomalous reflection and refraction, enhanced or suppressed backscattering, nonlinear effects, enhanced plasmonic effects (metals), engineered dielectric resonances (dielectrics), tunability, reconfigurability, and strong chiral properties. In the microwave regime, metamaterial- or metasurface-based devices can be used to design efficient antennas, RFIDs, invisibility cloaks, frequency selective surfaces (FSS), and sensors [21]. In terahertz (THz) and optical regime, several applications have been demonstrated which include retro-reflectors for visible backscattering communication (BackCom) [22], flat optical elements [23], absorbers [24, 25, 26], filters [27], modulators, mirrors, optical vortex generators [28, 29], beam splitters, and holograms [30, 31, 32]. Figure 9.4 shows a schematic summary of metamaterials and metasurfaces classification, fabrication, novel phenomena arising from them and their important applications.

In the context of communication technologies, metamaterials and metasurfaces are future candidates for the implementation of backscattering communication systems in the microwave to millimeter and optical regime. There have been recent proposals to utilize a transformative wireless concept that will be capable of sensing and controlling the radio environment by coating objects with the reconfigurable metasurfaces [33]. This type of smart radio environments can implement future wireless communication systems with uninterrupted connectivity without generating new EM signals but recycling existing (ambient) radio waves in the environment [34]. Researchers have proposed two types of intelligent reconfigurable metasurfaces applied to wireless networks. The first type of metasurfaces can be embedded into, e.g. building walls, and the wireless network operators can control their functionality via a software controller with a goal to improving the network coverage [35] or energy efficiency [36], etc. The second type of metasurfaces can be embedded into objects, e.g. smart dress with sensors for patient health monitoring and will be able to backscatter the EM waves generated by cellular base stations to communicate their sensed data to mobile phones. These novel functionalities can enable

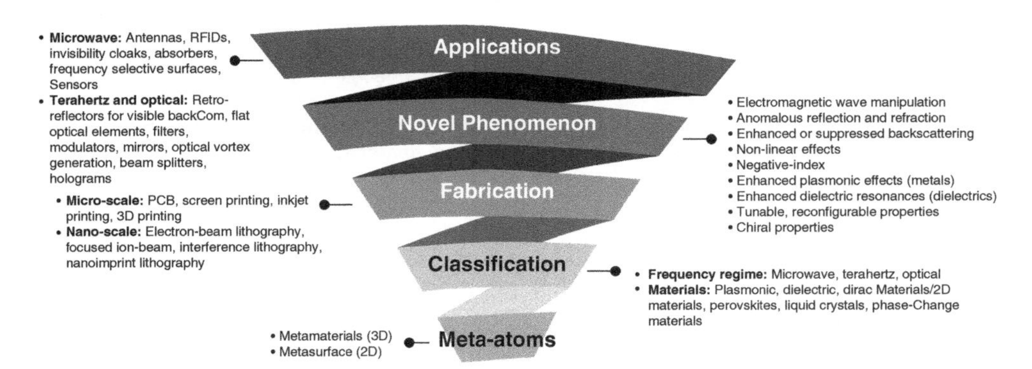


**Figure 9.4** Schematic illustration of metamaterials and metasurfaces classification, fabrication, novel phenomena arising from them and their applications.

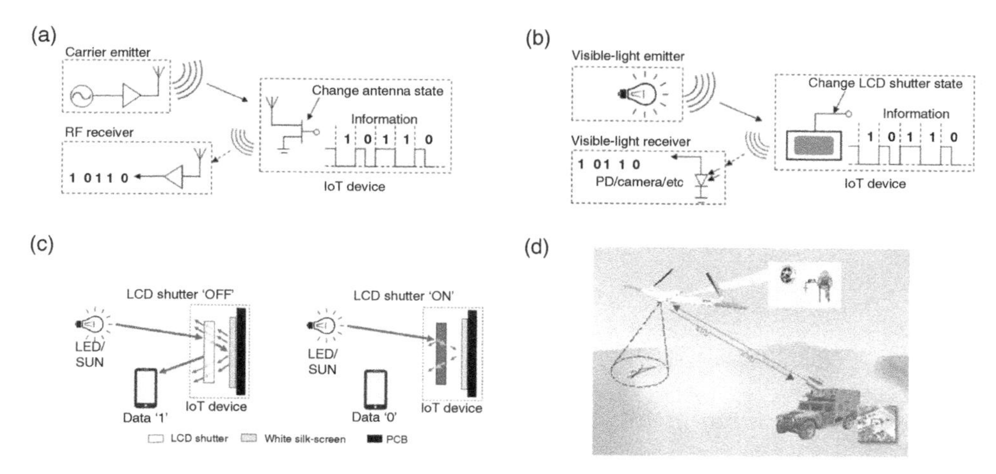


**Figure 9.5** (a) System architecture of a typical RF backscatter communication system. (b) System architecture of a typical VLBCS. (c) Concept illustration of a typical VLBCS. (d) A retroreflector-based visible light backscatter communication incorporated into an airborne reconnaissance concept significantly reduces the parasitic payload requirements for the onboard communications system. Sources: (a)–(c) Yun and Jang [46]. © 2016, Korean Institute of Electromagnetic Engineering and Science. (d) Gilbreath et al. [16]. © 2001, SPIE.

wireless network operators to offer new services without emitting additional EM power. A comprehensive survey of monostatic, bistatic, and ambient backscatter communication systems (BackCom) is presented here [37].

There is another emerging class of backscatter communications known as *visible light backscatter communication system (VLBCS)*, which is proposed to enable efficient data transmissions in RF limited environments, e.g. in critical infrastructure, hospitals, or on planes. In general, the principles of VLBCSs are similar to backscatter RF systems, hence the optical regime backscattering surfaces (known as retroreflectors) are required for practical implementation. The authors in [38, 39] have demonstrated backscatter transmitters that can send data to the backscatter receiver by backscattering visible light. Figure 9.5a–c shows a system architecture and overall concept of VLBCS in comparison to RF backscattering communication system. A VLBCS typically uses standard retroreflector based on bulky optics or some special fabrics. However, we note that the design efficiency of such VLBCSs can be significantly improved by using metasurface-based thin retroreflectors, recently demonstrated in [22]. This type of VLBCS will be extremely useful in scenarios where RF spectrum cannot be used for communications. A typical scenario has been demonstrated in Figure 9.5d.

Another important application of metasurfaces in future communication technology includes the efficient generation, and reception of orbital angular momentum (OAM) multiplexed communication signals [41]. Figure 9.6 shows the concept demonstration of OAM-multiplexing for high-speed and secure data transmission to moving platforms in the free space. The layout and concept of the seven-mode OAM multiplexer/demultiplexer is shown in Figure 9.7a. Here, seven single-mode fibers (SMFs) are connected to a fiber array followed by a microlens array to generate seven collimated Gaussian beams which

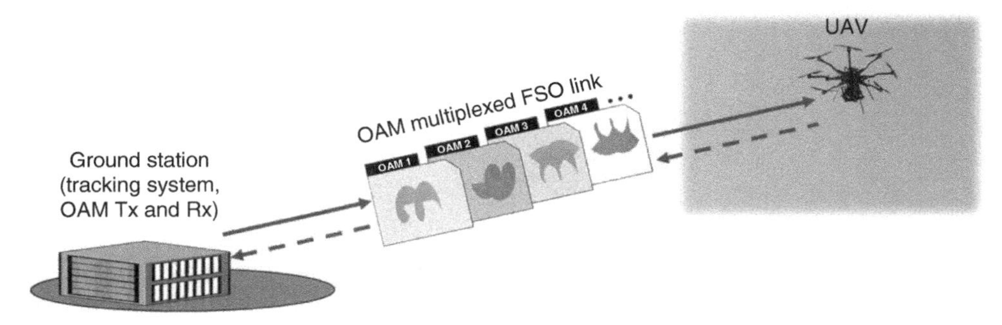


**Figure 9.6** Concept of an OAM multiplexing-based free-space optical communication link between an unmanned aerial vehicle (UAV) and a ground station. Source: Li et al. [25]. Licensed under Creative Commons Attribution 4.0.

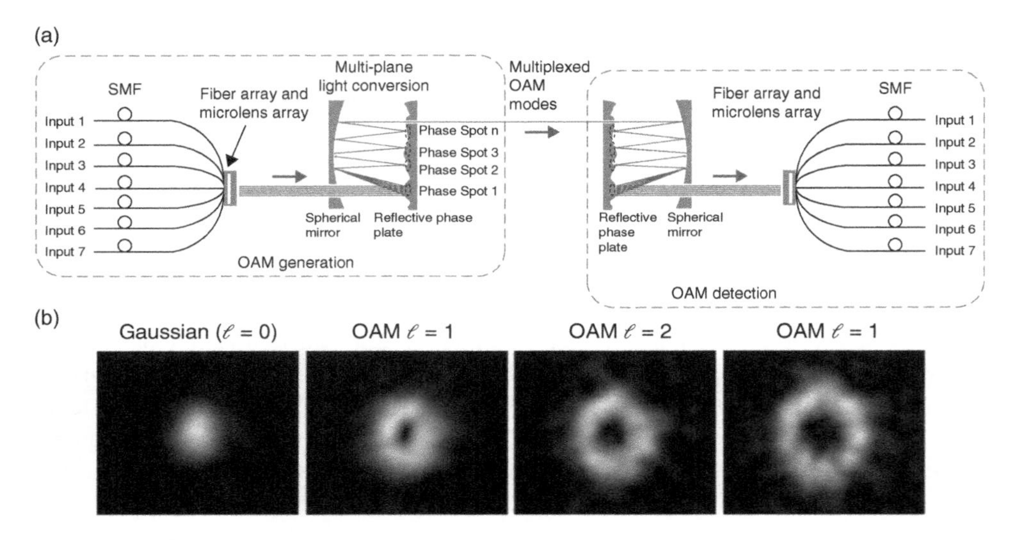


**Figure 9.7** Concept of an OAM multiplexing. (a) Schematic of the seven-mode OAM (de)multiplexer. (b) Intensity profiles of the generated Gaussian beam (i.e., $l = 0$) and OAM beams (i.e., $l = 1, 2, 3$).Source: Li et al. [25]. Licensed under Creative Commons Attribution 4.0.

can propagate in free space. The multiple OAM beams could be converted back to multiple Gaussian beams by using the multiplexer reversely. The retrieved beams are then sorted to corresponding SMF outputs. The field intensity profiles of the generated Gaussian beam and OAM beams are shown in Figure 9.7b. One can use gradient metasurfaces for efficient generation of OAM multiplexed beams; some examples are discussed in the next section. Finally, Figure 9.8 demonstrates various EM wave manipulation or control functionalities that can be implemented using metamaterials and metasurfaces.

In the rest of this chapter, we first present the basic physics of meta-atoms used to construct metasurfaces. Then, multiple examples of metasurface-based devices in optical regime are presented.

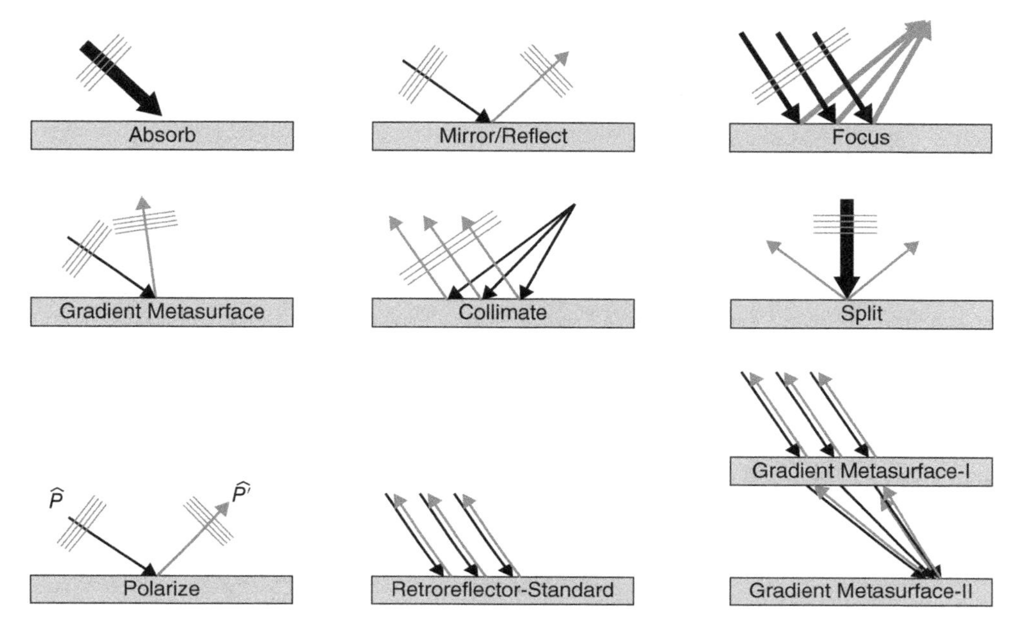


**Figure 9.8** Schematic description of typical wave manipulation operations that can be achieved using metasurfaces. Two gradient metasurfaces can be arranged using methodology proposed in [22] to design a retroreflector, which can serve as enabling platform for metasurface-based future visible light backscatter communication systems. Source: Modified from Arbabi et al. [7].

## 9.2 Meta-Atoms for Optical Frequencies

A light wave is about five thousand times the size of an atom in the conventional dielectric materials like glass. When it propagates through such dielectric materials, it is unable to see their tiny atoms individually. Hence, it preserves its overall properties, while propagating, as the medium looks more and less continuous to it. However, the multitude of such tiny atoms can scatter the wave. The same is true if an EM wave of much longer wavelength propagates through a certain region filled with closely spaced tiny (compared to wavelength) resonators. Such subwavelength scale resonators can be artificially engineered and their careful arrangement in 2D or 3D arrays can scatter the EM wave to exhibit intriguing phenomena. These tiny resonators are generally called meta-atoms, i.e. beyond natural atom. The term metamaterial is generally used when meta-atoms are arranged in 3D, while the subwavelength thick structures designed/fabricated by arranging meta-atoms in 2D are named as metasurfaces. The array of meta-atoms can give a unique way of tailoring various characteristics of the wave (like amplitude, phase, and polarization) at a miniaturized scale, which otherwise is almost impossible to achieve through naturally occurring materials. This section covers three basic all-dielectric meta-atoms (i.e., nano half-wave plates, step-indexed nanowaveguides [NSIW], and supercells) for the optical frequencies. In contrast to the conventional bulky materials in which wave's phasefronts are tailored by the optical constants of a material, meta-atoms provide a versatile platform of controlling phasefronts through their geometrical parameters. Such versatile control offers a unique degree of freedom to realize on-chip optical devices. The detailed design, functioning, and underlying mechanism of these meta-atoms are given in the following section.

### 9.2.1 Nano Half-Wave Plates

A nanoelement of high refractive index material can have the ability to control phase-fronts of the incident EM wave. A rectangular nanoelement can behave as a half-wave plate and due to its nanoscale dimensions (for optical frequencies) it can be referred as a nano half-wave plate (NHWP). Once a circularly polarized light is impinged on NHWP, it transmits/reflects both copolarized and cross-polarized components. Interestingly, the cross-polarized component acquires an additional phase, which is twice the orientation of NHWP as shown in Figure 9.9. Such unique control gives freedom to optimize a discrete set of NHWPs to achieve a complete phase control of (0–$2\pi$) along with uniform transmission for the transmitted/reflected light. Hence, an array of such NHWPs can be employed to achieve the desired phase mask at the aperture plane. Like half-wave plate, the incident circularly polarized light is transmitted as a copolarized light without any phase change, while the cross-polarized transmission acquires a relative phase which is twice the angle of rotation ($\theta$) as shown in Figure 9.9. Such phase control through varying the geometrical parameters of the antenna is known as Pancharatnam–Berry-phase (PB phase).

Consider a plane wave that is normally incident on an NHWP, and $T_{11}$ and $T_{22}$ represent complex transmission coefficients when the polarization of the incident light is along the two principle axes of the NHWP. When the NHWP is rotated at an angle $\theta$ (with the $x$-axis as depicted in Figure 9.9), the transmission coefficients of such rotated system can be extracted using Jones Calculus. Let the Jones vectors for the left and right circularly polarized are

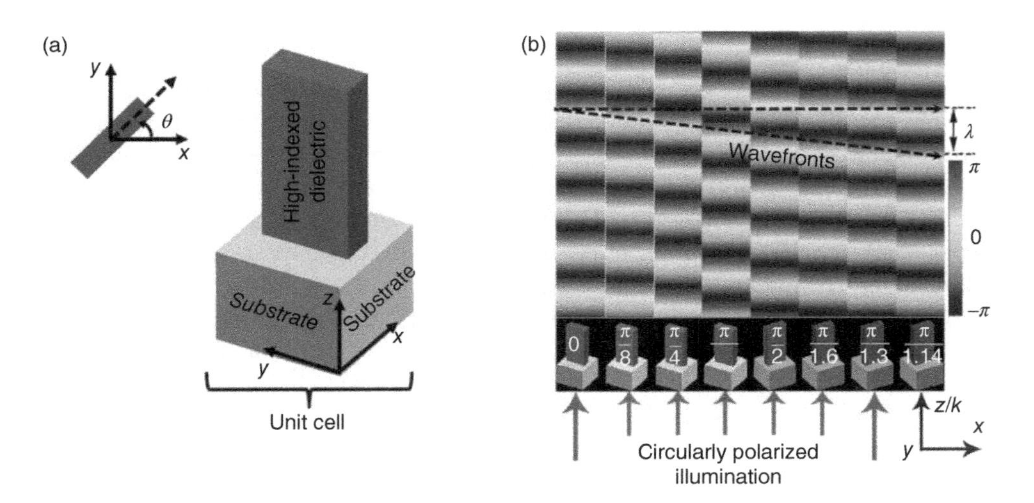


**Figure 9.9** (a) Schematic diagram of NHWP on a dielectric substrate. NHWP is the unit element of the PB-metasurfaces. Inset shows the top view the NHWP. (b) The phasefronts, of circular cross-polarized transmitted light, obtained through full-wave numerical simulations.

represented as

$$\vec{E}_{RCP} = \frac{1}{\sqrt{2}}\begin{bmatrix}1\\-i\end{bmatrix}, \qquad \vec{E}_{LCP} = \frac{1}{\sqrt{2}}\begin{bmatrix}1\\i\end{bmatrix} \tag{9.1}$$

and the transmitted light can be corelated with the incident light as

$$\vec{E}_t = \begin{bmatrix}T_{11} & T_{12}\\T_{21} & T_{22}\end{bmatrix} = \vec{E}_i \tag{9.2}$$

where $\begin{bmatrix}T_{11} & T_{12}\\T_{21} & T_{22}\end{bmatrix}$ is the transmission matrix which reduces to $\begin{bmatrix}T_{11} & 0\\0 & T_{22}\end{bmatrix}$ for the half-wave plate. For the rotated half-wave plate, the transmission matrix becomes

$$T' = R(-\theta) \cdot T \cdot R(\theta) \tag{9.3}$$

where $R(\theta) = \begin{bmatrix}\cos\theta & \sin\theta\\-\sin\theta & \cos\theta\end{bmatrix}$ represents the rotation matrix. After inserting the expressions of $T$, $R(\theta)$, and $R(-\theta)$, the generalized transmission matrix of the rotated half-wave plate becomes

$$\begin{aligned} T' &= \begin{bmatrix}\cos\theta & -\sin\theta\\ \sin\theta & \cos\theta\end{bmatrix}\begin{bmatrix}T_{11} & 0\\0 & T_{22}\end{bmatrix}\begin{bmatrix}\cos\theta & \sin\theta\\-\sin\theta & \cos\theta\end{bmatrix} \\ &= \begin{bmatrix}T_{11}\cos^2\theta + T_{22}\sin^2\theta & (T_{11}-T_{22})\sin\theta\cos\theta\\(T_{11}-T_{22})\sin\theta\cos\theta & T_{22}\cos^2\theta + T_{11}\sin^2\theta\end{bmatrix} \end{aligned} \tag{9.4}$$

When a left circularly polarized light beam is impinged on such rotated NHWP, the expression of the transmitted light takes the following form

$$\begin{aligned} \vec{E}_t &= T' \cdot \vec{E}_i = T' \cdot \vec{E}_{LCP} = T' \cdot \frac{1}{\sqrt{2}}\begin{bmatrix}1\\i\end{bmatrix} \\ &= \frac{1}{\sqrt{2}}\begin{bmatrix}T_{11}\cos^2\theta + T_{22}\sin^2\theta & (T_{11}-T_{22})\sin\theta\cos\theta\\(T_{11}-T_{22})\sin\theta\cos\theta & T_{22}\cos^2\theta + T_{11}\sin^2\theta\end{bmatrix}\begin{bmatrix}1\\i\end{bmatrix} \\ &= \left(\frac{T_{11}+T_{22}}{2}\right)\frac{1}{\sqrt{2}}\begin{bmatrix}1\\i\end{bmatrix} + \left(\frac{T_{11}-T_{22}}{2}\right)e^{i2\theta}\frac{1}{\sqrt{2}}\begin{bmatrix}1\\-i\end{bmatrix} \\ &= \left(\frac{T_{11}+T_{22}}{2}\right)\vec{E}_{LCP} + \left(\frac{T_{11}-T_{22}}{2}\right)e^{i2\theta}\vec{E}_{RCP} \end{aligned} \tag{9.5}$$

Equation 9.5 illustrates that the cross-polarized component of the transmitted electric field that acquires an additional phase of $2\theta$, while the scattered co-polarized component does not get such phase control. The underlying mechanism (for high-indexed all-dielectric NHWPs) behind the high transmission and the phase delay of 180° between $E_x$ and $E_y$ can be explained through antiferromagnetic resonance modes as depicted in Figure 9.10.

These results are plotted for hydrogenated amorphous silicon (a-Si:H) at 633 nm and show that the antiferromagnetic resonance modes are excited in a-Si : H-based NHWPs with even and odd vertical antiparallel magnetic dipole (MDs) and circular electric displacement currents (DCs). The straight arrows inside the NHWP represent the electric field vectors, whereas the circular arrows represent the magnetic field vectors. The red arrows at the top and bottom of the NHWP illustrate the directions of the transmitted and incident electric field vectors, respectively. When an $x$-polarized light is impinged on

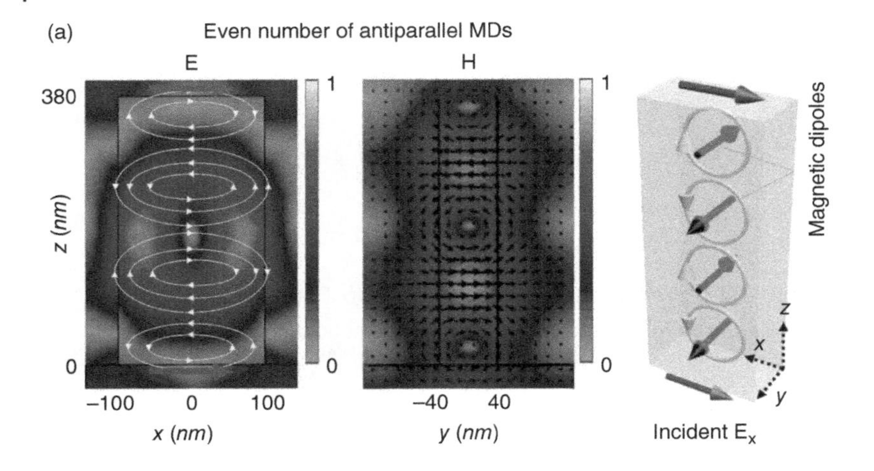


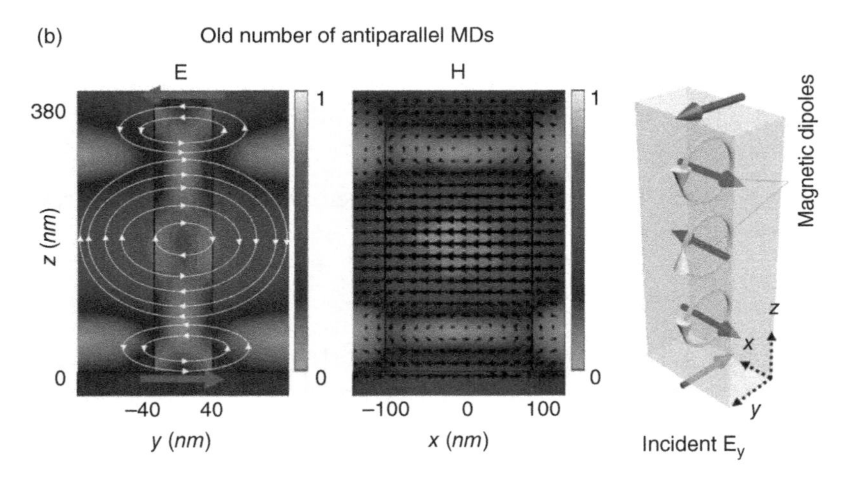


**Figure 9.10** Simulated results combined with artistic description to illustrate antiferromagnetic resonances in high-index hydrogenated silicon NHWPs under *x* and *y* directed linearly polarized illuminations. Source: Ansari et al. [5]. Copyright 2020 RSC. © 2020, The Royal Society of Chemistry.

an NHWP, the four antiparallel MDs are excited while the induced electric field has four electric DCs with varying directions. However, under *y*-polarized illumination, three electric DCs and antiparallel MDs are excited inside NHWP. These odd number of electric DCs result in opposite directions of electric field vectors for the input and output wave as shown in Figure 9.10b. It illustrates that when a circularly polarized light is impinged on an NHWP, the *x* (*y*) component of the electric field is retained (reversed) on the transmission side. Thus, enabling a phase delay of 180° that is required to achieve high efficiency of orthogonal polarization conversion. This simultaneous excitation of electric and magnetic resonances results in high transmission and 180°, phase delay between the orthogonal electric field components.

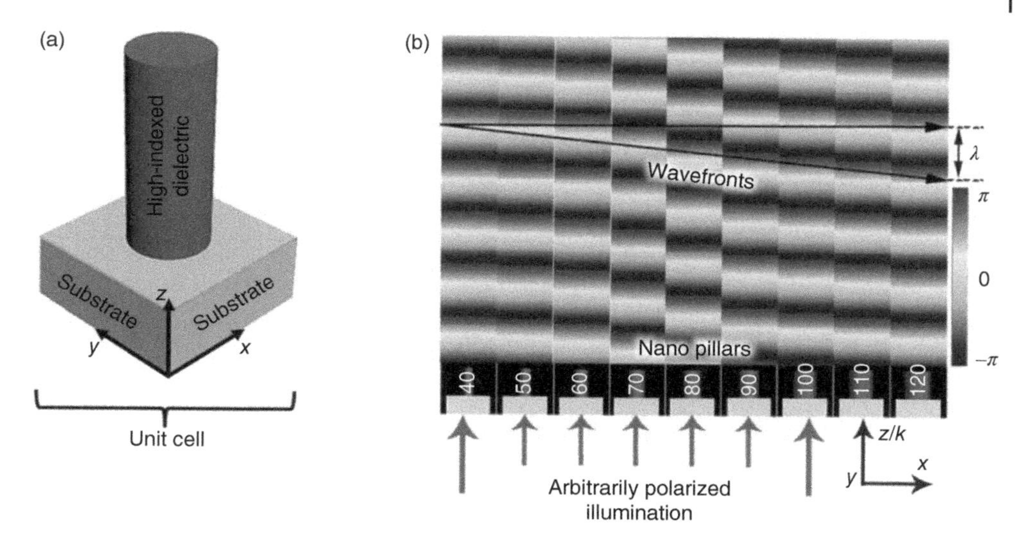


**Figure 9.11** (a) Schematic diagram of step-indexed nano-waveguide (SINW) on a dielectric substrate. SINW is the unit cell of polarization-insensitive metasurfaces. (b) The phasefronts of the transmitted light obtained through full-wave numerical simulations. Source: Adopted from Mahmood et al. [31], Copyright 2018 RSC.

## 9.2.2 Step-Indexed Nano-Waveguides

This section covers the working principle of a polarization-insensitive all-dielectric meta-atom which consists of a high-indexed nanopillar on top of the transparent dielectric substrate as shown in Figure 9.11.

In such polarization-insensitive meta-atoms, the complete phase control of $(0 - 2\pi)$ is achieved by meticulously varying the diameters of the nanocylinders. The underlying mechanism of achieving such phase control over the transmitted light can be explained through step-indexed waveguide (SIW) theory. Like SIWs, the field strongly confines in the high-indexed nanocylinder because of the significant difference between the refractive indices of nanopillar and the surrounding medium (air). Due to their resemblance with SIW and nanofeatured dimensions at the optical frequencies, such meta-atoms can be named as step-indexed nanowaveguides. Figure 9.12 illustrates the geometrical description and index profile of SINW.

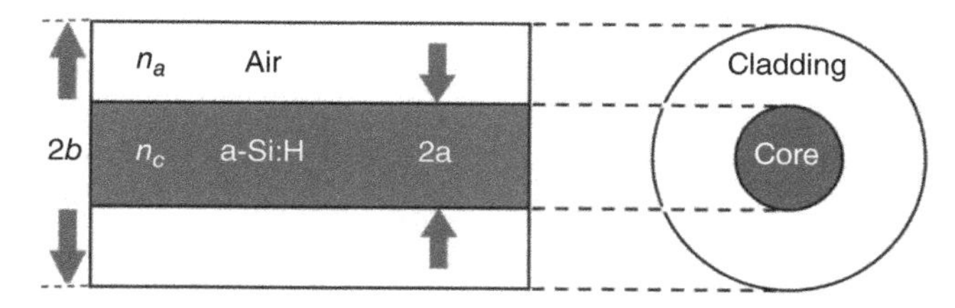


**Figure 9.12** The schematic illustration of the SINW consisting of core with radius "$a$" and cladding with "$b$". Source: Modified from Mahmood et al. [44]. Copyright 2018 RSC.

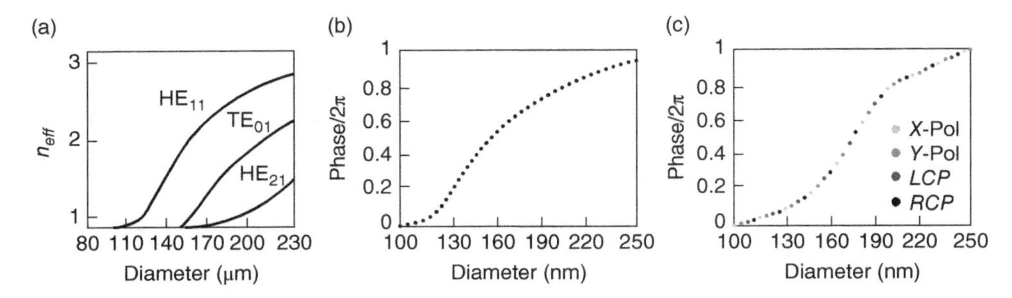


**Figure 9.13** (a) Effective refractive index as a function of diameter for some low-order modes of an SINW. (b) The calculated phase as a function of diameter through SINW theory. (c) The transmitted phase plot achieved from the full-wave EM simulations for different polarization states of the incident light. Source: Mahmood et al. [31]. Copyright 2018 RSC. © 2018, The Royal Society of Chemistry.

The core (having radius of "$a$") consists of a nanocylinder of high-index material $n_c$, while the cladding consists of air with the refractive index of $n_a$. The normalized propagation constant $\beta$ is correlated to the effective refractive index $n_{eff}$ and its value varies between $n_c$ and $n_a$. The relationship between $n_{eff}$ and $\beta$ is defined as:

$$n_{eff} = \frac{\beta}{k_0} \tag{9.6}$$

The graphical relationship between the $\beta$ and the diameter $D$ of SINW is plotted in Figure 9.13a. It shows that $HE_{11}$ is the dominant mode, and it will always propagate inside the SINW. The relationship between the phase $\Phi$ imparted by SINW and $n_{eff}$ is given as:

$$\Phi = \frac{2\pi}{\lambda} n_{eff} \tag{9.7}$$

These expressions and Figure 9.13a represent that by varying the diameter of an SINW, the $n_{eff}$ of the dominant mode is changed that results in a specific phase response $\Phi$ for the transmitted light. This phase response of the SINW is shown in Figure 9.13b and shows good agreement with the full-wave EM simulation results of Figure 9.13c. It should be noted that the smallest diameter of the SINW is limited by the maximum aspect ratio (that can be handled by the fabrication process), while the largest diameter should be less than the periodicity of a unit cell to meet the Nyquist sampling criteria ($U < \frac{\lambda_d}{2NA}$).

### 9.2.3 Supercell

Apart from NHWPs and SINWs, various other design strategies of meta-atoms have been presented by the researchers to effectively tailor the wavefronts of the incident light. One of those interesting strategies include the design of a super cell, in which the fundamental building block consists of a combination of resonators as illustrated in Figure 9.14. Such supercells provide an extra degree of freedom in design and allow the implementation of novel phenomena which otherwise are difficult to achieve through a single nanoresontor.

In this regard, authors in [31] presented a unique strategy of designing a super meta-cell that consisted of a combination of anisotropic NHWPs. In the proposed all-dielectric super

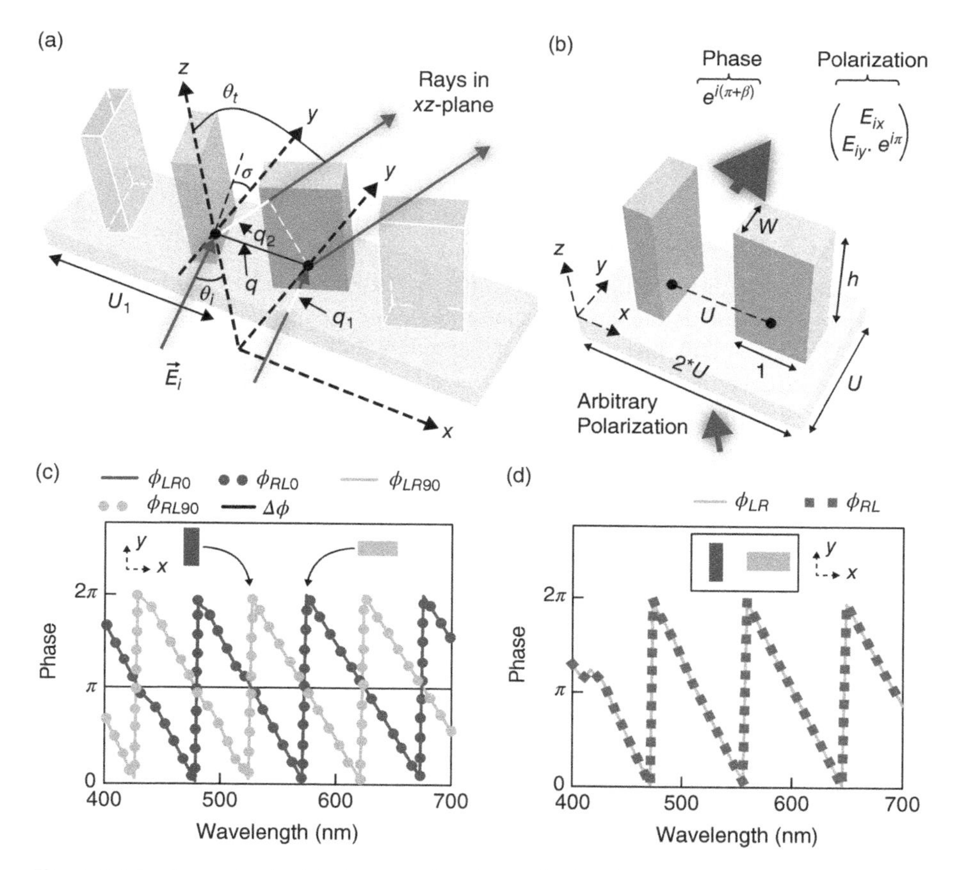


**Figure 9.14** Principle of achieving broadband polarization-insensitive phase control though super meta-cell containing combination of NHWPs. Source: Ansari et al. [6]. Licensed under CC BY 4.0.

cell, they effectively combined and modified the geometric (PB) and propagation phase modulation techniques to achieve an unaltered complete phase control for the incident light of any polarization state. Such polarization-insensitive complete phase control is impossible to achieve through any of these modulation techniques individually. Figure 9.14a shows two adjacent nonidentical NHWPs that are separated by a distance $q$. The transmitted electric fields through such pair of nonidentical NHWPs result in a phase delay of $\Delta\varphi = \frac{2\pi q}{U_1} + \alpha$, where $\alpha$ and $U_1$ represent the propagation phase shift (because of dissimilar parallel NHWPs) and the global displacement, respectively. This phase delay is a combined consequence of dissimilar dimensions of NHWPs and their displacement $q$. Figure 9.14a also depicts that these nonidentical NHWPs are either perpendicular or parallel to each other.

Keeping the collective impact of perpendicular and parallel NHWPs, the total transmitted electric field under an illumination of arbitrary polarization state can be written with the help of Jones calculus as

$$\begin{bmatrix} E_{tx} \\ E_{ty} \end{bmatrix} = \begin{bmatrix} \cos\sigma & -\sin\sigma \\ \sin\sigma & \cos\sigma \end{bmatrix} e^{i\left(\frac{2\pi q}{U_1}+\alpha+\beta\right)} \begin{bmatrix} 1 & 0 \\ 0 & e^{i\frac{2\pi q}{U_1}} \end{bmatrix} \begin{bmatrix} \cos\sigma & \sin\sigma \\ -\sin\sigma & \cos\sigma \end{bmatrix} \begin{bmatrix} E_{ix} \\ E_{iy} \end{bmatrix} \tag{9.8}$$

where $\beta$ shows the propagation phase shift due to nonidentical NHWPs, while $\sigma$ represents the angle between $y$-axis and longer axis of the NHWPs as shown in Figure 9.14a. The polarization and the phase of the transmitted light can be controlled by the parameters $q$, $U_1$, $\sigma$, $\alpha$, and $\beta$. By setting $U_1 = 2q$ and $\sigma = 0$, the geometry of the supercell is modified like in Figure 9.14b, and the Equation (9.8) is reduced to,

$$\begin{bmatrix} E_{tx} \\ E_{ty} \end{bmatrix} = e^{i(\pi+\alpha+\beta)} \begin{bmatrix} 1 & 0 \\ 0 & e^{i\pi} \end{bmatrix} \begin{bmatrix} E_{ix} \\ E_{iy} \end{bmatrix} = e^{i\gamma} \begin{bmatrix} 1 & 0 \\ 0 & -1 \end{bmatrix} \begin{bmatrix} E_{ix} \\ E_{iy} \end{bmatrix} \tag{9.9}$$

As shown in figure, $U$ denotes the displacement between two adjacent nonidentical NHWPs, and $2 * U$ represents the global displacement. For the incident light with linear horizontal polarization (LHP), linear vertical polarization (LVP), left circular polarization (LCP), and right circular polarization (RCP), the expressions for the transmitted light (using Equation (9.9)) are given as

$$\begin{bmatrix} E_{tx} \\ E_{ty} \end{bmatrix} = e^{i\gamma} \begin{bmatrix} 1 & 0 \\ 0 & -1 \end{bmatrix} \begin{bmatrix} 1 \\ 0 \end{bmatrix} = e^{i\gamma} \begin{bmatrix} 1 \\ 0 \end{bmatrix} \tag{9.10}$$

$$\begin{bmatrix} E_{tx} \\ E_{ty} \end{bmatrix} = e^{i\gamma} \begin{bmatrix} 1 & 0 \\ 0 & -1 \end{bmatrix} \begin{bmatrix} 0 \\ 1 \end{bmatrix} = -e^{i\gamma} \begin{bmatrix} 0 \\ 1 \end{bmatrix} \tag{9.11}$$

$$r\begin{bmatrix} E_{tx} \\ E_{ty} \end{bmatrix} = e^{i\gamma} \begin{bmatrix} 1 & 0 \\ 0 & -1 \end{bmatrix} \begin{bmatrix} 1 \\ -i \end{bmatrix} = e^{i\gamma} \begin{bmatrix} 1 \\ i \end{bmatrix} \tag{9.12}$$

$$\begin{bmatrix} E_{tx} \\ E_{ty} \end{bmatrix} = e^{i\gamma} \begin{bmatrix} 1 & 0 \\ 0 & -1 \end{bmatrix} \begin{bmatrix} 1 \\ i \end{bmatrix} = e^{i\gamma} \begin{bmatrix} 1 \\ -i \end{bmatrix} \tag{9.13}$$

The simulation results of distinct NHWP-based supercell are illustrated in Figures 9.14c and d, which verify the polarization-insensitive response of the meta-cell. Apart from this demonstration, other authors have also presented the concept of supercell for their metasurfaces design. For instance, Zhang et al. [45] proposed a chiral supercell to achieve a giant circular asymmetric transmission. Similarly, Ma et al. [46] employed a supercell-based design to demonstrate giant spin-selective asymmetric transmission.

## 9.3 Applications

This section covers a few highlighted applications, like metalensing, holography, light twisting, and EM wave absorption, implemented through metasurfaces employing the abovementioned meta-atoms.

A lens is a fundamental optical device to concentrate the incident light efficiently, and it has countless applications in various scientific fields like medicine, biology, optics, and physics, etc. The conventional glass-based lenses are limited in functionalities and are mostly bulky, imposing a major hurdle in compactness of the optical systems. The discovery of metamaterials undoubtedly opened up avenues to explore novel methods of designing compact lenses with novel functionalities. In this regard, an ultrathin multi-focus metalens (based upon NHWP meta-atoms) have been reported to concentrate a single inside beam at three different focal planes [23] as shown in Figure 9.15. The focal points in the real focal planes are labeled as F1, F2, and F3. The second real focal point corresponds to LCP

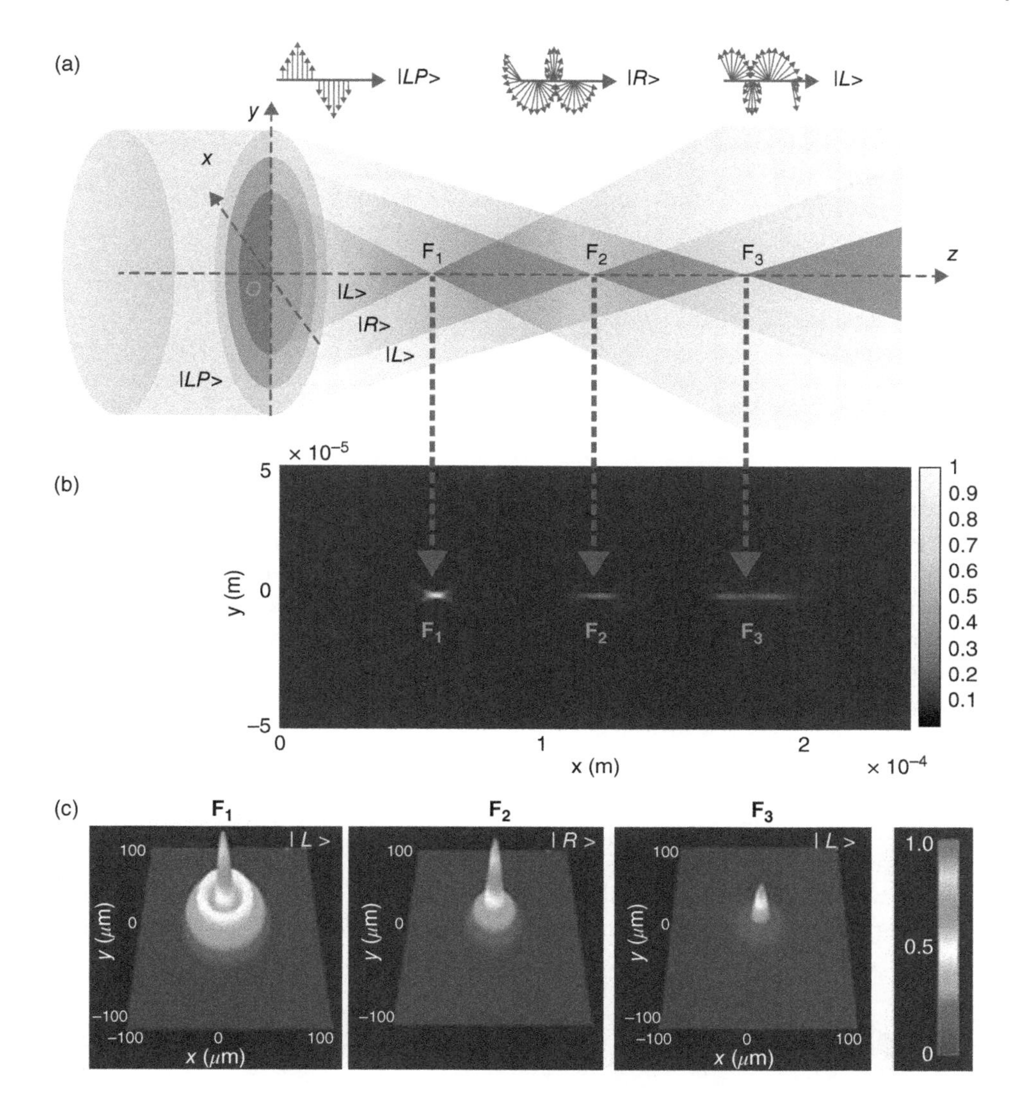


**Figure 9.15** (a, b) Schematic diagram illustrating the functioning of the multi-focus metalens. (b) Focal plane field profiles of numerically simulated structures. Source: Chen et al. [11]. Licensed under CC BY 4.0.

illumination, while the first and third real focal points correspond to the RCP illumination. In contrast to the conventional multi-focus diffractive lenses, the polarization and the position of the imaging planes can be controlled by varying the spin of the incident beam. Such mutli-focus meta-lens are vital for applications in optical free-space communications, imaging systems, and laser printing, etc.

Other than multi-focus metalens, an ultrathin step-zoom metalens with dual field of view has been presented by Zheng et al. [47]. The proposed zoom metalens consisted of silicon

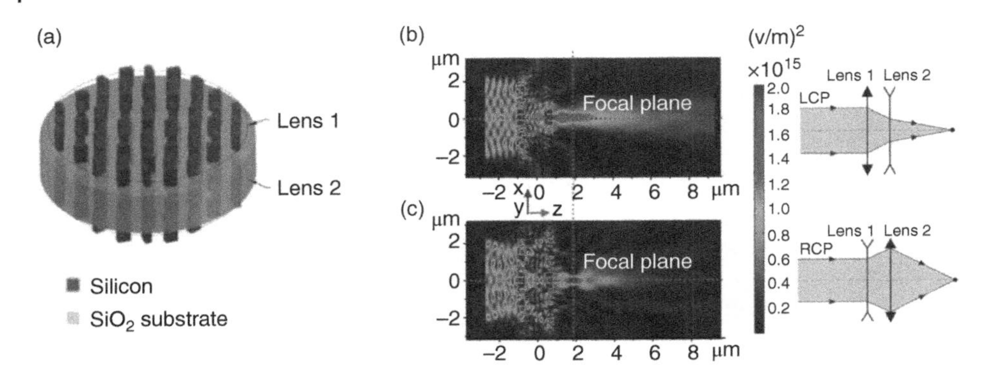


**Figure 9.16** (a) Schematic illustration of the step-zoom metalens, with dual field of view, designed on both sides of the substrate. (b, c) Numerical simulation results of dual FOV step zoom metalens under normal circularly polarized illumination. Source: Zheng et al. [48]. Copyright 2017 OSA. © 2017, The Optical Society.

NHWPs of spatially varying orientations on both sides of the glass substrate as shown in Figure 9.16a. Such compact step-zoom metalens can focus an incident circular polarized light with spin-dependent field of views without altering the imaging plane. The field profiles for circularly polarized illumination of different handedness are given in Figure 9.16b and c, respectively.

Other unique structures have also been employed to realize novel lensing functionalities. For example, authors in [48] presented a nanofeatured logarithmic spiral structure to realize helicity-controlled focusing as shown in Figure 9.17.

Other than the planar structures, 3D metalenses have been demonstrated to achieve novel functionalities like shaping the wave trajectory in 3D [49]. The freedom to control the EM wave characteristics through metamaterials also enabled to shape and twist light in a desired way. It has been reported that a unique phase mask can be generated by merging the phase profiles of spherical lens and spiral phase plate to realize a nanofeatured chip for

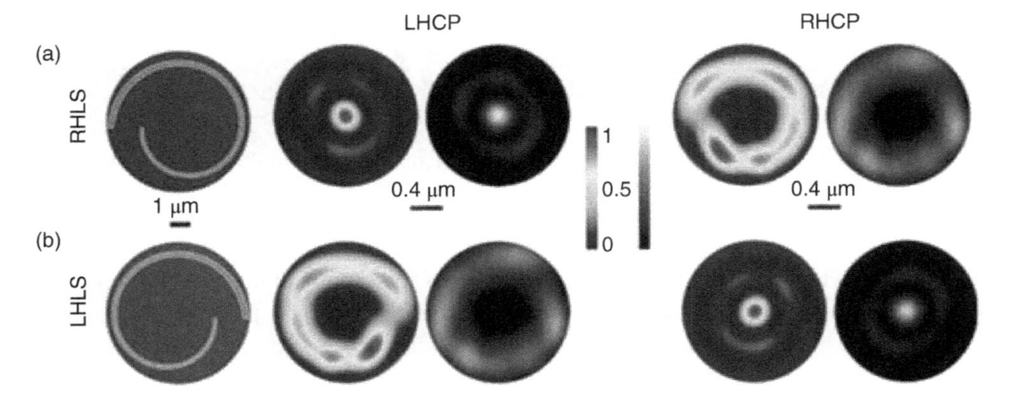


**Figure 9.17** (a, b) SEM images of the fabricated left- and right-handed structures and *xy*-plane simulated and measured intensity profiles, at the focal plane, under LCP and RCP illuminations. Source: Reprinted using a Creative Common License 4.0 from [48].

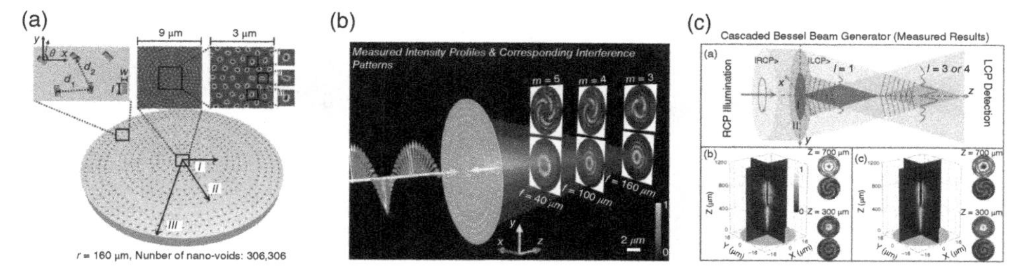


**Figure 9.18** (a) Schematic diagram of multi-focus vortex plate along with SEM images of the fabricated structure. (b) Measured intensity profiles and interference patterns at three different focal planes for a multi-focus vortex plate. Sources: Mehmood et al. [33] and Mei et al. [36]. © 2016, John Wiley and Sons.

generating focused optical vortices [50]. A schematic perspective of the proposed design is given in Figure 9.18a which consists of inverted NHWPs following the Babinet's principle. Instead of extruded NHWPs, they are milled into a 60-nm-thin gold film. Such Babinet design was adopted to achieve higher signal-to-noise ratio. The orientation angle ($\theta$) of each NHWP was found through the following expressions:

$$\theta_1(r) = \pm\frac{k_0}{2}\left(\sqrt{r^2+f^2} - |f|\right) \tag{9.14}$$

$$\theta_2(r) = \pm\frac{k_0}{2}\cdot m \cdot \tan^{-1}\left(\frac{y}{x}\right) \tag{9.15}$$

$$\theta_t(r) = \theta_1(r) + \theta_2(r) \tag{9.16}$$

where $\theta_t(r)$ represents the total rotation angle of each NHWP at a specific position $r$ as depicted in Figure 9.18a.

The measured results of the fabricated device are given in Figure 9.18b, which depicts distinct topological charges (of $m = 5, 4 and 3$) along with their corresponding interference patterns at each of three focal planes ($f$ = 40, 100 and 160 $\mu$m). The phenomenon of integrating distinct phase profiles has been extended to realize compact, highly efficient, and polarization-insensitive phase plates through an array of SINWs [44].

The concept is further extended to demonstrate the design of compact devices for the generation of zero-order and higher-order Bessel beams. The conventional optical component to generate the zero-order Bessel beam is known as axicon, which is a specialized glass of conical shape. However, the conventional axicons have serious limitations in terms of their functionalities. For example, it is not possible to design an axicon with a numerical aperture greater than 0.75 due to the design constraints of its conical structure. Moreover, an axicon alone cannot generate higher order Bessel beams carrying OAM. In contrast, metasurface-based axicon (named as metaaxicon) does not have such limitations. Hence, apart from on-chip realization, metaaxicons give more degrees of freedom in terms of compactness, design flexibility, and functionalities. In this regard, a multifunctional metasurfaces based upon intruded NHWPs in a 60-nm-thick Au film has been reported to realize higher order metaxiocons. The schematic description and the measured results of the two concentric metaxicons (designed for different topological charges) are given in Figure 9.18c. The ring-shaped intensity profiles having different radii appeared (as expected) in the measurements as both the metaaxicons scatter twisted beams of different topological charges.

**Figure 9.19** (a) Schematic description of the proposed meta-atom to generate beams with OAM mode $m = 1$. (b) Fabricated metasurface and the experimental setup. (c) Measurement results of the fabricated metasurface for the topological charge of $m = 1$. Source: Akram et al. [1]. Copyright 2019 Wiley. © 2019, John Wiley and Sons.

Other than such plasmonic metaaxicons, all-dielectric metasurfaces consisting of SINWs have also been presented to demonstrate polarization-insensitive highly efficient zero-order and higher-order Bessel beams [28].

The twisting of EM beam has been demonstrated for other frequency regimes as well. For example, a bilayer metasurface was presented for the microwave frequencies [43] to generate the optical vortices (OVs) with topological charges of $m = 1$ and 2. Figure 9.19a shows the schematic diagram of the meta-atom used for the said metasurface.

The proposed subwavelength resonator consisted of bilayer $z$-shaped copper (Cu) patterns printed on both sides of the Rogers RO4350B substrate. The Cu arms of the structure form a specific angle with the central strip that was optimized to obtain ultrahigh high polarization conversion efficiency ( 85%) along with achieving the required phase control in transmission. The optimized meta-atoms were then carefully arranged in an array, according to the phase mask's requirement for a specific topological charge, to realize the twisted beam. The fabricated structure is shown in the inset of Figure 9.19b, and the experimental setup to measure the response of structure is given in Figure 9.19b. The horn antennas (at 10 GHz) were used to impinge the EM wave onto the sample and to capture the transmitted field. The post-processed measurements for the twisted beam of topological charge $m = 1$ are given in Figure 9.19c. Similarly, the Bessel beams of zero and first order were also generated at 10 GHz through layered metasurface [51]. Apart from using metasurfaces, researchers have also reported other methods of generating twisted beams as well. For example, a single metadevice consisting of bilaterally symmetric grating was presented to generate optical beams in which values of topological charges can be dynamically controlled [52].

Another excellent demonstration of control over EM wave propagation through metamaterials can be provided via meta-holograms. Holography is a lensless photography in which, conventionally, an interference pattern is recorded by using a reference beam and a reflected beam (from the object). The recorded pattern is called a hologram that contains the information of phase and amplitude of the object. The hologram can be reconstructed by illuminating the interference pattern with a reference beam. However, for the computer-generated holograms (CGH), the recording of a hologram does not require an object. Instead, it can be generated using iterative numerical algorithms. Metasurfaces, having the ability of controlling the wavefront of the EM wave at a subwavelength scale, provide a promising platform to realize compact and high-resolution holograms. Benefiting from NHWP's ability of polarization-controlled transmission, an all-dielectric helicity

(a)

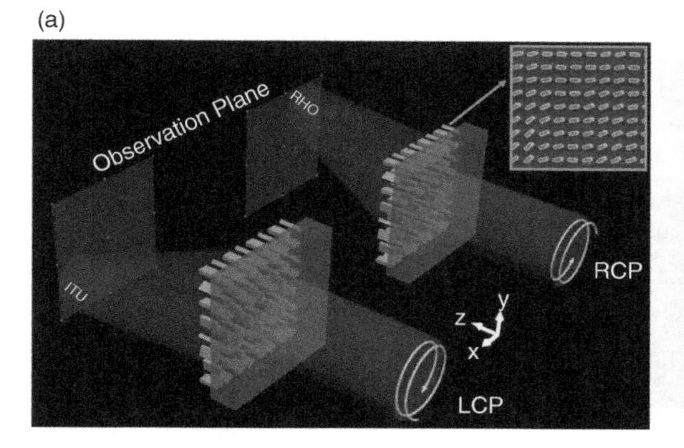


(b)

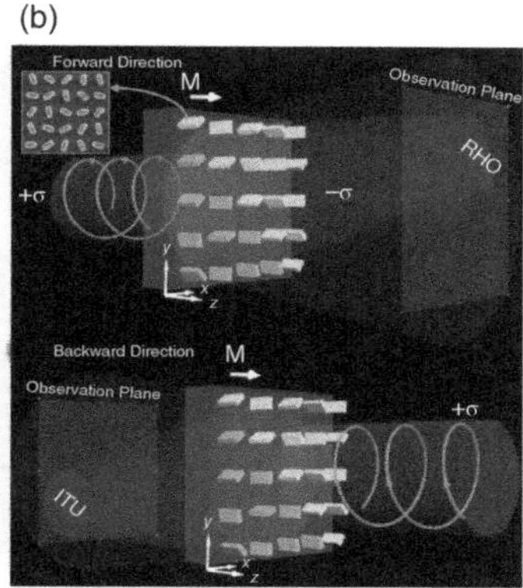


**Figure 9.20** Spin-encoded meta-holograms under circularly polarized illuminations of opposite handedness. Sources: (a) Ansari et al. [4]. © 2019, John Wiley and Sons. (b) Modified from Ansari et al. [5]. 2019 Wiley, Copyright 2020 RSC.

multiplexed meta-hologram has been presented for the visible wavelengths. The proposed meta-hologram demonstrated spin-controlled characteristics along with high transmission efficiency and image fidelity [30]. Two distinct off-axis holographic images of "ITU" and "RHO" can be reconstructed by illuminating the structure by left and right circularly polarized light respectively as depicted in Figure 9.20a.

This work was further extended to realize the direction-multiplexed meta-holograms. The authors used a novel approach of using monolayer 2D array of all-dielectric supercells to record the distinct holograms in forward and backward directions [32]. The graphical abstract of the direction multiplexed hologram results is depicted in Figure 9.20b. When a circularly polarized light was impinged in the forward direction ($+z$-direction), an image of "RHO" was reconstructed. However, the same metasurface reconstructed an image of "ITU" at the opposite side when illuminated by the similar handedness from the backward direction ($-z$-direction). Similar high-indexed NHWPs have also been used to realize efficient transmission-type holograms by combining the magnetic resonance and the geometric phase [53]. A Fourier hologram was designed for the normally incident circularly polarized light as given in Figure 9.21a. The reflected light forms a holographic image of a parrot in the far field at a plane vertical to the optical axis of incident beam. The required phase mask was obtained by the Gerchberg–Saxton (GS) algorithm. The measured holographic images exhibited high fidelity for these Fourier meta-holograms. They also presented image holograms that can produce the desired image above the surface of the structure and can be seen through a naked eye (for a bigger sample) or through an optical microscope. Figure 9.21b illustrates the working principle of an image hologram in which the reconstructed hologram was explicitly observed without any receiving screen under the white light.

Recently, a novel design strategy (of using 2D array of NHWP-based meta supercells) was used to realize a highly efficient, broadband, and polarization-insensitive meta-hologram [31]. The simulated results of the designed meta-holograms under illuminations of different polarizations (i.e., linear horizontally polarized [LHP], linear vertically

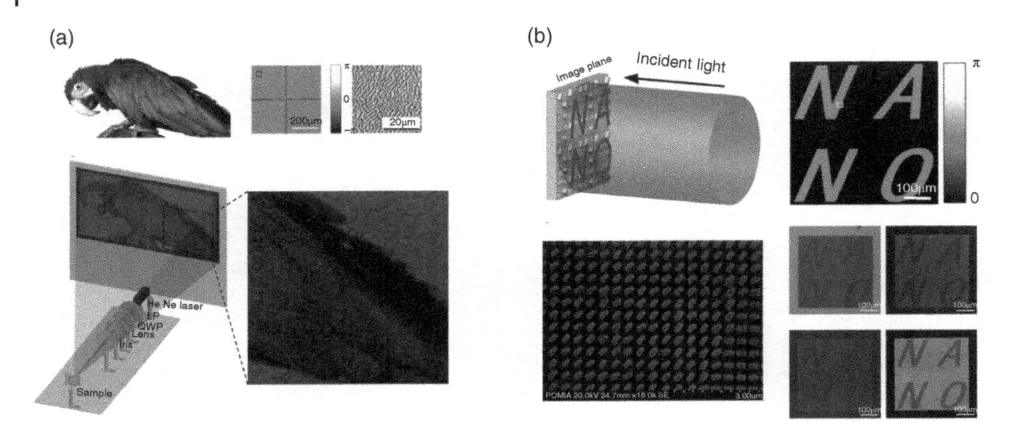


**Figure 9.21** (a) Phase mask obtained through GS algorithm, characterization setup, and measured results of the Fourier meta-hologram. (b) Graphical representation of the working principle of image meta-hologram, its employed phase mask, SEM image of the fabricated sample, and the measured results. Source: Li et al. [26]. © 2017, American Chemical Society.

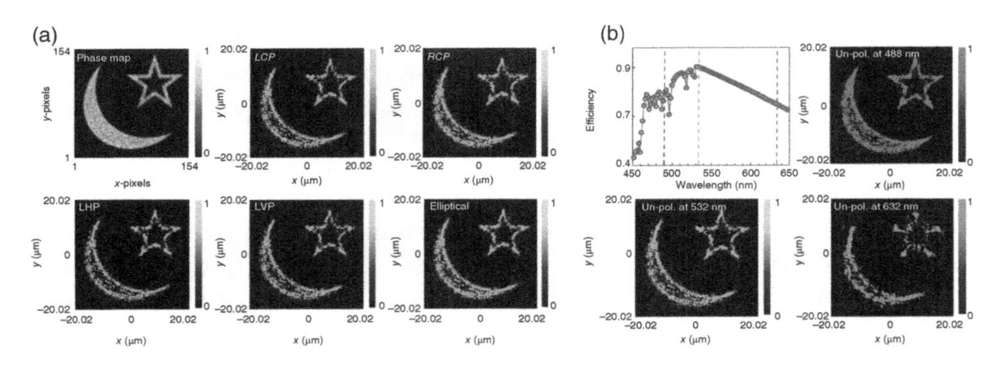


**Figure 9.22** (a) Polarization-insensitive response of the meta-hologram designed through NHWP-based supercells. (b) Broadband response of the meta-hologram, designed through NHWP-based supercells, under unpolarised illuminations. Source: Ansari et al. [6]. Licensed under CC BY 4.0.

polarized [LVP], left circularly polarized [LCP], and right circularly polarized [LCP], elliptically polarized) are given in Figure 9.22a.

The simulated results for the unpolarized blue, green, and red illuminations are shown in Figures 9.22b, which confirms the broadband capability of the proposed metasurface. Similar NHWP-based metasurfaces have also been used to demonstrate several other interesting applications like full-space spot generation that can be vital for many practical scenarios such as motion detection, facial recognition, and augmented reality [54].

Some demonstrations covering metamaterial-based absorbers are worthwhile to mention. EM wave absorbers play a pivotal role in a wide range of applications like thermal imaging, EM shielding, solar energy absorption, and thermal camouflaging. These applications require either broadband absorbers or single/multiple band absorbers. Metasurfaces provide an ideal platform to realize such EM absorbers through ultrathin 2D arrays of meta-atoms. In this regard, Rana et al. [24] numerically presented unique tungsten-based

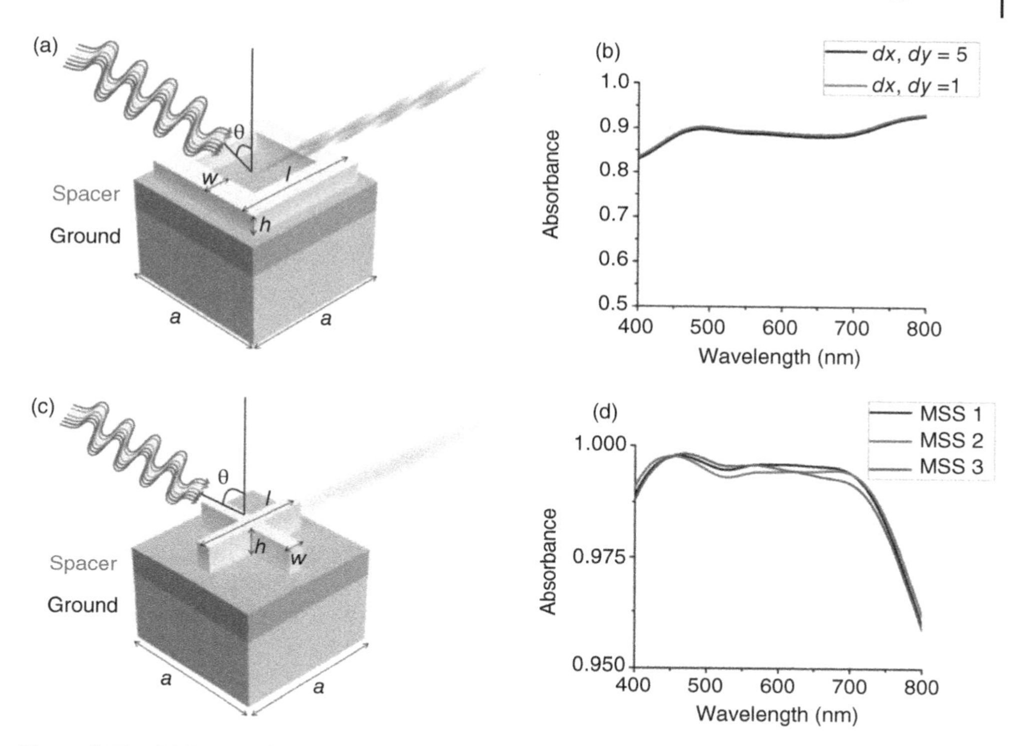


**Figure 9.23** (a) Square-shaped design for tungsten absorber. (b) Simulated absorption spectra of the square-shaped design. (c) Cross-shaped design for tungsten absorber. (d) Simulated absorption spectra of the cross-shaped design. Source: Rana et al. [39]. Licensed under CC BY 4.0.

optical meta-absorbers that can absorb the entire visible regime with an average absorbance of more than 98%. Their structures consisted of three layers on the substrate; the top layer of tungsten meta-atoms, a spacer layer of $SiO_2$ and then a thin layer of tungsten as a ground plane as shown in Figure 9.23a and c. The FDTD simulation results of the square-shaped and cross-shaped nanostructures exhibited desired broadband absorptance (covering the entire visible regime) as depicted in Figure 9.23b and d, respectively.

Similarly, Kim et al. [25] has presented a chromium (Cr) based meta-absorber for the visible light. Their Cr-based meta-absorber also had three layers as shown in Figure 9.24a; a top layer of ring-shaped Cr nanostructures, a dielectric layer of $SiO_2$, and then a thin bottom layer of Cr. The overall thickness of three layers was 250 nm, while a silicon substrate was chosen to make it compatible for the photonics integration. The SEM image of the fabricated device is shown in Figure 9.24b. The simulated and measured absorption spectra are depicted in Figure 9.24c, which exhibit an excellent absorption of more than 98% over the wavelength ranges of 500–710 nm.

Apart from using metasurfaces, thin dielectric films on top of the metal layer can also provide a cost-effective and simple way of obtaining wavelength-selective EM-absorption. Such absorption can be used to achieve enticing applications like color filtering. For example, Rana et al. [27] presented a unique design methodology of achieving color filters for cyan, magenta, and yellow (CMY) colors with enhanced color purity along with angle and polarization insensitiveness.

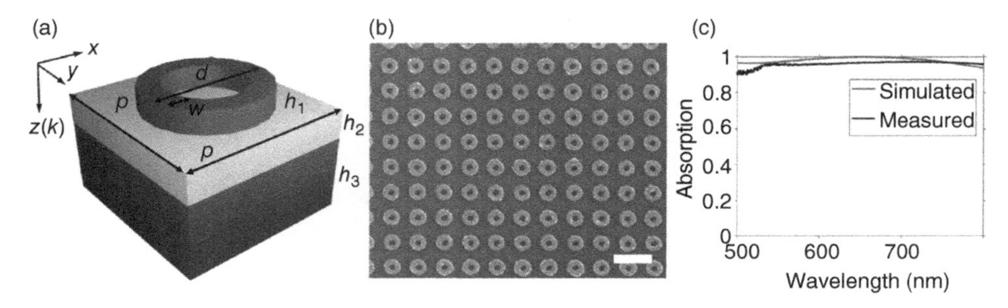


**Figure 9.24** (a) Schematic depiction of Cr-based meta-atom. (b) SEM image of the fabricated sample. (c) Simulated and measured absorption spectra of the meta-absorber containing array of meta-atoms. Source: Kim et al. [23]. © 2018, Walter de Gruyter.

Apart from their applications in the visible regime, EM absorbers are significantly important for other EM spectrums as well. For example, they are vital for the terahertz (THz) regime where it is hard to find naturally occurring materials with high absorption coefficients. In this regard, Zubair et al. [26] employed Cayley tree-based fractal nanoresonators to present a broadband, single-layered, and polarization-insensitive THz meta-absorber. They used meta-supercell consisting of different fractal orders to realize broadband THz absorption. Such broadband THz EM meta-absorber can be useful for applications like metasurface-based microbolometer sensor, in which broadband EM absorbers are required to be coupled with microbolometer sensors to have a complete THz detector.

## 9.4 Summary

In this chapter, we have presented an overview of EM wave manipulation strategies using metamaterials and metasurfaces in the optical regime. We first introduced the basic building blocks of meta-paradigm known as *meta-atoms* that can be arranged in 3D or 2D array configurations to design metamaterials or metasurfaces, respectively. Several examples of metasurface-based devices have been presented including the OAM carrying beam generators, holograms, and absorbers. Similar strategies can be utilized to design novel devices like gradient metasurface-based retroreflectors which can be used in futuristic visible light backscattering communication systems. It is to be noted that the basic design principles of metasurfaces are scaleable in nature, i.e. the design strategies utilized in optical regime can be easily rescaled to THz or microwave regime by rescaling the geometry and material parameters.

## References

**1** Zubair, M., Mughal, M.J., and Naqvi, Q.A. (2012). *Electromagnetic Fields and Waves in Fractional Dimensional Space*. Springer Science & Business Media.

**2** Caloz, C. and Itoh, T. (2005). *Electromagnetic Metamaterials: Transmission Line Theory and Microwave Applications*. Wiley.

3 Cui, T.J., Smith, D.R., and Liu, R. (2010). *Metamaterials*. Springer.
4 Lee, D., Ki Kim, H., and Lim, S. (2017). Textile metamaterial absorber using screen printed chanel logo. *Microwave and Optical Technology Letters* 59 (6): 1424–1427.
5 İbili, H., Karaosmanoğlu, B., and Ergül, Ö. (2018). Demonstration of negative refractive index with low-cost inkjet-printed microwave metamaterials. *Microwave and Optical Technology Letters* 60 (1): 187–191.
6 Walther, M., Ortner, A., Meier, A., et al. (2009). Terahertz metamaterials fabricated by inkjet printing. *Applied Physics Letters* 95 (25): 251107.
7 Jiang, W., Yan, L., Ma, H., et al. (2018). Electromagnetic wave absorption and compressive behavior of a three-dimensional metamaterial absorber based on 3D printed honeycomb. *Scientific Reports* 8 (1): 4817.
8 Shalaev, V.M., Cai, W., Chettiar, U.K., et al. (2005). Negative index of refraction in optical metamaterials. *Optics Letters* 30 (24): 3356–3358.
9 Enkrich, C., Pérez-Willard, F., Gerthsen, D., et al. (2005). Focused-ion-beam nanofabrication of near-infrared magnetic metamaterials. *Advanced Materials* 17 (21): 2547–2549.
10 Brueck, S.R.J. (2005). Optical and interferometric lithography–nanotechnology enablers. *Proceedings of the IEEE* 93 (10): 1704–1721.
11 Chou, S.Y., Krauss, P.R., and Renstrom, P.J. (1996). Nanoimprint lithography. *Journal of Vacuum Science & Technology B: Microelectronics and Nanometer Structures Processing, Measurement, and Phenomena* 14 (6): 4129–4133.
12 Zubair, M., Francavilla, M.A., Righero, M., Vecchi, G., and Dal Negro, L. (2015). Fast analysis of electrically large plasmonic arrays with aperiodic spiral order. *2015 IEEE International Conference on Computational Electromagnetics*, Hong Kong, China (2–5 February 2015), 53–55. IEEE.
13 Zubair, M., Francavilla, M.A., Zheng, D., Vipiana, F., and Vecchi, G. (2016). Dual-surface electric field integral equation solution of large complex problems. *IEEE Transactions on Antennas and Propagation* 64 (6): 2577–2582.
14 Vahabzadeh, Y., Achouri, K., and Caloz, C. (2016). Simulation of metasurfaces in finite difference techniques. *IEEE Transactions on Antennas and Propagation* 64 (11): 4753–4759.
15 Vahabzadeh, Y., Chamanara, N., Achouri, K., and Caloz, C. (2018). Computational analysis of metasurfaces. *IEEE Journal on Multiscale and Multiphysics Computational Techniques* 3: 37–49.
16 Alu, A. (2011). First-principles homogenization theory for periodic metamaterials. *Physical Review B* 84 (7): 075153.
17 Zubair, M., Sin Ang, Y., Ooi, K.J.A., and Ang, L.K. (2018). Fractional Fresnel coefficients for optical absorption in femtosecond laser-induced rough metal surfaces. *Journal of Applied Physics* 124 (16): 163101.
18 Naqvi, Q.A. and Zubair, M. (2016). On cylindrical model of electrostatic potential in fractional dimensional space. *Optik* 127 (6): 3243–3247.
19 Zubair, M. and Ang, L.K. (2016). Fractional-dimensional child-langmuir law for a rough cathode. *Physics of Plasmas* 23 (7): 072118.
20 Mughal, M.J. and Zubair, M. (2011). Fractional space solutions of antenna radiation problems: an application to hertzian dipole. *2011 IEEE 19th Signal Processing and*

*Communications Applications Conference (SIU)*, Antalya, Turkey (20–22 April 2011), 62–65. IEEE.

**21** Javed, A., Arif, A., Zubair, M., Mehmood, M.Q., and Riaz, K. (2020). A low-cost multiple complementary split-ring resonator based microwave sensor for contactless dielectric characterization of liquids. *IEEE Sensors Journal* 20 (19): 1132611334.

**22** Arbabi, A., Arbabi, E., xHorie, E., Kamali, S.M., and Faraon, A. (2017). Planar metasurface retroreflector. *Nature Photonics* 11 (7): 415.

**23** Chen, X., Chen, M., Mehmood, M.Q., et al. (2015). Longitudinal multifoci metalens for circularly polarized light. *Advanced Optical Materials* 3 (9): 1201–1206.

**24** Rana, A.S., Mehmood, M.Q., Jeong, H., Kim, I., and Rho, I. (2018). Tungsten-based ultrathin absorber for visible regime. *Scientific Reports* 8 (1): 1–8.

**25** Kim, I., So, S., Rana, A.S., Mehmood, M.Q., and Rho, J. (2018). Thermally robust ring-shaped chromium perfect absorber of visible light. *Nanophotonics* 7 (11): 1827–1833.

**26** Zubair, A., Zubair, M., Danner, A., and Mehmood, M.Q. (2020). Engineering multimodal spectrum of cayley tree fractal meta-resonator supercells for ultrabroadband terahertz light absorption. *Nanophotonics* 9 (3): 633–644.

**27** Rana, A.S., Zubair, M., Anwar, M.S., Saleem, M., and Mehmood, M.Q. (2020). Engineering the absorption spectra of thin film multilayer absorbers for enhanced color purity in cmy color filters. *Optical Materials Express* 10 (2): 268–281.

**28** Mahmood, N., Jeong, H., Kim, I., et al. (2019). Twisted non-diffracting beams through all dielectric meta-axicons. *Nanoscale* 11 (43): 20571–20578.

**29** Mei, S., Huang, K., Liu, H., et al. (2016). On-chip discrimination of orbital angular momentum of light with plasmonic nanoslits. *Nanoscale* 8 (4): 2227–2233.

**30** Ansari, M.A., Kim, I., Lee, D., et al. (2019). A spin-encoded all-dielectric metahologram for visible light. *Laser & Photonics Reviews* 13 (5): 1900065.

**31** Ansari, M.A., Tauqeer, T., Zubair, M., and Mehmood, M.Q. (2020). Breaking polarisation-bandwidth trade-off in dielectric metasurface for unpolarised white light. *Nanophotonics*, 9 (4): 963–971.

**32** Ansari, M.A., Kim, I., Rukhlenko, I.D., et al. (2020). Engineering spin and antiferromagnetic resonances to realize an efficient direction-multiplexed visible meta-hologram. *Nanoscale Horizons* 5 (1): 57–64.

**33** Di Renzo, M., Debbah, M., Phan-Huy, D.-T., et al. (2019). Smart radio environments empowered by reconfigurable AI meta-surfaces: an idea whose time has come. *EURASIP Journal on Wireless Communications and Networking* (1): 1–20.

**34** Basar, E., Di Renzo, M., De Rosny, J., et al. (2019). Wireless communications through reconfigurable intelligent surfaces. *IEEE Access* 7 : 116753–116773.

**35** Liaskos, C., Nie, S., Tsioliaridou, A., et al. (2018). A new wireless communication paradigm through software-controlled metasurfaces. *IEEE Communications Magazine* 56 (9): 162–169.

**36** Huang, C., Zappone, A., Alexandropoulos, G.C., Debbah, M., and Yuen, C. (2019). Reconfigurable intelligent surfaces for energy efficiency in wireless communication. *IEEE Transactions on Wireless Communications* 18 (8): 4157–4170.

**37** Van Huynh, N., Hoang, D.T., Lu, X., et al. (2018). Ambient backscatter communications: a contemporary survey. *IEEE Communications Surveys & Tutorials* 20 (4): 2889–2922.

**38** A. Liu, J. Li, G. Shen, et al. (2018). Enabling low-power duplex visible light communication. *arXiv preprint arXiv:1801.09812*.

**39** Yun, J. and Jang, J. (2016). Ambient light backscatter communication for iot applications. *Journal of Electromagnetic Engineering and Science* 16 (4): 214–218.

**40** Gilbreath, G.C., Rabinovich, W.S., Meehan, T.J., et al. (2001). Large-aperture multiple quantum well modulating retroreflector for free-space optical data transfer on unmanned aerial vehicles. *Optical Engineering* 40 (7): 1348 –1356. https://doi.org/10.1117/1.1383783.

**41** Li, L., Zhang, R., Zhao, Z., et al. (2017). High-capacity free-space optical communications between a ground transmitter and a ground receiver via a uav using multiplexing of multiple orbital-angular-momentum beams. *Scientific Reports* 7 (1): 1–12.

**42** Mei, S., Mehmood, M.Q., Hussain, S., et al. (2016) Flat helical nanosieves. *Advanced Functional Materials* 26 (29): 5255–5262.

**43** Akram, M.R., Mehmood, M.Q., Bai, X. et al. (2019). High efficiency ultrathin transmissive metasurfaces. *Advanced Optical Materials* 7 (11): 1801628.

**44** N. Mahmood, I. Kim, M.Q. Mehmood, et al. (2018). Polarisation insensitive multifunctional metasurfaces based on all-dielectric nanowaveguides. *Nanoscale* 10 (38): 18323–18330.

**45** Zhang, F., Pu, M., Li, X., et al. (2017). All-dielectric metasurfaces for simultaneous giant circular asymmetric transmission and wavefront shaping based on asymmetric photonic spin–orbit interactions. *Advanced Functional Materials* 27 (47): 1704295.

**46** Ma, D., Li, Z., Zhang, Y., et al. (2019). Giant spin-selective asymmetric transmission in multipolar-modulated metasurfaces. *Optics Letters* 44 (15): 3805–3808.

**47** Zheng, G., Wu, W., Li, Z., et al. (2017). Dual field-of-view step-zoom metalens. *Optics Letters* 42 (7): 1261–1264.

**48** Mehmood, M.Q., Liu, H., Huang, K., et al. (2015). Broadband spin-controlled focusing via logarithmic-spiral nanoslits of varying width. *Laser & Photonics Reviews* 9 (6): 674–681.

**49** Jiang, W.X., Ge, S., Han, T., et al. (2016). Shaping 3D path of electromagnetic waves using gradient-refractive-index metamaterials. *Advanced Science*, 3 (8): 1600022.

**50** Mehmood, M.Q., Mei, S., Hussain, S., et al. (2016). Visible-frequency metasurface for structuring and spatially multiplexing optical vortices. *Advanced Materials* 28 (13): 2533–2539.

**51** Akram, M.R., Mehmood, M.Q., Tauqeer, T., et al. (2019). Highly efficient generation of bessel beams with polarization insensitive metasurfaces. *Optics Express* 27 (7): 9467–9480.

**52** Huang, Kun, Liu, Hong, Restuccia, Sara, et al. (2018). Spiniform phase-encoded metagratings entangling arbitrary rational-order orbital angular momentum. *Light: Science & Applications* 7 (3): 17156–17156.

**53** Li,Z., Kim, I., Zhang, L., et al. (2017). Dielectric meta-holograms enabled with dual magnetic resonances in visible light. *ACS Nano* 11 (9): 9382–9389.

**54** Li, Z., Dai, Q., Mehmood, M.Q., et al. (2018). Full-space cloud of random points with a scrambling metasurface. *Light: Science & Applications* 7 (1): 1–8.

# 10

# Conclusion

*Qammer H. Abbasi[1], Hasan T. Abbas[1], Akram Alomainy[2], and Muhammad Ali Imran[1]*

[1] *School of Engineering, University of Glasgow, Glasgow, United Kingdom*
[2] *School of Electronic Engineering and Computer Science, Queen Mary University of London, London, United Kingdom*

The emergence of sensor networks and Internet of Things (IoT) has propelled the backscatter communication and RF sensing into mainstream research, with various potential applications most significantly in healthcare. Backscatter communication, in particular, enables an energy-efficient method technology for wireless networks that can be deployed in a large scale and can sustain themselves in terms of power. In this book, we have curated a list of chapters that touch upon some of the challenges in the development of backscatter technology and some of its use cases in the remote healthcare industry. We briefly discuss some of the hot research topics that have surfaced.

## 10.1 Future Hot Topics

### 10.1.1 Signal Modulation

In communication engineering, modulation refers to a process in which one or more properties of a carrier signal is varied based on a message signal. While one of the main inspirations behind the use of backscatter communication is the benefit of doing away with some of the costly RF infrastructures, achieving a reliable communication link is possible only due to the signal modulation schemes incorporated into the communication system. However, the power constraints of the network pose one of the biggest challenges in the selection of a particular modulation scheme.

### 10.1.2 Channel Estimation

One of the biggest challenges in the development of backscatter communication-enabled sensor networks is the availability of accurate channel state information (CSI). In real life, an accurate estimation of the CSI can be made possible through time-consuming and power hungry training signals. However, for networks like the IoT, this is not possible due to the power constraints.

*Backscattering and RF Sensing for Future Wireless Communication*, First Edition.
Edited by Qammer H. Abbasi, Hasan T. Abbas, Akram Alomainy, and Muhammad Ali Imran.

### 10.1.3 Low-Throughput

The inherent, low-power, and simplistic design of the backscatter communication end up in the prevalence of signal collisions as the one sensor may not be able to sense the transmission of another sensor. Due to the high likelihood of collisions, the overall sensor networks throughput is substantially reduced. Due to the increase in the network performance, there is work that needs to be done through which the channel capacity can be improved. One of the ways through which this can be achieved is the use of multi-antenna communication systems that permits more than one simultaneous transmission of signals.

### 10.1.4 Network Security

Another drawback of the uncomplicated nature of backscatter communication is the security vulnerability of the network. There are two ways in which the reliability of the communication system can be jeopardized. First, information meant for a particular receiver can be decoded by an eavesdropping entity, and second, a receiver can be prevented from decoding the information altogether through a jamming attack.

### 10.1.5 Energy Management

Some of the aforementioned challenges can be addressed by increasing the transmit powe.r of the signal, which may also increase the power of the backscattered signals. However, there is a delicate difference between the signal power and the prolonged self-sustaining capability of the sensor network. In this regard, efficient management of the energy resources is critical not only in the longevity of the network that may be battery operated but more importantly boosting the channel capacity and reliability of the network. Machine learning and artificial intelligence-based resource allocation and management have attracted a lot of interest in this regard, which may yield the most optimal parameters of the network.

## 10.2 Concluding Remarks

The congestion of the conventional sub-6 GHz radio frequency spectrum is a big concern especially in light of the imminent deployment of large-scale sensor networks. Software-controlled metasurfaces have brought backscatter communication to the mainstream as the propagation of communication signals in a given environment can be tailored as per the channel requirements. In the past few years, metasurfaces have found numerous applications due to their ability to control the propagation of electromagnetic waves. In communication systems, they have been used for beam focusing. This book enables the readers with the fundamental theories of metasurfaces, its applications in sensing for health and well-being monitoring, and backscatter communications. Throughout the chapters, the book presents open issues and challenges, addressing which can further advance the field.

# Index

Active relay 1
Amplitude shift keying 20
Angle of arrival 168
Anisotropic 78, 80, 118, 190
Aperture efficiency 4, 64, 80
Aperture reflector 4
Artificial impedance boundaries 72

Babinet's theory 48
Backscatter communication 19
Bandpass 50
Bandstop 50
Beam-steering 9, 54
Binary frequency shift keying 10
Birefringent media 94
Bit error rate 33
Bluetooth low energy 23
Boundary conditions 42, 46, 67, 71
Brewster effect 86
BST capacitor 12

Channel equalization 19
Channel state information 205
Chirp spread spectrum 20, 26, 27, 29, 31, 33
Circular polarization 16, 55, 91, 101, 148, 192
Coding sequence 8, 49
Conductivity 42
Constellation 25
Continuous wave 167
Cross-polarization conversion 59, 101
Cross-reflection 102
C2-symmetry 95
Cumulative distribution function 128

Demand response 122
Digital coding 49
Digital-to-analog convertor 28

Eigenvalue 46, 109
Electromagnetic bandgap 43
Electromagnetic wave 20, 35, 65, 80, 133, 159, 180
Equivalence principle 41, 73
Equivalent circuit 51, 138

Fabry–Pérot 46, 66
Fano resonances 48
Fast Fourier transform 27
Feature extraction 162, 168
Fermi velocity 45
Field-effect transistor 12
Field programmable gate array 12
Finite element 60, 181
FR-4 105, 112, 143
Free space wavenumber 10
Frequency-selective surface 2, 50
Frequency shift keying 10, 22

Gait approximation 168
Gradient metasurface 41, 77, 117, 184
Grating lobe condition 52

Half power beamwidth 63
Half-wave plate 95, 186
Heart rate monitoring 174
HFSS 48, 60, 62
Hologram 72, 182, 200
Human activity detection 170
Huygens' principle 42
Huygens source 77, 81
Huygens unit cell 77

Independent component analysis 163
Input impedance 137, 145
Intelligent healthcare 157
Interelement distance 4
Internet of things 1, 19, 119, 133, 152, 175, 205
IQ backscatter 25

Line-of-sight 34
Low impedance surface 103

*Backscattering and RF Sensing for Future Wireless Communication,* First Edition.
Edited by Qammer H. Abbasi, Hasan T. Abbas, Akram Alomainy, and Muhammad Ali Imran.

Machine learning 12, 144, 163, 203
Magnetic resonance imaging 72
Matching circuit 25, 138
Maxwell's equations 44, 87
Metamaterial 15, 41, 65, 72, 100, 180
Metasurface 14, 41, 50, 99, 100, 180
Micro-Doppler signature 162
Microwave 1, 17, 39, 65, 81, 117, 152, 176, 182, 200
Millimeter-wave 85
Mobile communication 9
Motion detection system 174
Multicarrier backscatter 34
Multiple-input-multiple-output 5

Narrowband IoT 19
Negative permittivity 42, 179
Noninvasive 6, 159
Non-line-of-sight 34

Omega-bianisotropic metasurface 63
Orbital angular momentum 44, 183
Orthogonal frequency division multiplexing 169
Orthogonality 26

Parabolic reflector 9
Particle swarm optimization 47
Passive infrared 157
Passive Lora 27
Perfect electric conductor 47, 100
Perfect lens 41
Permeability 7, 44, 112, 180
Phased array 9, 16, 77, 81
Phase shift keying 24
Phasor data concentrators 125
Phasor measurement units 122
Physiological monitoring radar system 167
P-I-N diode 25
Planar boundary 41
Plasmonic metasurface 44
Polarizability 76
Polarization extinction ratio 103
Polyimide 107
Power transmission coefficient 140
Principal component analysis 163
Printed circuit board 12, 53
Programmable metamaterial 15

Quadrature amplitude modulation 25
Quantization 7, 54
Quarter-wave plate 94

Radar cross-section 16, 50
Radiation cavities 61
Radio frequency identification 152, 175
Random optimization algorithm 8
Received signal strength indicator 143
Reconfigurable intelligent surface 2, 120, 202
Reflection coefficient circle 21
Right-hand circularly polarized 55, 92

Satellite antennas 50
Signal modulation scheme 19, 205
Signal-to-noise ratio 1, 195
Single-pole multi-throw switch 21
Sleep Apnea 173
Snell's law 3, 10, 97
Spectrogram 163, 169
Spherical wave 3
Split-ring resonator 42, 202
Supervisory control and data acquisition systems 122
Surface admittance 43
Surface impedance 3, 45, 65, 81
Symmetrical square loop 52

Tag antenna 20, 138
Tangential electric field 62
Temperature sensor 1, 146
Transmitarray antenna 53
Transport control protocol 128
Transverse electric 45
Transverse electromagnetic 48
Two-dimensional 3, 82, 120, 161, 180
2D metasurface 44–46

Ultra-high frequency 133
Up-chirp 28
User diagram protocol 128

Vector signal analyzer 31
Vehicle-to-grid 127
Virtual aperture 79

Wavefront transformation 55, 71
Wave propagation 44, 196
Wearable sensor 66, 147, 171
Wi-Fi 38, 121
Wilkinson power splitter 25
Wireless communication 1, 14, 38, 71, 85, 119, 131, 202

X-band 16, 102